The Hard-Won Pos

of S. J. Becker

CARTESIAN TENSORS:
With Applications to Mechanics,
Fluid Mechanics and Elasticity

ELLIS HORWOOD SERIES IN MATHEMATICS AND ITS APPLICATIONS

Series Editor: Professor G. M. BELL, Chelsea College, University of London

The works in this series will survey recent research, and introduce new areas and up-to-date mathematical methods. Undergraduate texts on established topics will stimulate student interest by including present-day applications, and the series can also include selected volumes of lecture notes on important topics which need quick and early publication.

In all three ways it is hoped to render a valuable service to those who learn, teach, develop and use mathematics.

MATHEMATICAL THEORY OF WAVE MOTION
G. R. BALDOCK and T. BRIDGEMAN, University of Liverpool.
MATHEMATICAL MODELS IN SOCIAL MANAGEMENT AND LIFE SCIENCES
D. N. BURGHES and A. D. WOOD, Cranfield Institute of Technology.
MODERN INTRODUCTION TO CLASSICAL MECHANICS AND CONTROL
D. N. BURGHES, Cranfield Institute of Technology and A. DOWNS, Sheffield University.
CONTROL AND OPTIMAL CONTROL
D. N. BURGHES, Cranfield Institute of Technology and A. GRAHAM, The Open University, Milton Keynes.
TEXTBOOK OF DYNAMICS
F. CHORLTON, University of Aston, Birmingham.
VECTOR AND TENSOR METHODS
F. CHORLTON, University of Aston, Birmingham.
TECHNIQUES IN OPERATIONAL RESEARCH
VOLUME 1: QUEUEING SYSTEMS
VOLUME 2: MODELS, SEARCH, RANDOMIZATION
B. CONOLLY, Chelsea College, University of London
MATHEMATICS FOR THE BIOSCIENCES
G. EASON, C. W. COLES, G. GETTINBY, University of Strathclyde.
HANDBOOK OF HYPERGEOMETRIC INTEGRALS: Theory, Applications, Tables, Computer Programs
H. EXTON, The Polytechnic, Preston.
MULTIPLE HYPERGEOMETRIC FUNCTIONS
H. EXTON, The Polytechnic, Preston
COMPUTATIONAL GEOMETRY FOR DESIGN AND MANUFACTURE
I. D. FAUX and M. J. PRATT, Cranfield Institute of Technology.
APPLIED LINEAR ALGEBRA
R. J. GOULT, Cranfield Institute of Technology.
MATRIX THEORY AND APPLICATIONS FOR ENGINEERS AND MATHEMATICIANS
A. GRAHAM, The Open University, Milton Keynes.
APPLIED FUNCTIONAL ANALYSIS
D. H. GRIFFEL, University of Bristol.
GENERALISED FUNCTIONS: Theory, Applications
R. F. HOSKINS, Cranfield Institute of Technology.
MECHANICS OF CONTINUOUS MEDIA
S. C. HUNTER, University of Sheffield.
GAME THEORY: Mathematical Models of Conflict
A. J. JONES, Royal Holloway College, University of London.
USING COMPUTERS
B. L. MEEK and S. FAIRTHORNE, Queen Elizabeth College, University of London.
SPECTRAL THEORY OF ORDINARY DIFFERENTIAL OPERATORS
E. MULLER-PFEIFFER, Technical High School, Ergurt.
SIMULATION CONCEPTS IN MATHEMATICAL MODELLING
F. OLIVEIRA-PINTO, Chelsea College, University of London.
ENVIRONMENTAL AERODYNAMICS
R. S. SCORER, Imperial College of Science and Technology, University of London.
APPLIED STATISTICAL TECHNIQUES
K. D. C. STOODLEY, T. LEWIS and C. L. S. STAINTON, University of Bradford.
LIQUIDS AND THEIR PROPERTIES: A Molecular and Macroscopic Treatise with Applications
H. N. V. TEMPERLEY, University College of Swansea, University of Wales and D. H. TREVENA, University of Wales, Aberystwyth.
GRAPH THEORY AND APPLICATIONS
H. N. V. TEMPERLEY, University College of Swansea.

CARTESIAN TENSORS:
With Applications to Mechanics, Fluid Mechanics and Elasticity

A. M. GOODBODY, B.Sc., Ph.D., A.R.C.S., D.I.C.
Department of Applied Mathematics and Theoretical Physics
University of Liverpool

ELLIS HORWOOD LIMITED
Publishers · Chichester

Halsted Press: a division of
JOHN WILEY & SONS
New York · Brisbane · Chichester · Toronto

First published in 1982 by

ELLIS HORWOOD LIMITED
Market Cross House, Cooper Street, Chichester, West Sussex, PO19 1EB, England

The publisher's colophon is reproduced from James Gillison's drawing of the ancient Market Cross, Chichester.

Distributors:

Australia, New Zealand, South-east Asia:
Jacaranda-Wiley Ltd., Jacaranda Press,
JOHN WILEY & SONS INC.,
G.P.O. Box 859, Brisbane, Queensland 40001, Australia

Canada:
JOHN WILEY & SONS CANADA LIMITED
22 Worcester Road, Rexdale, Ontario, Canada.

Europe, Africa:
JOHN WILEY & SONS LIMITED
Baffins Lane, Chichester, West Sussex, England.

North and South America and the rest of the world:
Halsted Press: a division of
JOHN WILEY & SONS
605 Third Avenue, New York, N.Y. 10016, U.S.A.

British Library Cataloguing in Publication Data
Goodbody, A. M.
Cartesian tensors: with applications to mechanics, fluid mechanics and elasticity. –
(The Ellis Horwood series in mathematics and its applications)
1. Calculus of tensors
I. Title
515'.63 QA433

Library of Congress Card No. 81–6884 AACR2

ISBN 0–85312–220–2 (Ellis Horwood Ltd., Publishers – Library Edn.)
ISBN 0–85312–377–2 (Ellis Horwood Ltd., Publishers – Student Edn.)
ISBN 0–470–27254–6 (Halsted Press)

Typeset in Great Britain by Activity, Salisbury.
Printed in the USA by The Maple-Vail Book Manufacturing Group, New York.

Contents

Chapter 1 Introductory Topics - Scalars and Vectors

1.1 Determinants 19
1.1.1 Introduction 19
1.1.2 Second-Order Determinants 19
1.1.3 Third-Order Determinants 20
1.1.4 Fourth and Higher Order Determinants 21
1.1.5 Properties of Third-Order Determinants 22
1.1.6 Minors and Cofactors 23
1.1.7 Further properties of Third-Order Determinants 23
Examples 1.1 25
Problems 1.1 25
1.2 Matrices 26
1.2.1 Introduction 26
1.2.2 Matrix addition 27
1.2.3 Multiplication by a Scalar 27
1.2.4 Matrix Multiplication 27
1.2.5 Properties of Matrix Multiplication 29
1.2.6 Zero and Unit Matrices 29
1.2.7 The Transpose Matrix. Symmetric and Skew-Symmetric Matrices 30
1.2.8 The Inverse Matrix 31
1.2.9 The Rank of a Matrix 31
Examples 1.2 32
Problems 1.2 33
1.3 Scalars 34
1.3.1 Introduction 34
1.3.2 Scalar Fields 34
Example 1.3. 35
Problems 1.3 -. 35

1.4 Vectors . . . 35
1.4.1 Introduction . . . 35
1.4.2 Definition of a Real Vector Space . . . 36
Examples 1.4 . . . 36
Problems 1.4 . . . 37
1.5 Vector Properties . . . 37
1.5.1 Properties of Vectors in R^3 . . . 37
1.5.2 The Scalar Product of Two Vectors . . . 37
1.5.3 The Vector Product of Two Vectors . . . 38
Example 1.5 . . . 39
Problems 1.5 . . . 39
1.6 The Geometrical Representation of Vectors . . . 40
1.6.1 Directed Line Segments . . . 40
1.6.2 The Parallelogram Law of Addition . . . 40
1.6.3 The Triangle Law of Addition . . . 41
1.6.4 The Geometrical Interpretation of the Scalar Product . . . 42
1.6.5 Direction Cosines . . . 44
1.6.6 The Geometrical Interpretation of the Vector Product . . . 45
Examples 1.6 . . . 48
Problems 1.6 . . . 49
1.7 Products of More than Two Vectors . . . 50
1.7.1 The Triple Scalar Product . . . 50
1.7.2 The Triple Vector Product . . . 50
1.7.3 Higher Products of Vectors . . . 51
1.7.4 Linear Dependence and Independence . . . 51
Examples 1.7 . . . 53
Problems 1.7 . . . 53
1.8 Vector Fields . . . 54
1.8.1 Definition . . . 54
1.8.2 Differentiation of Scalar Fields . . . 54
Examples 1.8 . . . 55
Problems 1.8 . . . 56
1.9 The Summation Convention . . . 56
1.9.1 Suffices . . . 56
1.9.2 The Kronecker Delta . . . 57
Examples 1.9 . . . 57
Problems 1.9 . . . 58
1.10 Transformations of Vectors . . . 58
1.10.1 The Invariance of Vectors – Transformation Properties . . . 58
1.10.2 Rotations . . . 59
1.10.3 The Rotation Matrix . . . 60
Examples 1.10 . . . 61
Problems 1.10 . . . 62

1.11 Rate of Change of a Vector in a Rotating Frame of Reference 63

Chapter 2 Second-Order Cartesian Tensors

2.1 Introduction 66
2.1.1 Linear Vector Functions 66
2.1.2 Dyads 66
Examples 2.1 67
Problems 2.1 67
2.2 Properties of Second-order Tensors. 68
2.2.1 Properties of Dyads 68
2.2.2 The Components of a Dyad 69
2.2.3 The Components of the Dot Product of a Tensor and a Vector 69
2.2.4 Properties of Second-Order Tensors 70
Examples 2.2 70
Problems 2.2 71
2.3 Further Properties of Second-Order Tensors 72
2.3.1 Pre- and Post-Multiplication of a Tensor and a Vector 72
2.3.2 The Transpose Tensor 72
Examples 2.3 73
Problems 2.3 73
2.4 Symmetric and Antisymmetric Tensors 74
Examples 2.4 76
Problems 2.4 76
2.5 Transformation Properties of Tensors 76
2.5.1 The Invariance of Tensors - Transformation Law 76
2.5.2 The Practical Transformation of Tensor Components 78
Example 2.5 78
Problems 2.5 79
2.6 The Scalar Invariants of a Second-Order Tensor 80
Examples 2.6 82
Problems 2.6 82
2.7 Inner Products 82
2.7.1 The Inner Product of Two Tensors - Contraction 82
2.7.2 The Double Inner Product of Two Tensors 83
Example 2.7 84
Problems 2.7 84
2.8 Vectors and Tensors with Complex Components 84
Example 2.8 85
Problems 2.8 85
2.9 The Eigenvector Problem 86
2.9.1 Statement of the Problem 86

2.9.2 Eigenvectors of the Hermitian Transpose Tensor 88
Example 2.9 89
Problems 2.9 90
2.10 The Eigenvector Problem: Degenerate Cases 91
2.10.1 Double Root of the Characteristic Equation 91
2.10.2 Treble Root of the Characteristic Equation 94
Examples 2.10 95
Problem 2.10 97
2.11 Hermitian Tensors 97
2.11.1 The Eigenvectors of Hermitian Tensors 97
2.11.2 Hermitian Tensors: The Degenerate Cases 98
Example 2.11 100
Problem 2.11 100
2.12 Real Symmetric Tensors 101
2.12.1 The Eigenvectors of Real Symmetric Tensors 101
2.12.2 Diagonalization of a Symmetric Tensor 101
Examples 2.12 102
Problems 2.12 103
2.13 The Cayley - Hamilton Theorem 103
Example 2.13 105
Problems 2.13 105
2.14 The Representation Theorem for Symmetric Tensors 106
2.15 Tensor Fields 108
2.15.1 Definition 108
2.15.2 Differentiation of Vector Fields 109
Examples 2.15 110
Problems 2.15 111

Chapter 3 Third-Order Cartesian Tensors
3.1 Tensors of the Third-Order 112
3.1.1 Definition 112
3.1.2 Products Involving Third-Order Tensors 113
3.1.3 The Transformation Law for Third-Order Tensors 113
Example 3.1 113
Problem 3.1 114
3.2 The Alternate Tensor 114
3.2.1 Definition 114
3.2.2 Applications of the Alternate Tensor 115
3.2.3 Differential Functions involving the Alternate Tensor 116
Examples 3.2 117
Problems 3.2 118
3.3 Cross Products of Tensors and Vectors 119
3.3.1 Definition 119

3.3.2 Rate of Change of a Tensor in a Rotating Frame of Reference 121
Examples 3.3 122
Problems 3.3 122
3.4 The Geometry of Rotations 123
3.4.1 General Rotations 123
3.4.2 Rotations about the Coordinate Axes 126
Examples 3.4 128
Problems 3.4 129
3.5 Integral Theorems in Tensor Form 130
3.5.1 The Tensor Form of Green's Theorem 130
3.5.2 Special Cases of Green's Theorem 130
3.5.3 The Tensor Form of Stokes's Theorem 131
3.5.4 Special Cases of Stokes's Theorem 131
Problems 3.5 132
3.6 Curvilinear Coordinates 132
3.6.1 Introduction 132
3.6.2 Cartesian Tensors Referred to Orthogonal Curvilinear Coordinates 134
3.6.3 Spherical Polar Coordinates 137
3.6.4 Cylindrical Polar Coordinates 140
Examples 3.6 141
Problems 3.6 142
3.7 Third-Order Tensor Fields 143
Examples 3.7 144

Chapter 4 Fourth-Order Cartesian Tensors
4.1 Tensors of the Fourth-Order - Definition 145
4.2 Isotropic Tensors. 145
4.2.1 Definition 145
4.2.2 The Most General Isotropic Tensor of the Second-Order 146
4.2.3 The Most General Isotropic Tensor of the Third-Order. 146
4.2.4 The Most General Isotropic Tensor of the Fourth-Order 147
4.2.5 Complex Isotropic Vectors 149
Examples 4.2 150
Problems 4.2 151

Chapter 5 The Inertia Tensor
5.1 The Inertia Tensor 152
5.1.1 Definition 152
5.1.2 Use of the Inertia Tensor - Kinetic Energy 153
5.1.3 The Moment of Inertia 155
5.1.4 Radius of Gyration 155
5.1.5 Momental Ellipsoid 156

Examples 5.1 156
Problems 5.1 158
5.2 Properties of the Inertia Tensor 159
5.2.1 The Components of the Inertia Tensor 159
5.2.2 The Parallel Axis Theorem 159
5.2.3 The Perpendicular Axes Theorem for Laminae 162
Examples 5.2 163
Problems 5.2 163
5.3 Calculation of the Inertia Tensor of Rectangular Bodies 164
5.3.1 Introduction 164
5.3.2 The Inertia Tensor of a Uniform Rectangular Block 165
5.3.3 Equimomental Systems 167
5.3.4 The Inertia Tensor of a Uniform Rectangular Block – An Alternative Approach 168
Examples 5.3 171
Problems 5.3 172
5.4 The Inertia Tensor of Non-Rectangular Bodies 173
5.4.1 The Inertia Tensor of a Uniform Parallelogram Lamina 173
5.4.2 The Inertia Tensor of a Uniform Solid Parallelepiped 174
5.4.3 The Inertia Tensor of a Uniform Triangular Lamina 175
5.4.4 The Inertia Tensor of a Uniform Solid Tetrahedron 177
Example 5.4 179
Problems 5.4 180
5.5 The Inertia Tensor of Bodies with Curved Boundaries 180
5.5.1 The Inertia Tensor of a Ring 180
5.5.2 The Inertia Tensor of a Circular Lamina 181
5.5.3 The Inertia Tensor of a Hollow Right Circular Cylinder 182
5.5.4 The Inertia Tensor of a Uniform Solid Right Circular Cylinder 184
5.5.5 The Inertia Tensor of a Uniform Elliptic Ring 185
5.5.6 The Inertia Tensor of an Elliptic Lamina 189
5.5.7 The Inertia Tensor of a Hollow Torus 190
5.5.8 The Inertia Tensor of a Solid Torus 192
5.5.9 The Inertia Tensor of a Spherical Shell 193
5.5.10 The Inertia Tensor of a Solid Sphere 194
Examples 5.5 195
Problems 5.5 197
5.6 Euler's Equations of Motion 199
Examples 5.6 201
Problems 5.6 203

Chapter 6 The Application of Cartesian Tensors to Fluid Mechanics
6.1 The Stress Tensor 204

6.1.1 Introduction . . . 204
6.1.2 The Stress Tensor . . . 205
6.1.3 Properties of the Stress Tensor . . . 207
6.1.4 The Appearance of the Stress Tensor in the Equation of Fluid Motion . . . 208
6.1.5 The Symmetry of the Stress Tensor . . . 209
6.1.6 Decomposition of the Stress Tensor . . . 210
Examples 6.1 . . . 211
Problems 6.1 . . . 217
6.2 Analysis of Fluid Motion . . . 219
6.2.1 The Rate of Strain Tensor . . . 219
6.2.2 The Contribution of the Rate of Strain Tensor to the Relative Velocity . . . 220
6.2.3 The Rate of Strain . . . 220
6.2.4 The Contribution of the Antisymmetric Tensor $\boldsymbol{\xi}$ to the Relative Velocity . . . 221
Examples 6.2 . . . 222
Problems 6.2 . . . 226
6.3 The Navier - Stokes Equations of Fluid Motion . . . 227
Chapter 7 The Application of Cartesian Tensors to Elasticity
7.1 Strain . . . 229
7.1.1 The Strain Tensor - Infinitesimal Strain . . . 229
7.1.2 The Contribution of the Strain Tensor to the Deformation . . 231
7.1.3 The Contribution of the Antisymmetric Tensor $\boldsymbol{\zeta}$ to the Deformation . . . 232
7.1.4 The Compatibility Equations . . . 232
Examples 7.1 . . . 233
Problems 7.1 . . . 235
7.2 The Relation between Stress and Strain for Elastic Solids . . . 236
7.2.1 Hooke's Law . . . 236
7.2.2 The Generalized Hooke's Law . . . 238
7.2.3 Simple Shear . . . 239
7.2.4 Isotropic Expansion . . . 239
7.2.5 The Extension of a Wire . . . 239
Examples 7.2 . . . 241
Problems 7.2 . . . 243
7.3 Strain Energy . . . 244
7.3.1 The Strain Energy Function . . . 244
7.3.2 Energy Stored in Dilatation and Distortion . . . 245
Example 7.3 . . . 246
Problems 7.3 . . . 246
7.4 The Displacement Equation . . . 247
Examples 7.4 . . . 248

Problems 7.4 252
7.5 The Rotating Disc Problem 253
7.6 Finite Deformation 257
7.6.1 Lagragian Finite Strain Components 257
7.6.2 The Eulerian Finite Strain Components 258
7.6.3 The Relation between the Eulerian Strain Tensor and the Rate - of - Strain Tensor 259
Examples 7.6 261
Problems 7.6 263

Appendix
A.1 General Tensors 266
A.1.1 Contravariant and Covariant Components 266
A.1.2 The Metric Tensor 268
A.1.3 The Quotient Theorem 269
A.2 Differentiation of General Tensors 270
A.2.1 The Christoffel Symbols 270
A.2.2 The Transformation Law for Christoffel Symbols 270
A.2.3 The Covariant Differentiation of Vectors 271
Examples A.1 272

Answers to Problems 276

Bibliography 293

Index 294

Preface

The basic material for this book is contained in a series of lectures given to Second Year Mathematics and Physics undergraduates at the University of Liverpool. It is the experience of the author that even the better students, who find the subject algebraically straightforward, experience difficulties in coming to terms with what a tensor actually is. To the reader who is familiar with the elements of vector algebra it may come as a surprise that he already knows what a tensor is. Vectors are tensors of order 1.

The enormous simplification of all tensor formulae effected by referring them to Cartesian rather than to curvilinear coordinates, and the fact that so much can be achieved by the use of Cartesian tensors alone, render the study of Cartesian tensors a worthwhile discipline in its own right. The object of this book is to introduce the reader to the theory and application of Cartesian tensors of order 2 and higher. Following an introductory chapter, in which the main results of vector algebra are stated, Chapter 2 on second-order Cartesian tensors, forms the theoretical backbone of the book. Third- and fourth-order tensors are treated briefly in Chapters 3 and 4. The final three chapters of the book are concerned with the main applications of Cartesian tensors to mechanics and continuum mechanics.

I am deeply indebted to Professor J. G. Oldroyd of the University of Liverpool for his critical reading of the text, for many hours of discussion, and for his encouragement to proceed with the writing of this book. I am also much indebted to Professor G. M. Bell, the Series Editor. He too read the text with extreme care, and as a result pointed out a number of errors and made numerous suggestions for improvements in the presentation. For help with the proof-reading I would like to express my grateful thanks to two of my colleagues, Dr. G. R. Baldock and Dr. T. Bridgeman. Between them they were responsible for bringing to light a number of errors hitherto undetected. Nevertheless those errors which remain are my responsibility and mine alone. Finally I should like to record my thanks to Miss Helen Wright for her unfailing patience and cheerfulness during the long job of typing the manuscript.

A. M. Goodbody
Liverpool, June 1981

List of Symbols

$A, B, C,$	Moments of Inertia
A, B, C, D, . . .	Points
$A, B, C, D, \ldots$	Matrices
$A,$	Plane surface area
$A',$	Transpose matrix of A
A_{ij}	Cofactor of element a_{ij}
$A^{-1},$	Inverse matrix of A
Adj A	Adjugate matrix of A
$\|A\|$, det A	Determinant of matrix A
A	Antisymmetric tensor,
$a, b, c, \ldots$	Length of sides etc.
$a, \ldots$	Radius of circle, sphere
$a_i, \ldots$	Components of vector **a**
$a_i, \ldots$	Covariant components of vector **a**
$a^i, \ldots$	Contravariant components of vector **a**
$a_{ij}, \ldots$	Components of matrix A, determinant $\|A\|$ or tensor **A**
a_{ijk}	Components of third order tensor **A**
(a_{ij})	Matrix whose (i, j) - element is a_{ij}
a, **b**, **c**	Vectors
d	Deviatoric Stress Tensor
$D, D_1, D_2,$	Determinants
$\mathscr{D}$	Domain
E	Alternate tensor
E	Young's Modulus
E	Complete Elliptic Integral of 2nd Kind
e	Rate-of-Strain tensor
$\mathbf{e}_1, \mathbf{e}_2, \mathbf{e}_3$	Orthonormal triad of vectors (basis)
e	Eccentricity of ellipse
F	Body Force per unit mass
$\mathbf{f}_s$	Total surface force

G	Centroid
G_{ik}	Components of Eulerian deformation tensor
g_{ik}	Components of Langrangian deformation tensor
g_{ik}	Components of metric tensor
g^{ik}	Components of associate tensor to g_{ik}
g	Acceleration due to gravity
$\mathbf{G}$	Eulerian deformation tensor
$\mathbf{g}$	Lagrangian deformation tensor
$\mathbf{H}$(P)	Angular momentum about P
h_i	Scale factor
$\mathbf{I}$(P)	Inertia tensor calculated at P
I_1, I_2, I_3	Scalar invariants of tensor
K	Complete elliptic integral of 1st Kind
K	Bulk modulus
k	Radius of gyration
M, m	Mass
$\mathbf{n}$	Normal vector
O	Origin
O	Null Matrix
p	Pressure
R	Rank of Matrix
R^3	Real Euclidean space of 3 dimensions
$\mathbf{r}$	The vector (x_1, x_2, x_3)
r	Radial spherical or cylindrical coordinate
$\bar{\mathbf{r}}$	Position vector of centroid
$\mathbf{S}$	Symmetric Tensor
S	Curved surface area
$\mathbf{T}$	Tensor
$\mathbf{T}'$	Tranpose of $\mathbf{T}$
T	Kinetic energy
U, U', U''	Strain energy functions
u^i	Curvilinear coordinates
$\mathbf{u}, \mathbf{v}$	Velocity
$v^j{}_{,k} v_{j,n}$	Covariant derivatives of v^j, v_j
V	Volume
$\mathscr{V}$	Real Vector space
x, y, z	Cartesian coordinates
x_1, x_2, x_3	Cartesian coordinates
x_1', x_2', x_3'	Cartesian coordinates with respect to rotated axes
$z = x + iy$	Complex number
z	Cylindrical polar coordinate
α, β, γ	Angles
$\mathbf{\Gamma}$	Moment of applied force

δ_{ij}	Kronecker delta
Δ	Determinant
$\boldsymbol{\epsilon}$	Infinitesimal strain tensor
$\bar{\boldsymbol{\epsilon}}$	Lagrangian strain tensor
ϵ_{ii}	Dilatation
ϵ_{ijk}	Components of Alternate tensor, **E**
$\boldsymbol{\eta}$	Eulerian Strain tensor
η_{ik}	Components of Eulerian strain tensor
θ	Angle (spherical or cylindrical polar coordinate)
Θ	Rotation matrix
θ_{ij}	Elements of rotation matrix
λ	Eigenvalue
λ, μ	Lamé constants
μ	Shear Modulus
μ	Moment of inertia
μ	Viscosity
ν	Poisson's Ratio
ν	Kinematic viscosity
$\boldsymbol{\xi}$	Vorticity tensor
ρ	Density
Σ	Summation sign
$\boldsymbol{\sigma}$	Stress tensor
σ	Surface density
$\boldsymbol{\Sigma}$	Stress (vector)
τ	Volume
ϕ	Scalar function
ϕ	Azimuthal angle (spherical polar coordinate)
ψ	Scalar function
$\boldsymbol{\omega}$	Vorticity
$\boldsymbol{\omega}$	Angular velocity
$\overline{\text{AB}}$	Vector between points A and B
$\mathbf{a.b}$	Scalar product of **a** and **b**
$\mathbf{a} \times \mathbf{b}$	Vector product of **a** and **b**
$\mathbf{a} \otimes \mathbf{b}$	Tensor product of **a** and **b**
$\mathbf{I}.\boldsymbol{\omega}$	Inner product of **I** and $\boldsymbol{\omega}$
$\mathbf{S}:\mathbf{T}$	Double inner product of **S** and **T**
$[\mathbf{x}, \mathbf{y}, \mathbf{z}]$	Triple scalar product $\mathbf{x} \times \mathbf{y}.\mathbf{z}$
$\lvert\mathbf{x}\rvert$	Modulus of $\mathbf{x} = \sqrt{(\mathbf{x}.\mathbf{x})}$
$\lvert\mathbf{T}\rvert$	Modulus of $\mathbf{T} = \sqrt{(\mathbf{T}:\mathbf{T})}$
$\mathbf{T}^{\mathrm{H}}$	Hermitian transpose of **T**
$\mathbf{1}$	Unit tensor
$\overline{x_i}, \bar{T}_{ik}$	Complex conjugate of x_i, T_{ik}
$\nabla\phi$	Gradient of ϕ

$\nabla . \mathbf{v}$	Divergence of **v**
$\nabla \times \mathbf{v}$	Curl of **v**
$\nabla^2 \phi$	Laplacian of ϕ
$\forall$	'For all'
$\frac{\mathrm{D}}{\mathrm{D}t}$	Derivative following the fluid motion
$\frac{\mathscr{D}}{\mathscr{D}\mathrm{t}}$	Convected derivative
$\int$	Integral
$\oint$	Line Integral
$[ij, k]$	Christoffel symbol of 1st kind
$\left\{ \begin{matrix} l \\ ij \end{matrix} \right\}$	Christoffel symbol of 2nd kind

CHAPTER 1

Introductory Topics-Scalars and Vectors

1.1 DETERMINANTS

1.1.1 Introduction

In this section the main properties of determinants are quoted without proof. For proofs of the theorems for determinants of arbitrary order the reader is referred to the books in the Bibliography. In the case of second or third-order determinants, which are the only ones to be employed in this book, the results are readily verifiable by direct calculation.

1.1.2 Second-Order Determinants

Consider the pair of simultaneous equations in the unknowns x_1, x_2:

$$\left.\begin{aligned} a_{11}x_1 + a_{12}x_2 &= b_1 , \\ a_{21}x_1 + a_{22}x_2 &= b_2 . \end{aligned}\right\} \tag{1.1}$$

If (x_1, x_2) are the Cartesian coordinates of a point in a plane, equations (1.1) represent two straight lines which either (i) intersect, (ii) are parallel or (iii) are coincident.

The solution of equations (1.1) may be written down formally by **Cramer's Rule**:

$$x_1 = \frac{b_1a_{22} - b_2a_{12}}{a_{11}a_{22} - a_{12}a_{21}} , \quad x_2 = \frac{b_2a_{11} - b_1a_{21}}{a_{11}a_{22} - a_{12}a_{21}} . \tag{1.2}$$

Let us write:

$$\begin{aligned} D &= a_{11}a_{22} - a_{12}a_{21} , \quad D_1 = b_1a_{22} - b_2a_{12} , \\ D_2 &= b_2a_{11} - b_1a_{21} , \end{aligned} \tag{1.3}$$

so that $$x_1 = D_1/D , \quad x_2 = D_2/D . \tag{1.4}$$

If $D \neq 0$, a unique solution exists for x_1 and x_2. This corresponds to the case (i), of two straight lines intersecting. The equations are then said to be **linearly**

independent (see section 1.7.4). If $D = 0$ while $D_1 \neq 0$ and/or $D_2 \neq 0$, no solution exists. This corresponds to the case (ii), of two parallel straight lines. The equations are then said to be **inconsistent**. If $D = D_1 = D_2 = 0$, infinitely many solutions for x_1, x_2 exist. This corresponds to case (iii) of two coincident straight lines. The equations are then said to be **linearly dependent** (see section 1.7.4).

The quantities D, D_1 and D_2 play an important role in the formal solution of equations (1.1). D is called the **determinant of the coefficients** of the simultaneous equations (1.1). It is written:

$$D = \begin{vmatrix} a_{11} & a_{12} \\ a_{21} & a_{22} \end{vmatrix} . \tag{1.5}$$

D_1 and D_2 are also determinants. They are written:

$$D_1 = \begin{vmatrix} b_1 & a_{12} \\ b_2 & a_{22} \end{vmatrix} , \quad D_2 = \begin{vmatrix} a_{11} & b_1 \\ a_{21} & b_2 \end{vmatrix} . \tag{1.6}$$

The numbers a_{11}, a_{12} etc. are called the **elements** of the determinant. The pairs of numbers a_{11}, a_{12} and a_{21}, a_{22} are called **rows** and the pairs of numbers a_{11}, a_{21} and a_{12}, a_{22} are called **columns**. The determinants D, D_1 and D_2 have two rows and two columns and are therefore called **second-order determinants**.

1.1.3 Third-Order Determinants

Consider the three simultaneous equations

$$\left.\begin{aligned} a_{11}x_1 + a_{12}x_2 + a_{13}x_3 &= b_1 , \\ a_{21}x_1 + a_{22}x_2 + a_{23}x_3 &= b_2 , \\ a_{31}x_1 + a_{32}x_2 + a_{33}x_3 &= b_3 \end{aligned}\right\} . \tag{1.7}$$

If (x_1, x_2, x_3) are the Cartesian coordinates of a point in space, equations (1.7) represent three planes. As in the case of two equations, the solution may be written down formally using Cramer's Rule:

$$x_1 = D_1/D , \quad x_2 = D_2/D , \quad x_3 = D_3/D , \tag{1.8}$$

where D is the determinant of the coefficients given by:

$$D = a_{11} \begin{vmatrix} a_{22} & a_{23} \\ a_{32} & a_{33} \end{vmatrix} - a_{12} \begin{vmatrix} a_{21} & a_{23} \\ a_{31} & a_{33} \end{vmatrix} + a_{13} \begin{vmatrix} a_{21} & a_{22} \\ a_{31} & a_{32} \end{vmatrix} . \tag{1.9}$$

D is called a **third-order determinant** and is written:

$$D = \begin{vmatrix} a_{11} & a_{12} & a_{13} \\ a_{21} & a_{22} & a_{23} \\ a_{31} & a_{32} & a_{33} \end{vmatrix} . \tag{1.10}$$

D_1 is a third-order determinant derived from D by replacing the first column with the column of right-hand sides.

Thus

$$D_1 = \begin{vmatrix} b_1 & a_{12} & a_{13} \\ b_2 & a_{22} & a_{23} \\ b_3 & a_{32} & a_{33} \end{vmatrix} . \tag{1.11}$$

D_2 and D_3 are defined in a similar way.

If $D \neq 0$, the planes represented by equations (1.7) intersect in a single point and the equations have a unique solution. If $D = 0$, however, there are no fewer than seven possibilities to consider†.

1.1.4 Fourth and Higher Order Determinants

A set of four linear simultaneous equations in four unknowns gives rise to a solution which may be written down formally in terms of fourth-order determinants. The method is of no great practical use, however, because of the amount of computation involved.

The fourth-order determinant

$$D = \begin{vmatrix} a_{11} & a_{12} & a_{13} & a_{14} \\ a_{21} & a_{22} & a_{23} & a_{24} \\ a_{31} & a_{32} & a_{33} & a_{34} \\ a_{41} & a_{42} & a_{43} & a_{44} \end{vmatrix} \tag{1.12}$$

is defined in terms of third-order determinants by the relation

$$D = a_{11} \begin{vmatrix} a_{22} & a_{23} & a_{24} \\ a_{32} & a_{33} & a_{34} \\ a_{42} & a_{43} & a_{44} \end{vmatrix} - a_{12} \begin{vmatrix} a_{21} & a_{23} & a_{24} \\ a_{31} & a_{33} & a_{34} \\ a_{41} & a_{43} & a_{44} \end{vmatrix}$$
$$+ a_{13} \begin{vmatrix} a_{21} & a_{22} & a_{24} \\ a_{31} & a_{32} & a_{34} \\ a_{41} & a_{42} & a_{44} \end{vmatrix} - a_{14} \begin{vmatrix} a_{21} & a_{22} & a_{23} \\ a_{31} & a_{32} & \bar{a}_{33} \\ a_{41} & a_{42} & a_{43} \end{vmatrix} . \tag{1.13}$$

†For a full treatment of all possible cases the reader is referred to *Advanced Engineering Mathematics* by Erwin Kreyszig.

Third-order determinants have already been defined in terms of second-order determinants (1.9).

Determinants of higher order are defined in a similar way. Thus the nth-order determinant

$$D = \begin{vmatrix} a_{11} & a_{12} & a_{13} & \dots & a_{1n} \\ a_{21} & a_{22} & a_{23} & \dots & a_{2n} \\ \dots & \dots & \dots & \dots & \dots \\ a_{n1} & a_{n2} & a_{n3} & \dots & a_{nn} \end{vmatrix} \tag{1.14}$$

is defined by an expansion of $(n-1)$th-order determinants:

$$D = a_{11} \begin{vmatrix} a_{22} & a_{23} & \dots & a_{2n} \\ a_{32} & a_{33} & \dots & a_{3n} \\ \dots & \dots & \dots & \dots \\ a_{n2} & a_{n3} & \dots & a_{nn} \end{vmatrix} - a_{12} \begin{vmatrix} a_{21} & a_{23} & \dots & a_{2n} \\ a_{31} & a_{33} & \dots & a_{3n} \\ \dots & \dots & \dots & \dots \\ a_{n1} & a_{n3} & \dots & a_{nn} \end{vmatrix}$$

$$+ \dots + (-1)^{n-1} a_{1n} \begin{vmatrix} a_{21} & a_{22} & \dots & a_{2n-1} \\ a_{31} & a_{32} & \dots & a_{3n-1} \\ \dots & \dots & \dots & \dots \\ a_{n1} & a_{n2} & \dots & a_{nn-1} \end{vmatrix} . \tag{1.15}$$

Each $(n-1)$th-order determinant is then defined by an expansion of $(n-1)$ determinants of order $(n-2)$, and so on until second-order determinants – defined by (1.3) and (1.5) – are obtained.

1.1.5 Properties of Third-Order Determinants

We shall now list the most important properties of third-order determinants. They may all be verified by direct calculation. Similar properties hold for determinants of higher order.

(1) The value of a determinant is unaltered if the rows and columns are interchanged,

i.e.
$$\begin{vmatrix} a_{11} & a_{12} & a_{13} \\ a_{21} & a_{22} & a_{23} \\ a_{31} & a_{32} & a_{33} \end{vmatrix} = \begin{vmatrix} a_{11} & a_{21} & a_{31} \\ a_{12} & a_{22} & a_{32} \\ a_{13} & a_{23} & a_{33} \end{vmatrix} . \tag{1.16}$$

(2) If any two rows are interchanged, the sign of the determinant is reversed,

i.e.
$$\begin{vmatrix} a_{21} & a_{22} & a_{23} \\ a_{11} & a_{12} & a_{13} \\ a_{31} & a_{32} & a_{33} \end{vmatrix} = -\begin{vmatrix} a_{11} & a_{12} & a_{13} \\ a_{21} & a_{22} & a_{23} \\ a_{31} & a_{32} & a_{33} \end{vmatrix} . \tag{1.17}$$

In view of property (1) the same is true if two columns are interchanged. In this and all the properties which follow the word 'row' may be replaced by 'column'.

(3) If two rows of a determinant are identical, the value of the determinant is zero,

i.e.
$$\begin{vmatrix} a_{11} & a_{12} & a_{13} \\ a_{11} & a_{12} & a_{13} \\ a_{31} & a_{32} & a_{33} \end{vmatrix} \equiv 0 . \tag{1.18}$$

1.1.6 Minors and Cofactors

The second-order determinant obtained from a third-order determinant by omitting the ith row and the jth column is called the **minor** of the element, a_{ij}, which belongs to row i and column j. The minor multiplied by $(-1)^{i+j}$ is called the **cofactor** of a_{ij} and is written A_{ij}. Thus, for the determinant

$$D = \begin{vmatrix} a_{11} & a_{12} & a_{13} \\ a_{21} & a_{22} & a_{23} \\ a_{31} & a_{32} & a_{33} \end{vmatrix} ,$$

$$A_{11} = \begin{vmatrix} a_{22} & a_{23} \\ a_{32} & a_{33} \end{vmatrix} , \quad A_{12} = \begin{vmatrix} a_{23} & a_{21} \\ a_{33} & a_{31} \end{vmatrix} ,$$

$$A_{13} = \begin{vmatrix} a_{21} & a_{22} \\ a_{31} & a_{32} \end{vmatrix} . \tag{1.19}$$

Hence the definition of a third-order determinant may be written:

$$D = a_{11}A_{11} + a_{12}A_{12} + a_{13}A_{13} . \tag{1.20}$$

The expression (1.20) is called an **expansion by cofactors.**

1.1.7 Further Properties of Third-Order Determinants

(4) The expression (1.20) is an expansion by the elements of the first row, with their respective cofactors. A determinant may be expanded by the elements of any row.

Thus $$D = a_{21}A_{21} + a_{22}A_{22} + a_{23}A_{23} \; . \qquad (1.21)$$

(5) If all the elements of one row of a determinant are multiplied by a factor k, the value of the determinant is multiplied by k.

Thus $$\begin{vmatrix} a_{11} & ka_{12} & a_{13} \\ a_{21} & ka_{22} & a_{23} \\ a_{31} & ka_{32} & a_{33} \end{vmatrix} \equiv k \begin{vmatrix} a_{11} & a_{12} & a_{13} \\ a_{21} & a_{22} & a_{23} \\ a_{31} & a_{32} & a_{33} \end{vmatrix} . \qquad (1.22)$$

(6) If two rows of a determinant are proportional, the value of the determinant is zero,

i.e. $$\begin{vmatrix} a_{11} & a_{12} & a_{13} \\ ka_{11} & ka_{12} & ka_{13} \\ a_{31} & a_{32} & a_{33} \end{vmatrix} \equiv 0 \; . \qquad (1.23)$$

(7) If each element of a row can be expressed as a binomial, the determinant may be written as the sum of two determinants,

thus $$\begin{vmatrix} (a_{11} + b_{11}) & (a_{12} + b_{12}) & (a_{13} + b_{13}) \\ a_{21} & a_{22} & a_{23} \\ a_{31} & a_{32} & a_{33} \end{vmatrix}$$

$$= \begin{vmatrix} a_{11} & a_{12} & a_{13} \\ a_{21} & a_{22} & a_{23} \\ a_{31} & a_{32} & a_{33} \end{vmatrix} + \begin{vmatrix} b_{11} & b_{12} & b_{13} \\ a_{21} & a_{22} & a_{23} \\ a_{31} & a_{32} & a_{33} \end{vmatrix} . \qquad (1.24)$$

(8) The value of a determinant is unaltered if to the elements of any row are added a constant multiple of the elements of any other row.

Thus $$\begin{vmatrix} (a_{11} + ka_{21}) & (a_{12} + ka_{22}) & (a_{13} + ka_{23}) \\ a_{21} & a_{22} & a_{23} \\ a_{31} & a_{32} & a_{33} \end{vmatrix} = \begin{vmatrix} a_{11} & a_{12} & a_{13} \\ a_{21} & a_{22} & a_{23} \\ a_{31} & a_{32} & a_{33} \end{vmatrix} . \qquad (1.25)$$

Properties (2), (5) and (8) are particularly useful in the practical evaluation of determinants. A determinant of high order may be evaluated by reducing its order one unit at a time until it is a second-order determinant. Properties (5) and (8) are used to reduce all the elements except one in a particular row to zero. Property (2) is then used to bring the non-zero element into the top left-hand corner of the determinant.

Examples 1.1

(1) $$\begin{vmatrix} 1 & 1 & 1 \\ a & b & c \\ d & e & f \end{vmatrix} = \begin{vmatrix} 1 & 0 & 0 \\ a & b-a & c-b \\ d & e-d & f-e \end{vmatrix} = \begin{vmatrix} b-a & c-b \\ e-d & f-e \end{vmatrix}$$

$$= (b-a)(f-e) - (c-b)(e-d) \ .$$

(2) $$\begin{vmatrix} 1 & 1 & 1 \\ p & q & r \\ p^2 & q^2 & r^2 \end{vmatrix} = \begin{vmatrix} 1 & 0 & 0 \\ p & q-p & r-p \\ p^2 & q^2-p^2 & r^2-p^2 \end{vmatrix} = (q-p)(r-p)\begin{vmatrix} 1 & 1 \\ q+p & r+p \end{vmatrix}$$

$$= (q-r)(r-p)(p-q) \ .$$

(3) $$\begin{vmatrix} p & 2p & 3p \\ q-r & 2(r-p) & 3(p-q) \\ p & 2q & 3r \end{vmatrix} = 6p \begin{vmatrix} 1 & 1 & 1 \\ q-r & r-p & p-q \\ p & q & r \end{vmatrix}$$

$$= 6p \begin{vmatrix} 1 & 0 & 0 \\ q-r & 2r-p-q & p+r-2q \\ p & q-p & r-p \end{vmatrix} = 12p\,(p^2 + q^2 + r^2 - qr - rp - pq) \ .$$

Problems 1.1

Evaluate the following determinants:

(1) $\begin{vmatrix} 1 & 0 & 0 \\ 6 & 2 & 2 \\ 1 & 4 & 5 \end{vmatrix}$, (2) $\begin{vmatrix} 1 & 9 & 8 \\ 0 & 7 & 6 \\ 0 & 4 & 2 \end{vmatrix}$, (3) $\begin{vmatrix} 1 & 1 & 1 \\ 2 & 1 & 2 \\ 3 & 4 & -5 \end{vmatrix}$,

(4) $\begin{vmatrix} 1 & 2 & 1 \\ 1 & 3 & 4 \\ 1 & 1 & 5 \end{vmatrix}$, (5) $\begin{vmatrix} 1 & 2 & 4 \\ 7 & 6 & 16 \\ 3 & 2 & 12 \end{vmatrix}$, (6) $\begin{vmatrix} 1 & 2 & 4 \\ 4 & 8 & 16 \\ 6 & 9 & 5 \end{vmatrix}$,

(7) $\begin{vmatrix} 2 & 1 & 4 \\ 3 & 3 & -2 \\ 6 & 2 & 1 \end{vmatrix}$, (8) $\begin{vmatrix} 3 & 6 & 7 \\ 1 & 2 & 4 \\ 2 & 1 & 9 \end{vmatrix}$, (9) $\begin{vmatrix} 3 & 2 & 1 & 9 \\ 1 & 4 & 6 & -7 \\ 2 & -8 & 4 & -3 \\ -1 & 0 & 2 & 4 \end{vmatrix}$,

(10) $$\begin{vmatrix} (b^2+c^2) & b^2 & c^2 \\ a^2 & (c^2+a^2) & c^2 \\ a^2 & b^2 & (a^2+b^2) \end{vmatrix} .$$

1.2 MATRICES

1.2.1 Introduction

Consider the pair of simultaneous equations:

$$\left.\begin{aligned} a_{11}x_1 + a_{12}x_2 &= b_1 , \\ a_{21}x_1 + a_{22}x_2 &= b_2 \end{aligned}\right\} . \tag{1.26}$$

The array $\begin{bmatrix} a_{11} & a_{12} \\ a_{21} & a_{22} \end{bmatrix}$ is called the **matrix of coefficients**. The matrix, which has two rows and two columns, is called a **(2 × 2) square matrix**. Another matrix may be formed by including the right-hand sides. Thus the matrix $\begin{bmatrix} a_{11} & a_{12} & b_1 \\ a_{21} & a_{22} & b_2 \end{bmatrix}$ with two rows and three columns, is a (2 × 3) matrix. The numbers a_{11}, a_{12}, etc. are the **elements** of the matrix.

A matrix with only one column is called a **column matrix** or **column vector** and is written:

$$A = \begin{bmatrix} a_1 \\ a_2 \\ a_3 \\ \cdots\cdots \\ a_m \end{bmatrix} \tag{1.27}$$

A is an $(m \times 1)$ column vector. A matrix with only one row is called a **row matrix** or **row vector** and is written:

$$B = [b_1, b_2, b_3, \ldots, b_n] . \tag{1.28}$$

B is a $(1 \times n)$ row vector.

The numbers of rows and columns of a matrix constitute the **dimensions** of the matrix. The element which occupies the position in the ith row and the jth column is designated a_{ij}, and the matrix of such elements as (a_{ij}). A matrix with the same number of rows and columns is called a **square matrix**. The elements of such a matrix for which the row and column numbers are equal constitute the **leading diagonal**. Thus the leading diagonal of the matrix of coefficients of equations (1.26) consists of the elements a_{11} and a_{22}.

1.2.2 Matrix Addition

Matrix addition is defined only for matrices of the same dimension. They are then said to be **compatible for addition**.

Thus if
$$A = \begin{bmatrix} a_{11} & a_{12} & a_{13} \\ a_{21} & a_{22} & a_{23} \end{bmatrix}, \quad B = \begin{bmatrix} b_{11} & b_{12} & b_{13} \\ b_{21} & b_{22} & b_{23} \end{bmatrix}, \tag{1.29}$$

then
$$A + B = \begin{bmatrix} a_{11} + b_{11} & a_{12} + b_{12} & a_{13} + b_{13} \\ a_{21} + b_{21} & a_{22} + b_{22} & a_{23} + b_{23} \end{bmatrix}. \tag{1.30}$$

1.2.3 Multiplication by a Scalar

If A is the matrix defined by (1.29), the matrix kA, where k is a scalar, is given by:

$$kA = \begin{bmatrix} ka_{11} & ka_{12} & ka_{13} \\ ka_{21} & ka_{22} & ka_{23} \end{bmatrix}. \tag{1.31}$$

If $k = -1$ the resulting matrix is written $-A$.

1.2.4 Matrix Multiplication

Consider the pair of simultaneous equations:

$$\left.\begin{aligned} y_1 &= a_{11}x_1 + a_{12}x_2 \;, \\ y_2 &= a_{21}x_1 + a_{22}x_2 \end{aligned}\right\}. \tag{1.32}$$

The equations may be represented symbolically by the matrix equation

$$\begin{bmatrix} y_1 \\ y_2 \end{bmatrix} = \begin{bmatrix} a_{11} & a_{12} \\ a_{21} & a_{22} \end{bmatrix} \begin{bmatrix} x_1 \\ x_2 \end{bmatrix}, \tag{1.33}$$

or
$$\mathbf{y} = A\mathbf{x} \tag{1.34}$$

where $A\mathbf{x}$ is the **product** of the (2×2) matrix A and the (2×1) column matrix $\mathbf{x}$. Thus, from (1.32),

$$A\mathbf{x} = \begin{bmatrix} a_{11}x_1 + a_{12}x_2 \\ a_{21}x_1 + a_{22}x_2 \end{bmatrix}. \tag{1.35}$$

Suppose now that z_1, z_2, z_3 are given by

$$\left.\begin{aligned} z_1 &= b_{11}y_1 + b_{12}y_2 \ , \\ z_2 &= b_{21}y_1 + b_{22}y_2 \ , \\ z_3 &= b_{31}y_1 + b_{32}y_2 \end{aligned}\right\} . \tag{1.36}$$

Equations (1.36) may be represented by the matrix equation

$$\begin{bmatrix} z_1 \\ z_2 \\ z_3 \end{bmatrix} = \begin{bmatrix} b_{11} & b_{12} \\ b_{21} & b_{22} \\ b_{31} & b_{32} \end{bmatrix} \begin{bmatrix} y_1 \\ y_2 \end{bmatrix} , \tag{1.37}$$

or $$\mathbf{z} = B\mathbf{y} \ , \tag{1.38}$$

where $B\mathbf{y}$ is the product of the (3×2) matrix B and the (2×1) column matrix $\mathbf{y}$. From (1.36) we see that $B\mathbf{y}$ is the matrix

$$\begin{bmatrix} b_{11}y_1 + b_{12}y_2 \\ b_{21}y_1 + b_{22}y_2 \\ b_{31}y_1 + b_{32}y_2 \end{bmatrix} . \tag{1.39}$$

We notice that the product of a (3×2) matrix and a (2×1) matrix is a (3×1) matrix.

Finally, let us attempt to write the (3×1) column matrix $\mathbf{z}$ in terms of the (2×1) column matrix $\mathbf{x}$. Symbolically we may do this by noting that

$$\mathbf{z} = B\mathbf{y} = B(A\mathbf{x}) \ . \tag{1.40}$$

Substituting the values of y_1, y_2 obtained from (1.32) into equations (1.36), we have:

$$\left.\begin{aligned} z_1 &= b_{11}(a_{11}x_1 + a_{12}x_2) + b_{12}(a_{21}x_1 + a_{22}x_2) \ , \\ z_2 &= b_{21}(a_{11}x_1 + a_{12}x_2) + b_{22}(a_{21}x_1 + a_{22}x_2) \ , \\ z_3 &= b_{31}(a_{11}x_1 + a_{12}x_2) + b_{32}(a_{21}x_1 + a_{22}x_2) \end{aligned}\right\} . \tag{1.41}$$

Equations (1.41) may be re-written:

$$\left.\begin{aligned} z_1 &= (b_{11}a_{11} + b_{12}a_{21})x_1 + (b_{11}a_{12} + b_{12}a_{22})x_2 \ , \\ z_2 &= (b_{21}a_{11} + b_{22}a_{21})x_1 + (b_{21}a_{12} + b_{22}a_{22})x_2 \ , \\ z_3 &= (b_{31}a_{11} + b_{32}a_{21})x_1 + (b_{31}a_{12} + b_{32}a_{22})x_2 \ , \end{aligned}\right\} \tag{1.42}$$

which correspond to the symbolic form

$$\mathbf{z} = (BA)\mathbf{x} \ . \tag{1.43}$$

Thus the product matrix BA is the (3×2) matrix

$$\begin{bmatrix} b_{11}a_{11} + b_{12}a_{21} & b_{11}a_{12} + b_{12}a_{22} \\ b_{21}a_{11} + b_{22}a_{21} & b_{21}a_{12} + b_{22}a_{22} \\ b_{31}a_{11} + b_{32}a_{21} & b_{31}a_{12} + b_{32}a_{22} \end{bmatrix} \tag{1.44}$$

The (i,k) element of the product matrix BA is the scalar product (see section 1.5.2) of the ith row of B and the kth column of A. For this reason the **pre-multiplier** (B) must have the same number of columns as the number of rows in the **post-multiplier** (A). The matrices are then said to be **compatible for multiplication.**

1.2.5 Properties of Matrix Multiplication

A comparison of (1.40) and (1.43) shows that $B(A\mathbf{x}) = (BA)\mathbf{x}$. This a particular case of the general result that **matrix multiplication is associative**. Thus, provided the matrices are compatible for multiplication,

$$(AB)C = A(BC) \tag{1.45}$$

for any three matrices A, B and C.

Matrix multiplication is also **distributive**, i.e.

$$A(B + C) = AB + AC \ ,$$

$$(A + B)C = AC + BC \ . \tag{1.46}$$

Matrix multiplication is not, however, commutative. Thus even when AB and BA exist

$$AB \neq BA \text{ in general} \ .$$

Every $(n \times n)$ square matrix A has a determinant which we shall write $|A|$. If A and B are square matrices of the same order, then

$$|AB| = |A|\,|B| \ . \tag{1.47}$$

This result may be extended to any number of matrices.

1.2.6 Zero and Unit Matrices

A matrix, all of whose elements are zero, is called a **zero matrix** or **null matrix**. It has the property that if O is a zero matrix compatible with a matrix A for addition,

then $$A + O = A \ . \tag{1.48}$$

If A is a matrix of dimensions $(m \times n)$ and O is a null matrix of dimensions $(n \times p)$, then

$$AO = P \,, \tag{1.49}$$

where P is a null matrix of dimensions $(m \times p)$.

A square matrix with ones on the leading diagonal and zeros everywhere else is called a **unit matrix**. Thus the (3×3) unit matrix is

$$\begin{bmatrix} 1 & 0 & 0 \\ 0 & 1 & 0 \\ 0 & 0 & 1 \end{bmatrix} . \tag{1.50}$$

A unit matrix has the property that if A is a unit matrix, then for any other matrix B,

$$AB = B, \quad BA = B \,, \tag{1.51}$$

provided only that the matrices are compatible for multiplication.

1.2.7 The Transpose Matrix, Symmetric and Skew-Symmetric Matrices

To any matrix there corresponds a transpose matrix, formed by interchanging the rows and columns. Thus if A is the (2×3) matrix $\begin{bmatrix} 6 & 1 & 2 \\ 3 & 4 & 5 \end{bmatrix}$, the transpose matrix A' is the (3×2) matrix $\begin{bmatrix} 6 & 3 \\ 1 & 4 \\ 2 & 5 \end{bmatrix}$.

A square matrix which is equal to its transpose is known as a **symmetric matrix**. Thus the matrix $\begin{bmatrix} 6 & 5 & 4 \\ 5 & 3 & 2 \\ 4 & 2 & 1 \end{bmatrix}$ is a symmetric matrix. A (3×3) symmetric matrix has only six independent elements rather than the nine independent elements in the general (3×3) matrix. A (4×4) symmetric matrix has only ten independent elements, and so on (see Problems 1.2, No. 5).

A square matrix which is equal to the negative of its transpose is known as **skew-symmetric matrix**. Thus the matrix $\begin{bmatrix} 0 & 3 & 2 \\ -3 & 0 & 1 \\ -2 & -1 & 0 \end{bmatrix}$ is a skew-symmetric matrix.

A (3×3) skew-symmetric matrix has zeros on the leading diagonal, and only three independent elements (see Problems 1.2, No. 5).

The following two theorems are easily proved.

Theorem 1.2.7(i) The transpose of the sum (or difference) of two matrices is equal to the sum (or difference) of the transposes, i.e. $(A \pm B)' = A' \pm B'$.

Theorem 1.2.7 (ii) Any square matrix may be expressed *uniquely* as the sum of a symmetric and a skew-symmetric matrix. Thus $A \equiv \frac{1}{2}(A + A') + \frac{1}{2}(A - A')$ where $\frac{1}{2}(A + A')$ is symmetric and $\frac{1}{2}(A - A')$ is skew-symmetric.

1.2.8 The Inverse Matrix

Suppose A is an nth order square matrix and $\mathbf{x}$, $\mathbf{y}$ are $(n \times 1)$ column vectors such that

$$A\mathbf{x} = \mathbf{y} \,. \tag{1.52}$$

Equation (1.52) represents n linear simultaneous equations in the unknowns $x_1, x_2, \ldots, x_n$, the elements of $\mathbf{x}$. Suppose the equations have a unique solution

$$\mathbf{x} = B\mathbf{y} \,. \tag{1.53}$$

B is said to be the **inverse** of the matrix A and is written A^{-1}.

Thus
$$\mathbf{x} = A^{-1}\mathbf{y} \,, \tag{1.54}$$

$$A\mathbf{x} = AA^{-1}\mathbf{y} \,, \tag{1.55}$$

or
$$AA^{-1}\mathbf{y} = \mathbf{y} \,. \tag{1.56}$$

Hence AA^{-1} is a unit matrix.

In practice A^{-1} may be found by solving equations (1.52), or from the formula:

$$A^{-1} = \frac{(\text{Adj } A)'}{|A|} \,, \tag{1.57}$$

where $|A|$ is the determinant of A and Adj A is the **adjugate matrix** whose (i,j) – element is the cofactor of a_{ij}, the (i,j) element of A,

i.e.
$$(\text{Adj } A)_{ij} = A_{ij} \,. \tag{1.58}$$

Clearly A^{-1} exists if and only if $|A|$ is non-zero. This is precisely the condition for equations (1.52) to have a unique solution. A matrix whose determinant is zero, and which therefore has no inverse, is said to be **singular.**

1.2.9 The Rank of a Matrix

A matrix formed by omitting any number of rows and/or columns from a matrix A is said to be a **submatrix** of A. As a matter of convention A is also said to be a submatrix of itself.

The **rank** R of a matrix A (rectangular or square) is defined to be the dimension of the largest non-singular square submatrix of A. It follows therefore that the rank of a rectangular matrix cannot exceed the smaller of the two dimensions, also that the rank of a square matrix is less than its dimension if and only if the matrix is singular. The only matrix (of any dimension) to have zero rank is the null matrix.

The concept of rank is closely connected with that of linear dependence to be discussed in section 1.7.4. It is vital to the treatment of the eigenvector problem which is dealt with in sections 2.9.1 *et seq.*

Examples 1.2

Let $A = \begin{bmatrix} 3 & 3 & 1 \\ 1 & -2 & 4 \end{bmatrix}$, $B = \begin{bmatrix} 2 & 2 \\ 4 & -1 \end{bmatrix}$, $C = \begin{bmatrix} 6 & 0 \\ -1 & -5 \\ 2 & 3 \end{bmatrix}$, $D = \begin{bmatrix} 1 & 7 & 2 \\ 8 & 2 & 4 \end{bmatrix}$.

(1) $A + D = \begin{bmatrix} 4 & 10 & 3 \\ 9 & 0 & 8 \end{bmatrix}$; $A - D = \begin{bmatrix} 2 & -4 & -1 \\ -7 & -4 & 0 \end{bmatrix}$.

$A \pm B, A \pm C$ etc. are not possible additions because the matrices are not compatible for addition, i.e. they have differing dimensions.

(2) $BA = \begin{bmatrix} 2 & 2 \\ 4 & -1 \end{bmatrix}\begin{bmatrix} 3 & 3 & 1 \\ 1 & -2 & 4 \end{bmatrix} = \begin{bmatrix} 8 & 2 & 10 \\ 11 & 14 & 0 \end{bmatrix}$;

$$CB = \begin{bmatrix} 6 & 0 \\ -1 & -5 \\ 2 & 3 \end{bmatrix}\begin{bmatrix} 2 & 2 \\ 4 & -1 \end{bmatrix} = \begin{bmatrix} 12 & 12 \\ -22 & 3 \\ 16 & 1 \end{bmatrix};$$

$$AC = \begin{bmatrix} 3 & 3 & 1 \\ 1 & -2 & 4 \end{bmatrix}\begin{bmatrix} 6 & 0 \\ -1 & -5 \\ 2 & 3 \end{bmatrix} = \begin{bmatrix} 17 & -12 \\ 16 & 22 \end{bmatrix};$$

$$CA = \begin{bmatrix} 6 & 0 \\ -1 & -5 \\ 2 & 3 \end{bmatrix}\begin{bmatrix} 3 & 3 & 1 \\ 1 & -2 & 4 \end{bmatrix} = \begin{bmatrix} 18 & 18 & 6 \\ -8 & 7 & -21 \\ 9 & 0 & 14 \end{bmatrix}.$$

(3) The matrix A has one (2×3) submatrix, A, three (2×2) submatrices $\begin{bmatrix} 3 & 1 \\ -2 & 4 \end{bmatrix}, \begin{bmatrix} 3 & 1 \\ 1 & 4 \end{bmatrix}, \begin{bmatrix} 3 & 3 \\ 1 & -2 \end{bmatrix}$, also two (1×3) submatrices, three (2×1) submatrices, six (1×2) submatrices (of which two are equal), and six (1×1) submatrices (of which there are two equal pairs).

(4) The rank of matrices A,B,C,D is 2 but the rank of matrices $E = \begin{bmatrix} -1 & 3 & -4 \\ -2 & 6 & -8 \end{bmatrix}$

and $F = \begin{bmatrix} 0 & 1 \\ 0 & 7 \end{bmatrix}$ is 1.

(5) The rank of a matrix is equal to that of its transpose.

Problems 1.2

(1) Express each of the sets of equations as a single matrix equation:

$$\text{(i)}\ \begin{aligned} a_1x + b_1y + c_1z &= d_1 \,, \\ a_2x + b_2y + c_2z &= d_2 \,, \\ a_3x + b_3y + c_3z &= d_3 \,, \end{aligned} \qquad \text{(ii)}\ \begin{aligned} x + 2y - z + 4 &= 0, \\ 2x - y + 3z &= 0 \,, \\ x + 2y - 3 &= 0 \,. \end{aligned}$$

(2) Let $A = \begin{bmatrix} 2 & 1 \\ 1 & 2 \end{bmatrix}$, $B = \begin{bmatrix} 3 & 2 \\ 1 & 4 \\ 0 & 2 \end{bmatrix}$, $C = \begin{bmatrix} -1 & 4 & 1 \\ 3 & -6 & -7 \end{bmatrix}$,

$$D = \begin{bmatrix} 1 & -1 & 2 \\ 0 & -4 & 3 \end{bmatrix}.$$

Write down the dimensions of the matrices, A, B, C, D. Find if possible (i) $A + B$; (ii) $B + C$; (iii) $C + D$; (iv) $C - D$; (v) AB; (vi) BC; (vii) CD; (viii) CB; (ix) BA.

(3) Using the matrices defined in Problem No. 2 demonstrate (i) that the distributive law holds, by computing AC, AD and $A(C + D)$; (ii) that the associative law holds by computing $B(AC)$ and $(BA)C$; (iii) that the commutative law does not hold by computing BD and DB.

(4) The equations

$$\left.\begin{aligned} x' &= x \cos\theta + y \sin\theta \,, \\ y' &= -x \sin\theta + y \cos\theta \,, \\ z' &= z \,, \end{aligned}\right\} \left.\begin{aligned} x'' &= x' \,, \\ y'' &= y' \cos\phi + z' \sin\phi \,, \\ z'' &= -y' \sin\phi + z' \cos\phi \,, \end{aligned}\right\}$$

represent a rotation of axes (x,y,z) by θ about the z-axis to new axes (x',y',z'), followed by a rotation of the (x',y',z') axes by ϕ about the x' axis. By multiplying two matrices express (x'',y'',z'') directly in terms of (x,y,z).

(5) Show that there are $\frac{1}{2}n(n + 1)$ independent elements in an $(n \times n)$ symmetric matrix and $\frac{1}{2}n(n - 1)$ independent elements in an $(n \times n)$ skew-symmetric matrix.

(6) Construct two (2×2) non-zero matrices A and B to demonstrate that the relation $AB = O$, where O is the (2×2) zero matrix, does *not* imply that either A or B is a zero matrix. If $A = \begin{bmatrix} a_{11} & a_{12} \\ a_{21} & a_{22} \end{bmatrix}$, $B = \begin{bmatrix} b_{11} & b_{12} \\ b_{21} & b_{22} \end{bmatrix}$, write down necessary and sufficient conditions for AB to be the zero matrix.

(7) Express each of the square matrices as the sum of a symmetric and a skew-symmetric matrix:

$$\text{(i)} \begin{bmatrix} 2 & 1 \\ 0 & -3 \end{bmatrix}; \quad \text{(ii)} \begin{bmatrix} 3 & 2 & -1 \\ 6 & -3 & -2 \\ -4 & 5 & 0 \end{bmatrix}.$$

(8) How many (2×2) submatrices has the matrix

$$A = \begin{bmatrix} 1 & 2 & 4 \\ 3 & 8 & 7 \\ 4 & 10 & 11 \end{bmatrix}$$? Determine the rank of A.

(9) Determine the ranks of the matrices:

$$\text{(i)} \begin{bmatrix} 2 & 0 & 8 \\ 3 & 1 & 4 \end{bmatrix}; \quad \text{(ii)} \begin{bmatrix} 0 & 1 & 4 \\ 0 & 2 & 7 \end{bmatrix}; \quad \text{(iii)} \begin{bmatrix} 0 & 1 & -4 \\ 0 & 2 & -8 \end{bmatrix};$$

$$\text{(iv)} \begin{bmatrix} 1 & 6 & 2 \\ 3 & 0 & 9 \\ 1 & 4 & 4 \end{bmatrix}; \quad \text{(v)} \begin{bmatrix} 2 & 0 & 8 \\ 3 & 1 & 4 \\ 1 & -1 & 12 \end{bmatrix}; \quad \text{(vi)} \begin{bmatrix} 1 & 2 & 4 \\ 3 & 8 & 14 \\ 4 & 10 & 18 \end{bmatrix};$$

$$\text{(vii)} \begin{bmatrix} -3 & 2 & 1 \\ 4 & -3 & 2 \\ 1 & 1 & 1 \end{bmatrix}; \quad \text{(vii)} \begin{bmatrix} 2 & 2 & 4 \\ 3 & 3 & 6 \\ 1 & 1 & 2 \end{bmatrix}.$$

1.3 SCALARS

1.3.1 Introduction

A **scalar** is a quantity which may be represented by a single number. Examples of scalar quantities are temperature, pressure, density, mass, and the distance between two points. Speed is also a scalar quantity but velocity, which depends on direction, is not.

A scalar quantity is independent of the units in which it is specified. Consider, for example, the boiling point of water at a given point in space. It will depend on certain physical parameters, such as pressure, but it is the same temperature whether measured on the Celsius ('Centigrade') or Fahrenheit scale.

Just as the temperature at a given point depends in no way on the choice of scale, it is also independent of the coordinate system (Cartesian, spherical polar etc.) used to specify its location in space. It is thus invariant with respect to coordinate transformations. This book is concerned with a class of invariants called **Cartesian tensors**, of which the simplest example is a scalar. In the language of tensors a scalar is a **tensor of order zero.**

1.3.2 Scalar Fields

Suppose a scalar ϕ is defined at every point P within a domain D by the relation $\phi = \phi(\mathrm{P})$. If, for every point P, there is a unique value $\phi(\mathrm{P})$, then $\phi = \phi(\mathrm{P})$ defines a scalar function of position called a **scalar field**.

The temperature, pressure and density distributions throughout a given substance are examples of scalar fields.

Example 1.3

(1) Let (x_1,x_2,x_3) be Cartesian coordinates and let $r^2 = x_1^2 + x_2^2 + x_3^2$. Consider the scalar field $\phi = 3r^2 + 2r$. Show that the derived scalar field $\partial\phi/\partial x_1$ has the same value at every point whether calculated using Cartesian or spherical polar coordinates.

We have in spherical polar coordinates:

$$\phi = 3r^2 + 2r \ ,$$

then $$\frac{\partial\phi}{\partial x_1} = \frac{\mathrm{d}\phi}{\mathrm{d}r}\frac{\partial r}{\partial x_1} = (6r+2)\frac{x_1}{r} \ .$$

In Cartesian coordinates:

$$\phi = 3(x_1^2 + x_2^2 + x_3^2) + 2\sqrt{(x_1^2 + x_2^2 + x_3^2)} \ ,$$

therefore $$\frac{\partial\phi}{\partial x_1} = 6x_1 + \frac{2x_1}{\sqrt{(x_1^2 + x_2^2 + x_3^2)}} = (6r+2)\frac{x_1}{r} \ .$$

Problems 1.3

(1) Let $\phi(x_1,x_2,x_3) = 3x_1^2 - 2x_2$ define a scalar field. Find the value of ϕ at (i) (0,0,0); (ii) (1,1,1); (iii) (1,1,0).

(2) Let $\phi(r) = 3r^2 - 2r$ define a scalar field. Find the value of ϕ at (i) (0,0,1); (ii) (1,−2,2); (iii) (2,0,−3).

(3) The pressure at a distance r from the centre of a spherically symmetrical pressure distribution is given by $p = k/r^3$ where k is a constant. Show that the pressure gradient $\mathrm{d}p/\mathrm{d}r$ has the same value at a given point whether calculated using Cartesian or spherical polar coordinates.

1.4 VECTORS

1.4.1 Introduction

Whereas scalars can be represented by a single number, **vectors** require as many numbers as the number of dimensions of the space in which they are defined. Displacement, velocity, acceleration, and force are all vector quantities. With each is associated a direction as well as a magnitude.

In this book we shall assume that all vectors are defined in the **real Euclidean space of 3 dimensions**. This is the space of the physical world, denoted by R^3. In R^3 vectors may be represented by three numbers, called **components**.

R^3 is an example of a **real vector space**. Elements of this space are called **points** and, once a Cartesian coordinate system has been chosen, with every point is associated a **position vector**. The position vectors are represented by ordered triples (x_1,x_2,x_3) where x_1,x_2,x_3 are the Cartesian coordinates of the point. Before proceeding further we shall give a formal definition of a real vector space.

1.4.2 Definition of a Real Vector Space

The elements of a real vector space have to satisfy certain axioms, four for addition and four for multiplication by a real scalar. The axioms, which are given below, are, of course, true for R^3 which was discussed in section 1.4.1.

Axioms for Addition

Let $\mathscr{V}$ be a real vector space and **x**, **y**, **z** any three elements of that space. Then there exists a unique vector **x** + **y** in $\mathscr{V}$ called the **sum** of **x** and **y**. It is required that

(1) Addition be **associative**,

i.e. $$\mathbf{x}+(\mathbf{y}+\mathbf{z}) = (\mathbf{x}+\mathbf{y})+\mathbf{z} \ . \tag{1.59}$$

(2) Addition be **commutative**,

i.e. $$\mathbf{x}+\mathbf{y} = \mathbf{y}+\mathbf{x} \ , \tag{1.60}$$

(3) There exists a vector **0** called the **zero vector**, such that $\mathbf{x}+\mathbf{0} = \mathbf{x}$ for all **x** in $\mathscr{V}$ (1.61)

(4) For each **x** in $\mathscr{V}$ there exists a vector $(-\mathbf{x})$ in $\mathscr{V}$ such that $\mathbf{x}+(-\mathbf{x}) = \mathbf{0}$. (1.62)

Axioms for Scalar Multiplication

For any vectors **x**, **y** in $\mathscr{V}$ and any real numbers α,β there exists a unique vector $\alpha\mathbf{x}$ in $\mathscr{V}$ called the **product** of α and **x**. It is required that

(5) $$\alpha(\mathbf{x}+\mathbf{y}) = \alpha\mathbf{x}+\alpha\mathbf{y} \ , \tag{1.63}$$

(6) $$(\alpha+\beta)\mathbf{x} = \alpha\mathbf{x}+\beta\mathbf{x} \ , \tag{1.64}$$

(7) $$(\alpha\beta)\mathbf{x} = \alpha(\beta\mathbf{x}) \ , \tag{1.65}$$

(8) $$1.\mathbf{x} = \mathbf{x} \ . \tag{1.66}$$

Examples 1.4

(1) The Cartesian plane R^2 is a real vector space of two dimensions. Elements of the vector space are points, which may be represented by ordered pairs (x_1,x_2). The zero vector is the position vector of the origin. The point whose position vector is $-\mathbf{x}$ is the reflection in the origin of the point whose position vector is **x**.

(2) The set of polynomials of degree n in x, with real coefficients, forms a real vector space of dimension n. Elements of the space are polynomials,

$$\sum_{r=0}^{n} a_r x^r = a_0 + a_1 x + a_2 x^2 + \ldots + a_n x^n \ ,$$

which may be represented by ordered **(n + 1)-tuples** $(a_0, a_1, \ldots a_n)$.

Problems 1.4

(1) Show that axioms (1) – (8) are satisfied by the vector spaces mentioned in Examples (1) and (2).

(2) Show that the set of complex numbers $z = x + iy$ forms a real vector space. Can you suggest a definition of a **complex vector space**?

1.5 VECTOR PROPERTIES

1.5.1 Properties of Vectors in R^3

We shall now set down the properties of vectors in R^3 which are required for the remaining chapters of this book.

The sum of two vectors $\mathbf{x} = (x_1, x_2, x_3)$ and $\mathbf{y} = (y_1, y_2, y_3)$ is defined by:

$$\mathbf{x} + \mathbf{y} = (x_1 + y_1, x_2 + y_2, x_3 + y_3) \ . \tag{1.67}$$

This definition satisfies axioms (1) and (2) of section 1.4.2.

The **zero vector** or **null vector** is a vector all of whose components are zero.

Thus $$\mathbf{0} = (0,0,0) \ . \tag{1.68}$$

This definition satisfies axiom (3).

To every vector $\mathbf{x} = (x_1, x_2, x_3)$ there corresponds a vector $-\mathbf{x} = (-x_1, -x_2, -x_3)$. This definition satisfies axiom (4).

For every vector $\mathbf{x} = (x_1, x_2, x_3)$ there exists for any scalar number k a vector $k\mathbf{x}$ with components (kx_1, kx_2, kx_3). This definition satisfies axioms (5) – (8).

1.5.2 The Scalar Product of Two Vectors

Let $\mathbf{x}$ and $\mathbf{y}$ be vectors with components (x_1, x_2, x_3) and (y_1, y_2, y_3) respectively. The **scalar product** of $\mathbf{x}$ and $\mathbf{y}$ is defined by

$$\mathbf{x}.\mathbf{y} = x_1y_1 + x_2y_2 + x_3y_3 \ . \tag{1.69}$$

The name of the product emphasizes the fact that it is a scalar, independent of choice of Cartesian axes. This fact, which is by no means obvious from the algebraic definition (1.69), will follow at once from the alternative definition of $\mathbf{x}.\mathbf{y}$ to be given in section 1.6.4. The invariance of the algebraic form (1.69) will be proved conclusively in Examples 1.10.

The scalar product is commutative and distributive,

i.e. $$\mathbf{x}.\mathbf{y} = \mathbf{y}.\mathbf{x} \ , \tag{1.70}$$

and $$\mathbf{x}.(\mathbf{y} + \mathbf{z}) = \mathbf{x}.\mathbf{y} + \mathbf{x}.\mathbf{z} \ , \tag{1.71}$$

where $\mathbf{z}$ is any third vector in R^3. These results follow at once from the definition (1.69). There is no associative law for the scalar product of two vectors. Two vectors $\mathbf{x}$, $\mathbf{y}$ which satisfy the relation $\mathbf{x}.\mathbf{y} = 0$ are said to be **orthogonal**.

The **modulus** or **length** of a vector is given by:

$$x = |\mathbf{x}| = +\sqrt{(\mathbf{x}.\mathbf{x})} = +\sqrt{(x_1^2 + x_2^2 + x_3^2)} \; . \tag{1.72}$$

The modulus is a scalar and is essentially positive. The scalar product of a vector with itself is generally written $\mathbf{x}^2$ rather than $\mathbf{x}.\mathbf{x}$.

Hence, since $x = \sqrt{(\mathbf{x}.\mathbf{x})} = \sqrt{\mathbf{x}^2}$,

therefore
$$x^2 = \mathbf{x}^2 \; . \tag{1.73}$$

A **unit vector** is a vector of unit modulus. We shall denote the unit vectors along the three Cartesian coordinate axes by $\mathbf{e}_1, \mathbf{e}_2, \mathbf{e}_3$.

Thus
$$\mathbf{e}_1 = (1,0,0), \; \mathbf{e}_2 = (0,1,0), \; \mathbf{e}_3 = (0,0,1) \; . \tag{1.74}$$

The set of vectors $\mathbf{e}_1, \mathbf{e}_2, \mathbf{e}_3$ is said to form an **orthogonal unit triad** of vectors. Such a set of vectors is also known as as basis, but basis is a more general term applied to any set of n vectors in R^n which are **linearly independent** – a term to be defined in section 1.7.4.

The introduction of the unit vectors $\mathbf{e}_1, \mathbf{e}_2, \mathbf{e}_3$ provides an alternative representation of a vector.

Thus
$$\begin{aligned} \mathbf{x} &= (x_1, x_2, x_3) \; , \\ &= (x_1, 0, 0) + (0, x_2, 0) + (0, 0, x_3) \; , \\ &= x_1\mathbf{e}_1 + x_2\mathbf{e}_2 + x_3\mathbf{e}_3 \; , \end{aligned}$$

or
$$\mathbf{x} = \sum_{i=1}^{3} x_i\mathbf{e}_i \; . \tag{1.75}$$

Note also that

$$\mathbf{x}.\mathbf{e}_1 = x_1, \qquad \mathbf{x}.\mathbf{e}_2 = x_2, \qquad \mathbf{x}.\mathbf{e}_3 = x_3 \; ,$$

or
$$\mathbf{x}.\mathbf{e}_i = x_i; \qquad i = 1, 2, 3. \tag{1.76}$$

Hence
$$\mathbf{x} = (\mathbf{x}.\mathbf{e}_1)\mathbf{e}_1 + (\mathbf{x}.\mathbf{e}_2)\mathbf{e}_2 + (\mathbf{x}.\mathbf{e}_3)\mathbf{e}_3 \; ,$$

or
$$\mathbf{x} = \sum_{i=1}^{3} (\mathbf{x}.\mathbf{e}_i)\mathbf{e}_i \; . \tag{1.77}$$

The shorthand notation used in (1.75) and (1.77) will be employed extensively throughout the remainder of the book, although we shall soon drop the summation symbol Σ (see section 1.9.1).

1.5.3 The Vector Product of Two Vectors

The vector product of two vectors $\mathbf{x}, \mathbf{y}$ with components (x_1, x_2, x_3) and (y_1, y_2, y_3) respectively is a vector defined by:

$$\mathbf{x} \times \mathbf{y} = (x_2y_3 - x_3y_2, x_3y_1 - x_1y_3, x_1y_2 - x_2y_1) \; . \tag{1.78}$$

This expression may be represented formally by the determinant

$$\begin{vmatrix} \mathbf{e}_1 & \mathbf{e}_2 & \mathbf{e}_3 \\ x_1 & x_2 & x_3 \\ y_1 & y_2 & y_3 \end{vmatrix} . \tag{1.79}$$

The determinant (1.79) differs from the third-order determinants defined in section 1.1.3 in that there are vectors in the first row. It may be expanded according to (1.9) to give a vector which is essentially that given in (1.78).

The vector product is distributive but not commutative,

i.e. $$\mathbf{x} \times (\mathbf{y} + \mathbf{z}) = (\mathbf{x} \times \mathbf{y}) + (\mathbf{x} \times \mathbf{z}) , \tag{1.80}$$

where $\mathbf{z}$ is any third vector in R^3. In place of the commutative law we have:

$$\mathbf{x} \times \mathbf{y} = -\mathbf{y} \times \mathbf{x} . \tag{1.81}$$

Again these results follow at once from the definition. There is no associative law for the vector product of two vectors.

Example 1.5

(1) Let $\mathbf{e}_1, \mathbf{e}_2, \mathbf{e}_3$ be unit vectors in the directions of the three coordinate axes as defined in section 1.5.2.

Then (i) $\mathbf{e}_2.\mathbf{e}_3 = \mathbf{e}_3.\mathbf{e}_1 = \mathbf{e}_1.\mathbf{e}_2 = 0$;

(ii) $\mathbf{e}_1^2 = \mathbf{e}_2^2 = \mathbf{e}_3^2 = 1$;

(iii) $\mathbf{e}_2 \times \mathbf{e}_3 = \mathbf{e}_1, \ \mathbf{e}_3 \times \mathbf{e}_2 = -\mathbf{e}_1$,

$\mathbf{e}_3 \times \mathbf{e}_1 = \mathbf{e}_2, \ \mathbf{e}_1 \times \mathbf{e}_3 = -\mathbf{e}_2$,

$\mathbf{e}_1 \times \mathbf{e}_2 = \mathbf{e}_3, \ \mathbf{e}_2 \times \mathbf{e}_1 = -\mathbf{e}_3$;

(iv) $\mathbf{e}_1 \times \mathbf{e}_1 = \mathbf{e}_2 \times \mathbf{e}_2 = \mathbf{e}_3 \times \mathbf{e}_3 = \mathbf{0}$.

Problems 1.5

(1) Let $\mathbf{x} = (1,0,1), \mathbf{y} = (2,3,-2), \mathbf{z} = (4,6,-4)$.

Calculate (i) $\mathbf{e}_1.\mathbf{x}$, $\mathbf{e}_2.\mathbf{y}$, $\mathbf{e}_3.\mathbf{z}$;

(ii) $\mathbf{e}_1 \times \mathbf{x}$, $\mathbf{e}_2 \times \mathbf{y}$, $\mathbf{e}_3 \times \mathbf{z}$.

(iii) By calculating $\mathbf{x.z}$, $\mathbf{y.z}$, $(\mathbf{x} + \mathbf{y}).\mathbf{z}$ etc. verify the truth of the distributive law for scalar and vector products. Verify also the commutative law for scalar products by computing $\mathbf{y.z}$ and $\mathbf{z.y}$, and show that $\mathbf{x} \times \mathbf{y} = -\mathbf{y} \times \mathbf{x}$.

(2) By computing a suitable scalar product, show that a zero scalar product does not imply that either of the vectors in the product is zero. Prove a similar result for vector products.

(3) Find y and z the moduli of **y** and **z** respectively. Verify that $(1/y)$**y** and $(1/z)$**z** are the same unit vector.
(4) Prove that, for any vector **a**, if **b** = $\mathbf{e}_1$ X **a** then **b.a** = 0. Prove also that, for any vectors **a** and **b**, if **c** = **a** X **b** then **c.a** = **c.b** = 0.

1.6 THE GEOMETRICAL REPRESENTATION OF VECTORS

1.6.1 Directed Line Segments

We have seen that the position vector **x** of a point A may be represented by an ordered triple (x_1, x_2, x_3) the elements of which are the coordinates of the point A. Alternatively the position vector may be represented by a **directed line segment**, namely the line drawn from the origin to the point, with an arrow directed towards the point.

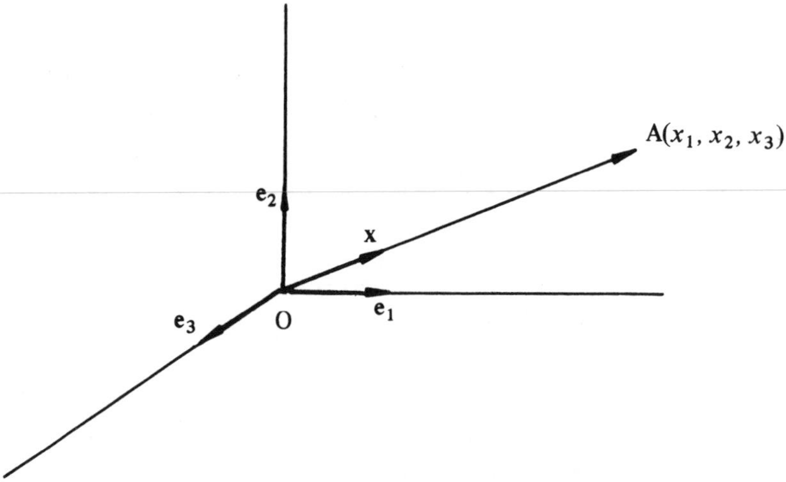

Fig. 1.1

In Fig. 1.1 the position vector of A is represented by the directed line segment OA, while unit vectors $\mathbf{e}_1, \mathbf{e}_2, \mathbf{e}_3$ along the coordinate axes are represented by directed line segments of unit length.

We shall now see how to add vectors represented in this way and how the definitions of scalar and vector products, framed in terms of directed line segments, are equivalent to the definitions already given.

1.6.2 The Parallelogram Law of Addition

Let **x** and **y** be the position vectors of the points A and B respectively. Let C be a third point such that BC is equal and parallel to OA. The directed line segment

BC is said to represent the **free vector x** and is written $\overline{BC}$. It is equal in magnitude and direction to the position vector **x**, but it is not tied to the origin.

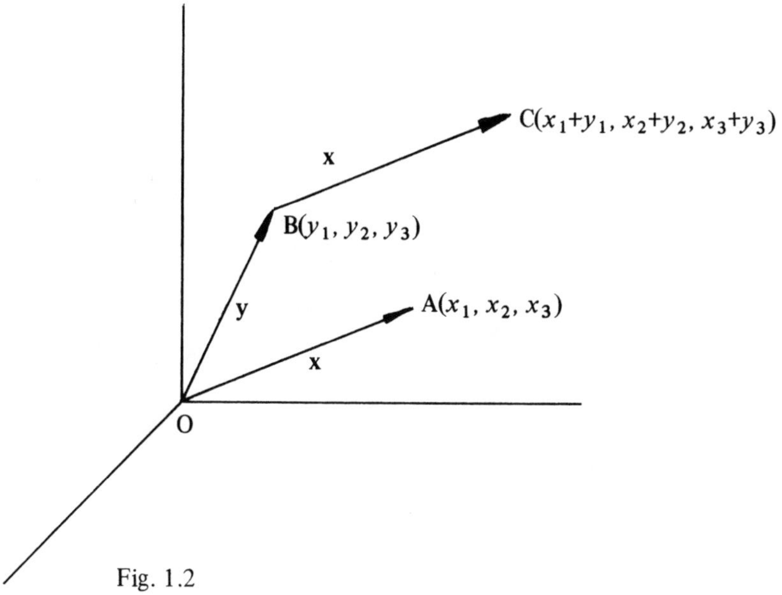

Fig. 1.2

The coordinates of C are $(x_1 + y_1, x_2 + y_2, x_3 + y_3)$. Hence the directed line segment OC represents the vector **x** + **y**. By constructing BC to be equal and parallel to OA we completed the parallelogram OBCA. Hence, this construction for finding the vector sum **x** + **y** is known as the **Parallelogram Law of Addition**.

1.6.3 The Triangle Law of Addition

The development of the construction described in section 1.6.2 for the addition of two position vectors drawn from the same origin suggests a construction for adding two vectors represented by directed line segments OB and BC arranged as indicated in Fig. 1.3. The construction is to complete the triangle OBC by drawing the third side, OC.

Thus $$\overline{OC} = \overline{OB} + \overline{BC} \ ,$$

or $$\mathbf{z} = \mathbf{x} + \mathbf{y} \ .$$

This construction is known as the **Triangle Law of Addition**.

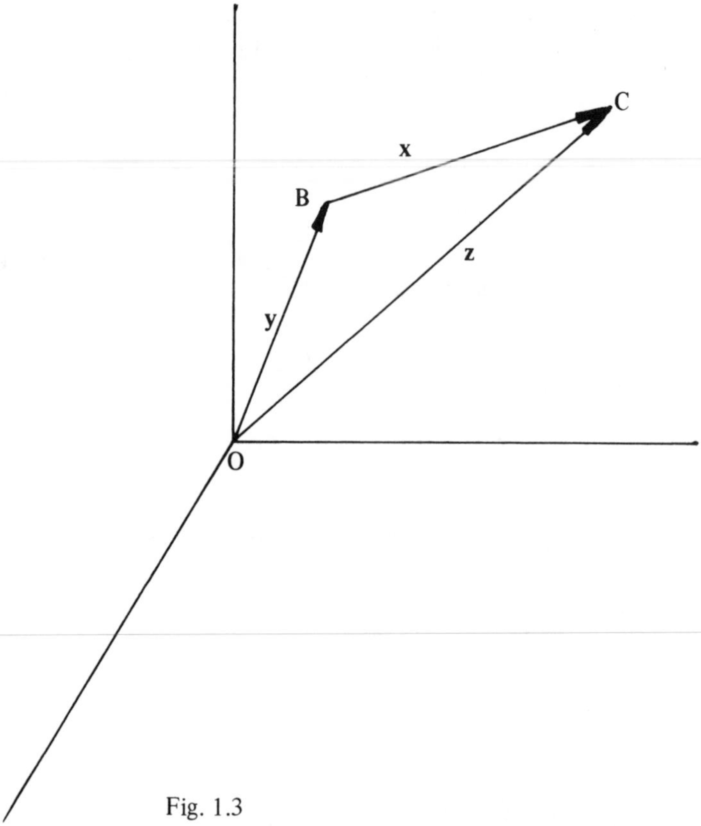

Fig. 1.3

1.6.4 The Geometrical Interpretation of the Scalar product

Let **x** and **y** be vectors represented by the directed line segments OA and OB respectively (Fig. 1.4). If θ is the angle between OA and OB the definition of the scalar product is:

$$\mathbf{x}.\mathbf{y} = |\mathbf{x}||\mathbf{y}| \cos\theta = xy \cos\theta \ . \tag{1.82}$$

According to this new definition of the scalar product, it is clearly commutative. A geometrical proof is required to show that the product is distributive. We first note, however, that the scalar product may be interpreted in terms of projections. Thus $x \cos\theta$ is the projection of **x** on **y**, and **x.y** is the projection of **x** on **y** multiplied by the modulus of **y**. Similarly **x.y** is also the projection of **y** on **x** multiplied by the modulus of **x**.

Theorem 1.6.4 The scalar product of two vectors is distributive, i.e. if **x**, **y** and **z** are any three vectors then

$$\mathbf{x}.(\mathbf{y}+\mathbf{z}) = \mathbf{x}.\mathbf{y} + \mathbf{x}.\mathbf{z} \ .$$

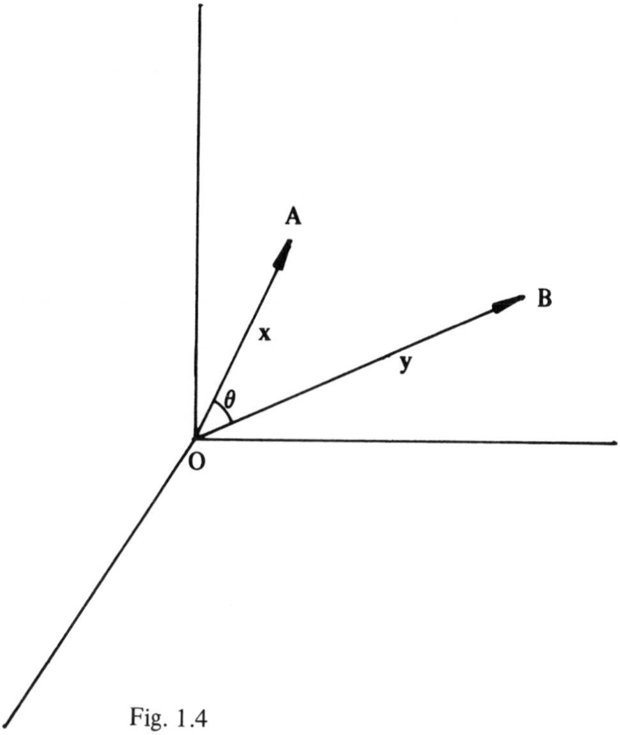

Fig. 1.4

Proof Let the directed line segments AB, BC, AC represent **y**, **z**, (**y** + **z**) respectively (Fig. 1.5). Let PQR be a line in the direction of the vector **x** such that PQ is the projection of **y** on **x**, QR is the projection of **z** on **x** and PR is the projection of (**y** + **z**) on **x**.

Then
$$\mathbf{x}.\mathbf{y} = \text{PQ}\,|\mathbf{x}|\ ,$$
$$\mathbf{x}.\mathbf{z} = \text{QR}\,|\mathbf{x}|\ ,$$
$$\mathbf{x}.(\mathbf{y}+\mathbf{z}) = \text{PR}\,|\mathbf{x}|\ ,$$
but
$$\text{PR} = \text{PQ} + \text{QR}\ ,$$
hence
$$\mathbf{x}.(\mathbf{y}+\mathbf{z}) = \mathbf{x}.\mathbf{y} + \mathbf{x}.\mathbf{z}\ .$$

This proves the theorem.

Using the distributive law, and writing **x** and **y** in terms of their components, we have:

$$\mathbf{x}.\mathbf{y} = (x_1\mathbf{e}_1 + x_2\mathbf{e}_2 + x_3\mathbf{e}_3).(y_1\mathbf{e}_1 + y_2\mathbf{e}_2 + y_3\mathbf{e}_3)\ , \qquad (1.83)$$
$$= x_1y_1 + x_2y_2 + x_3y_3\ . \qquad (1.84)$$

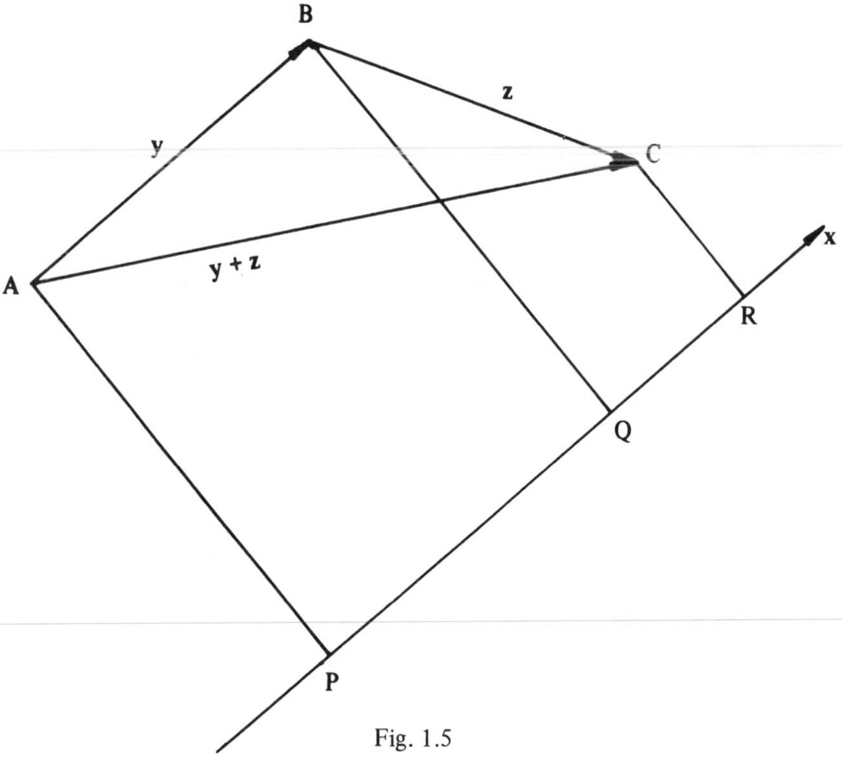

Fig. 1.5

We have thus recovered the algebraic form of the definition given in section 1.5.2.

A useful feature of the definition of **x.y** given in (1.82) is that it is a **basis free** definition, that is, it is in terms of quantities (two lengths and an angle) which are independent of any coordinate system. It follows therefore that **x.y** is a scalar.

1.6.5 Direction Cosines

Let the vector **x** (Fig. 1.6) be inclined at angles $\alpha_1, \alpha_2, \alpha_3$ to the x_1, x_2, x_3 axes respectively.

Now $\mathbf{x.e}_1 = x\cos\alpha_1 = x_1$, (1.85)

therefore $\cos\alpha_1 = x_1/x$ (1.86)

The quantities $\cos\alpha_1$, $\cos\alpha_2$, $\cos\alpha_3$ are called the **direction cosines** of the vector **x**.

Also $\mathbf{x} = (x_1, x_2, x_3)$,

hence $\mathbf{x} = (x\cos\alpha_1, x\cos\alpha_2, x\cos\alpha_3)$. (1.87)

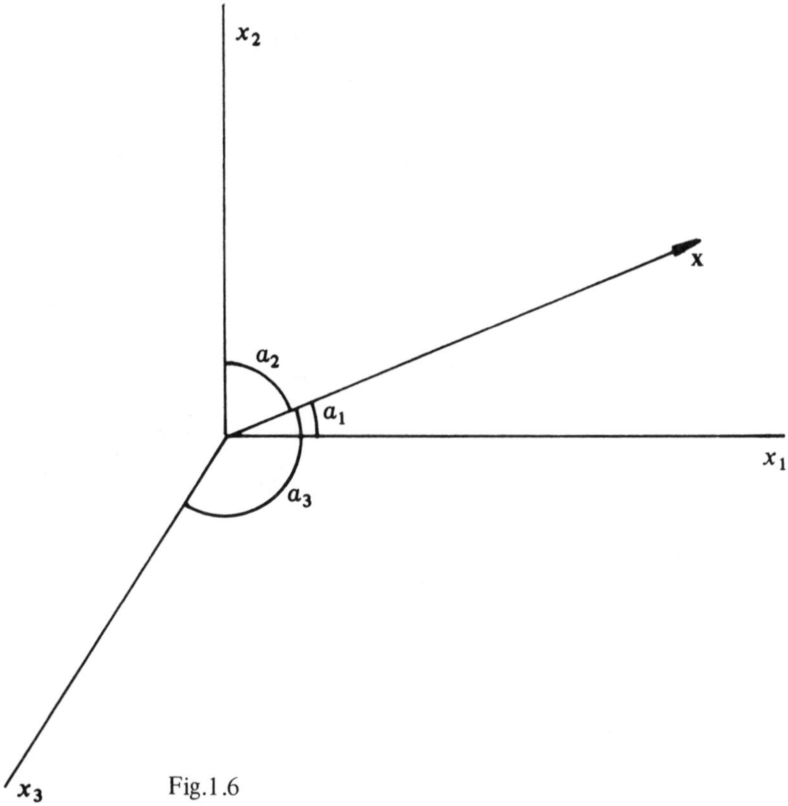

Fig.1.6

A unit vector $\check{\mathbf{x}}$ in the direction of $\mathbf{x}$ is $\mathbf{x}/|\mathbf{x}|$,

Therefore $\check{\mathbf{x}} = (\cos\alpha_1, \cos\alpha_2, \cos\alpha_3)$, (1.88)

hence $\cos^2\alpha_1 + \cos^2\alpha_2 + \cos^2\alpha_3 = 1$. (1.89)

We have shown that (i) the unit vector in the direction of $\mathbf{x}$ has as its components the direction cosines of $\mathbf{x}$, and (ii) that the sum of the squares of the direction cosines is unity.

Any vectors which have the same direction cosines are equally inclined to the three coordinate axes and are therefore parallel.

1.6.6 The Geometrical Interpretation of the Vector Product

The vectors $\mathbf{x}$ and $\mathbf{y}$ are represented by directed line segments. The vector product $\mathbf{x} \times \mathbf{y}$ is a vector represented by a directed line segment whose direction and length are defined as follows. The length of $\mathbf{x} \times \mathbf{y}$ is given by:

$$|\mathbf{x} \times \mathbf{y}| = xy \sin\theta \ , \tag{1.90}$$

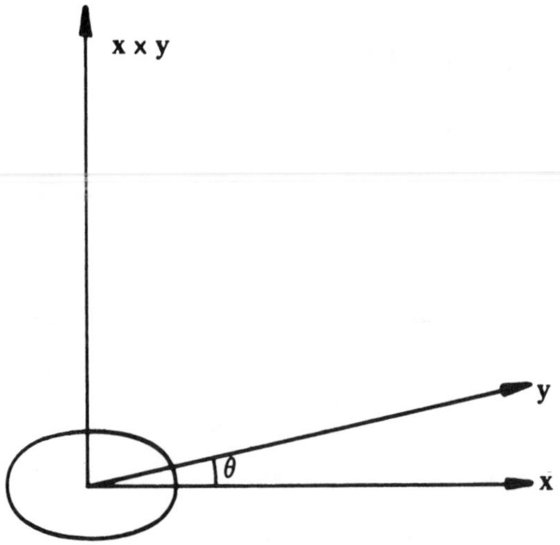

Fig. 1.7

where θ is the acute angle between **x** and **y**. The direction of $\mathbf{x} \times \mathbf{y}$ is perpendicular to the plane of **x** and **y** in the direction given by the **right hand screw rule** going from **x** to **y**.

The result (1.81) follows at once from the definition, but the distributive law (1.80) requires a geometric proof. Before embarking on this proof it is useful to think of the vector product $\mathbf{x} \times \mathbf{y}$ as the result of the operator $(\mathbf{x} \times)$ operating on the vector **y**. The effect on **y** is to rotate the vector through an angle $\pi/2$ in the plane of **y** and the perpendicular to the plane of **x** and **y** (see Fig. 1.7), and to magnify it by the factor $x \sin \theta$.

Theorem 1.6.6 (i) The vector product of two vectors is distributive, i.e. if **x**, **y** and **z** are any three vectors then

$$\mathbf{x} \times (\mathbf{y} + \mathbf{z}) = (\mathbf{x} \times \mathbf{y}) + (\mathbf{x} \times \mathbf{z}) \ .$$

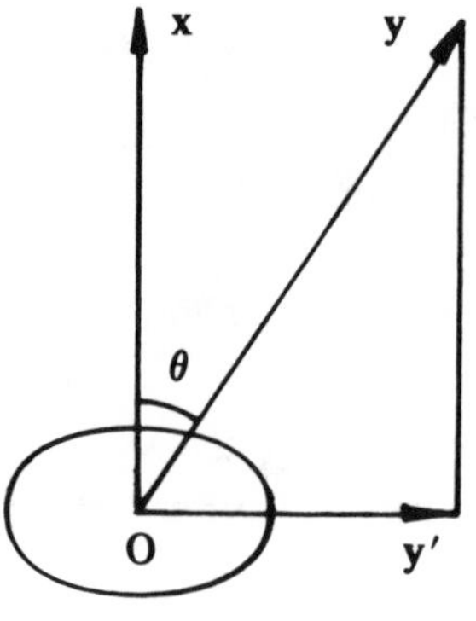

Fig. 1.8

Lemma. We prove first that if $\mathbf{y}'$ is the vector projection of $\mathbf{y}$ onto a plane perpendicular to $\mathbf{x}$ (Fig. 1.8) then

$$\mathbf{x} \times \mathbf{y} = \mathbf{x} \times \mathbf{y}' \ . \tag{1.91}$$

The result is immediately obvious because both vectors are of magnitude $xy \sin \theta$ in the direction perpendicular to the plane of $\mathbf{x}$, $\mathbf{y}$ and $\mathbf{y}'$.

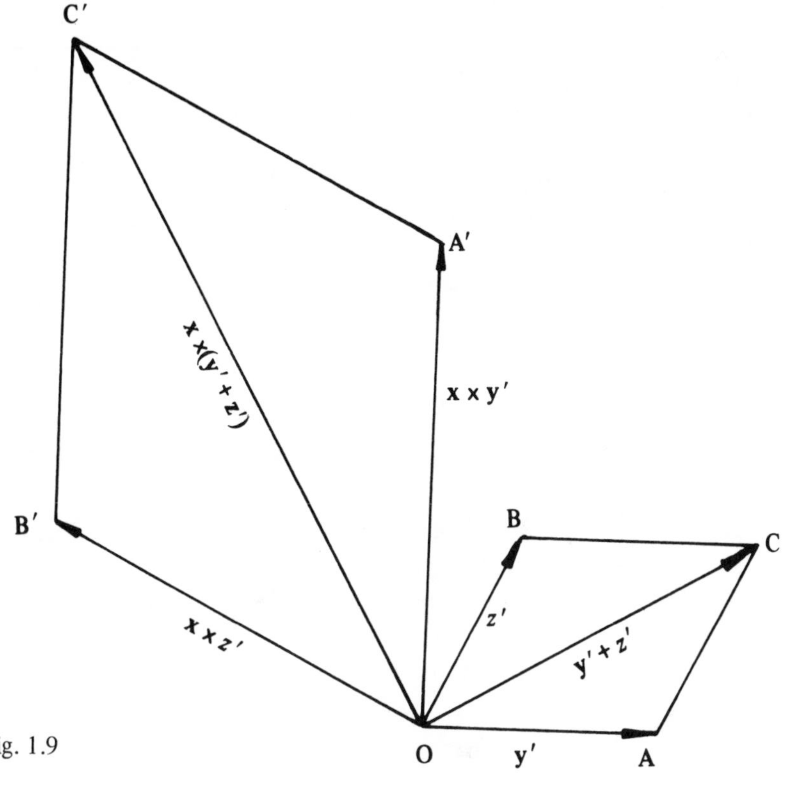

Fig. 1.9

Proof. The diagram in Fig. 1.9 is in the plane perpendicular to the vector $\mathbf{x}$ (it is a plan view of Fig. 1.8).

The effect of the operation $(\mathbf{x} \times)$ on the vector $\mathbf{y}'$ represented by OA, is to rotate the vector by $\pi/2$ and to increase its magnitude by the factor $|\mathbf{x}|$. Thus OA′, which represents $\mathbf{x} \times \mathbf{y}'$, is perpendicular to OA and

$$\text{OA}' = |\mathbf{x}|\, \text{OA} \ .$$

Similarly

$$\text{OB}' = |\mathbf{x}|\, \text{OB} \ ,$$

$$\text{OC}' = |\mathbf{x}|\, \text{OC} \ ,$$

and OB′, OC′ are perpendicular respectively to OB and OC. Thus the parallelogram OACB representing $\mathbf{y}'$, $\mathbf{z}'$ and $\mathbf{y}' + \mathbf{z}'$ has been rotated by the angle $\pi/2$ and increased in size linearly by the factor $|\mathbf{x}|$. But OC′ represents $\mathbf{x} \times (\mathbf{y}' + \mathbf{z}')$. By the parallelogram law of addition, however, it also represents $\mathbf{x} \times \mathbf{y}' + \mathbf{x} \times \mathbf{z}'$.

Hence $\quad \mathbf{x} \times (\mathbf{y}' + \mathbf{z}') = \mathbf{x} \times \mathbf{y}' + \mathbf{x} \times \mathbf{z}'$.

Employing the result of the lemma we have therefore:

$$\mathbf{x} \times (\mathbf{y} + \mathbf{z}) = \mathbf{x} \times \mathbf{y} + \mathbf{x} \times \mathbf{z} \; .$$

This proves the theorem.

It has already been pointed out (Example 1.5, No. 1) that $\mathbf{e}_2 \times \mathbf{e}_3 = \mathbf{e}_1$ etc. Using these results together with the distributive law, we find that

$$\mathbf{x} \times \mathbf{y} = (x_2 y_3 - x_3 y_2)\mathbf{e}_1 + (x_3 y_1 - x_1 y_3)\mathbf{e}_2 + (x_1 y_2 - x_2 y_1)\mathbf{e}_3 \; , \qquad (1.92)$$

which is the algebraic form of the definition, already given in (1.78).

Examples 1.6

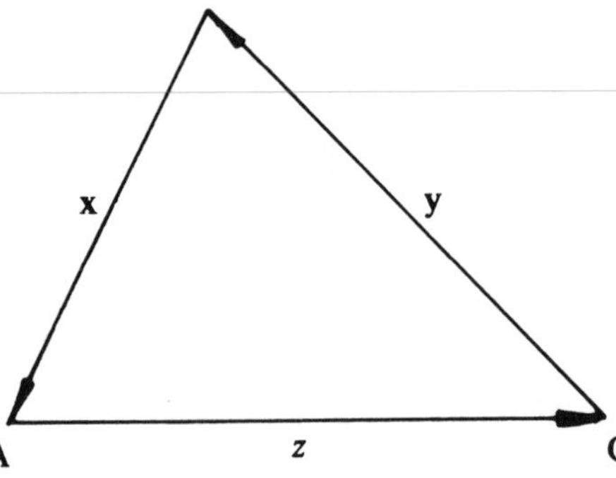

Fig.1.10

The vector sum of three vectors drawn continuously round a triangle is zero (Fig. 1.10).

Thus $\mathbf{x} + \mathbf{y} + \mathbf{z} = \mathbf{0}$.

(2)

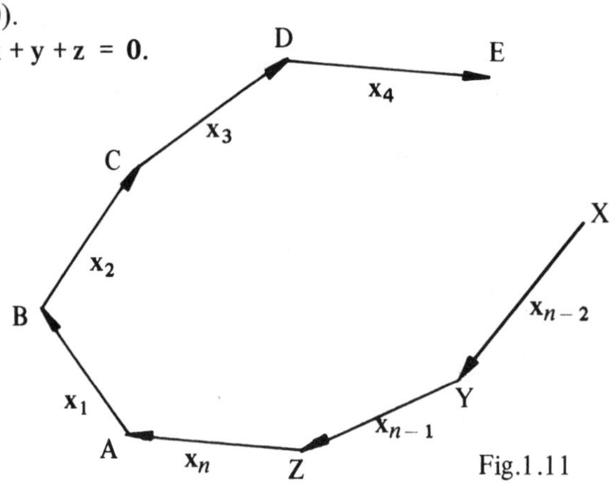

Fig.1.11

The result of Example 1 may be extended to any *closed* polygon (Fig. 1.11).

Thus $\mathbf{x}_1 + \mathbf{x}_2 + \ldots + \mathbf{x}_{n-1} + \mathbf{x}_n = \mathbf{0}$.

(3)

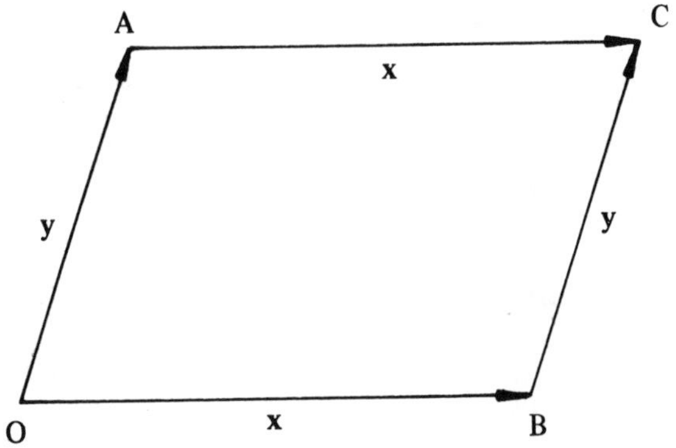

Fig. 1.12

The points A and B (Fig. 1.12) have position vectors **x** and **y** respectively with respect to an origin O. The position vector of a third point, C, is $\mathbf{x} + \mathbf{y}$. OABC is therefore a parallelogram since $\overline{AC} = \mathbf{x}$ and $\overline{BC} = \mathbf{y}$.

The area of triangle OAB $= \frac{1}{2}|\mathbf{x} \times \mathbf{y}|$ and the area of the parallelogram OABC is $|\mathbf{x} \times \mathbf{y}|$. The vector $\frac{1}{2}(\mathbf{x} \times \mathbf{y})$ is called the **vector area** of triangle OAB.

(4) Let **n** be a unit vector. The direction cosines of **n** are $\mathbf{n}.\mathbf{e}_1$, $\mathbf{n}.\mathbf{e}_2$, $\mathbf{n}.\mathbf{e}_3$.

Problems 1.6

(1) The vertices of a triangle have position vectors **a**, **b**, **c** with respect to an origin O. Show that the area of the triangle is equal to $\frac{1}{2}|\mathbf{b} \times \mathbf{c} + \mathbf{c} \times \mathbf{a} + \mathbf{a} \times \mathbf{b}|$.

(2) Find the angle between the vectors (i) (1, 2, 3), (1, 0, 1); (ii) (1, 2, 3), (−1, 0, 1); (iii) (1, 2, 3), (−1, 0, −1); (iv) (1, 2, 3), (1, 0, −1).

(3) Use vector methods to show that the angle subtended by a diameter of a sphere at a point on the surface is a right angle.

(4) Write down the direction cosines of the vector $\mathbf{a} = (3, 4, 5)$. Hence write down the unit vector in the direction of **a**.

(5) Prove that for any vectors **y**, **z**:

$$\mathbf{y}.(\mathbf{y} \times \mathbf{z}) = \mathbf{z}.(\mathbf{y} \times \mathbf{z}) = 0 .$$

Prove further that a vector perpendicular to $(\mathbf{y} \times \mathbf{z})$ is in the plane of **y** and **z**. Hence prove that a vector perpendicular to $(\mathbf{y} \times \mathbf{z})$ and also perpendicular to a third vector **x** has the form

$$k\{(\mathbf{x}.\mathbf{z})\mathbf{y} - (\mathbf{x}.\mathbf{y})\mathbf{z}\} .$$

[You may assume **x** is not perpendicular to **y** or **z**].

1.7 PRODUCTS OF MORE THAN TWO VECTORS

1.7.1 The triple Scalar Product

The **Triple Scalar Product** [**x**, **y**, **z**] of the three vectors **x**, **y**, **z** is defined to be

$$[\mathbf{x}, \mathbf{y}, \mathbf{z}] = (\mathbf{x} \times \mathbf{y}).\mathbf{z} \ . \tag{1.93}$$

By writing the vectors in terms of their components and using the determinantal representation of **x** X **y** given by (1.79), it may be shown that

$$[\mathbf{x}, \mathbf{y}, \mathbf{z}] = \begin{vmatrix} x_1 & x_2 & x_3 \\ y_1 & y_2 & y_3 \\ z_1 & z_2 & z_3 \end{vmatrix} \tag{1.94}$$

We may then make use of the properties of determinants set out in sections 1.1.5 – 1.1.7 to deduce further properties of the triple scalar product,

namely $(\mathbf{x} \times \mathbf{y}).\mathbf{z} = \mathbf{x}.(\mathbf{y} \times \mathbf{z})$,

and $[\mathbf{x}, \mathbf{y}, \mathbf{z}] = [\mathbf{y}, \mathbf{z}, \mathbf{x}] = [\mathbf{z}, \mathbf{x}, \mathbf{y}]$. (1.95)

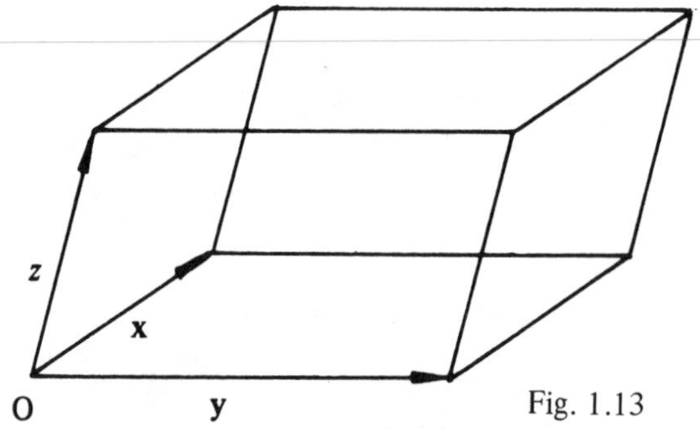

Fig. 1.13

If the vectors **x**, **y**, **z** are represented by directed line segments (Fig. 1.13), the triple scalar product [**x**, **y**, **z**] is the volume of the parallelepiped whose sides are equal and parallel to the line segments. It follows therefore that [**x**, **y**, **z**] = 0 if and only if the vectors **x**, **y** and **z** are coplanar.

The geometrical interpretation of [**x**, **y**, **z**], being independent of any co-ordinate system, or **basis free**, (see section 1.6.4) provides us with a ready insight as to why the triple scalar product is a scalar invariant under coordinate transformations.

1.7.2 The Triple Vector Product

There are three **triple vector products** of the vectors **x**, **y**, **z** namely **x** X (**y** X **z**), **y** X (**z** X **x**) and **z** X (**x** X **y**). It is sufficient to quote the expansion for **x** X (**y** X **z**),

namely $\mathbf{x} \times (\mathbf{y} \times \mathbf{z}) = \mathbf{y}(\mathbf{x}.\mathbf{z}) - \mathbf{z}(\mathbf{x}.\mathbf{y})$, (1.96) from which it may be deduced that

$$(\mathbf{x} \times \mathbf{y}) \times \mathbf{z} = \mathbf{y}(\mathbf{x}.\mathbf{z}) - \mathbf{x}(\mathbf{y}.\mathbf{z}) \ . \tag{1.97}$$

The proof of (1.96) is given in books on vector algebra. The following line of reasoning, however, which is not a proof, gives some insight into the nature of the triple vector product.

The geometric definition of the vector product of two vectors tells us that $\mathbf{x} \times (\mathbf{y} \times \mathbf{z})$ is a vector perpendicular to $\mathbf{x}$ and to $(\mathbf{y} \times \mathbf{z})$. Hence $\mathbf{x} \times (\mathbf{y} \times \mathbf{z}) = k[(\mathbf{z}.\mathbf{x})\mathbf{y} - (\mathbf{x}.\mathbf{y})\mathbf{z}]$ where k is a factor to be determined (see Problems 1.6, No. 5). Now take for example $\mathbf{x} = \mathbf{y} = \mathbf{e}_1, \mathbf{z} = \mathbf{e}_2$ and we find $k = 1$ to recover (1.96).

The way to remember expansions (1.96) and (1.97) is as follows. First write down the middle vector and multiply it by the scalar product of the other two. Then subtract the other vector within the bracket multiplied by the scalar product of the other two. This rule is valid for both (1.96) and (1.97).

1.7.3 Higher Products of Vectors

The expansions for triple scalar and triple vector products lead to expansions for higher products. We have, for example:

$$(\mathbf{x} \times \mathbf{y}).(\mathbf{z} \times \mathbf{w}) = (\mathbf{y}.\mathbf{w})(\mathbf{x}.\mathbf{z}) - (\mathbf{y}.\mathbf{z})(\mathbf{x}.\mathbf{w}) \ , \tag{1.98}$$

$$(\mathbf{x} \times \mathbf{y}) \times (\mathbf{z} \times \mathbf{w}) = [\mathbf{x}, \mathbf{y}, \mathbf{w}]\mathbf{z} - [\mathbf{x}, \mathbf{y}, \mathbf{z}]\mathbf{w} \ , \tag{1.99}$$

$$(\mathbf{x} \times \mathbf{y}) \times (\mathbf{z} \times \mathbf{w}) = [\mathbf{z}, \mathbf{w}, \mathbf{x}]\mathbf{y} - [\mathbf{z}, \mathbf{w}, \mathbf{y}]\mathbf{x} \ . \tag{1.100}$$

From (1.99) and (1.100) we deduce the result:

$$[\mathbf{x}, \mathbf{y}, \mathbf{z}]\mathbf{w} = [\mathbf{w}, \mathbf{y}, \mathbf{z}]\mathbf{x} + [\mathbf{x}, \mathbf{w}, \mathbf{z}]\mathbf{y} + [\mathbf{x}, \mathbf{y}, \mathbf{w}]\mathbf{z} \ . \tag{1.101}$$

1.7.4 Linear Dependence and Independence

Let $\mathbf{x}_1, \mathbf{x}_2, \ldots, \mathbf{x}_n$ be n ($\geqslant 2$) vectors. If there exists a set of numbers $\alpha_1, \alpha_2, \ldots, \alpha_n$ *not all zero*, such that

$$\alpha_1 \mathbf{x}_1 + \alpha_2 \mathbf{x}_2 + \ldots + \alpha_n \mathbf{x}_n = \mathbf{0} \ , \tag{1.102}$$

the vectors $\mathbf{x}_1, \mathbf{x}_2, \ldots, \mathbf{x}_n$ are said to be **linearly dependent**. If no such set of numbers exists, the vectors are said to be **linearly independent**.

If $n = 2$ we have $\alpha_1 \mathbf{x}_1 + \alpha_2 \mathbf{x}_2 = \mathbf{0}$ (1.103)

as the condition for linear dependence. In general (1.103) is satisfied by

$$\mathbf{x}_1 = -(\alpha_2/\alpha_1)\mathbf{x}_2 \ , \tag{1.104}$$

where all quantities are non-zero. If $\mathbf{x}_1, \mathbf{x}_2$ are represented by directed line segments, the condition (1.104) is satisfied if the two vectors are parallel. The condition (1.103) for linear dependence is also satisfied if (say) $\mathbf{x}_1 = \mathbf{0}, \alpha_2 = 0$, $\alpha_1 \neq 0, \mathbf{x}_2 \neq \mathbf{0}$. This, however, is a degenerate case.

If $n = 3$ we have $$\alpha_1 \mathbf{x}_1 + \alpha_2 \mathbf{x}_2 + \alpha_3 \mathbf{x}_3 = \mathbf{0} \tag{1.105}$$

as the condition for linear dependence. If all quantities are non-zero, condition (1.105) is equivalent to

$$[\mathbf{x}_1, \mathbf{x}_2, \mathbf{x}_3] = 0 \,, \tag{1.106}$$

which, as we have already seen (section 1.7.1) is the necessary and sufficient condition for the vectors $\mathbf{x}_1, \mathbf{x}_2, \mathbf{x}_3$ to be coplanar. Degenerate cases occur when one or two vectors are zero or when two or three vectors are parallel.

Finally, when $n = 4$,

we have $$\alpha_1\mathbf{x}_1 + \alpha_2\mathbf{x}_2 + \alpha_3\mathbf{x}_3 + \alpha_4\mathbf{x}_4 = \mathbf{0} \,. \tag{1.107}$$

Four vectors in R^3 are *always* linearly dependent. It follows that a set of numbers $\alpha_1, \alpha_2, \alpha_3, \alpha_4$, satisfying (1.107) always exists. Reference to equation (1.101) shows that such a set of numbers is

$$\alpha_1 = -[\mathbf{x}_2, \mathbf{x}_3, \mathbf{x}_4], \alpha_2 = [\mathbf{x}_3, \mathbf{x}_4, \mathbf{x}_1] \,,$$

$$\alpha_3 = -[\mathbf{x}_4, \mathbf{x}_1, \mathbf{x}_2], \alpha_4 = [\mathbf{x}_1, \mathbf{x}_2, \mathbf{x}_3] \,. \tag{1.108}$$

Again, degenerate cases occur when, for example, three vectors are coplanar, two vectors are parallel, or one vector is zero. We observe, however, that any vector in R^3 may be expressed as a linear combination of three linearly independent vectors (see Problems 1.7, No. 4).

It was hinted in section 1.2.9 that the concept of rank was closely bound up with that of linear dependence and independence. We are now in a position to formalize this connection by thinking of the rows (or columns) of a matrix as components of a vector. We may then talk meaningfully of the linear dependence or otherwise of the rows of a matrix. Equation (1.94) expresses the connection between the vectors and their components. Clearly the rows of the matrix are linearly independent if and only if its determinant is non-zero. In this case the rank of the matrix is 3. This result may be extended by the following theorem.

Theorem 1.7.4 The rank of a (3×3) matrix is equal to the number of linearly independent rows (or columns).

Proof. Let us denote the determinant of the matrix by the triple scalar product of vectors as in (1.94).

Thus $$\Delta = \begin{vmatrix} x_1 & x_2 & x_3 \\ y_1 & y_2 & y_3 \\ z_1 & z_2 & z_3 \end{vmatrix} = [\mathbf{x}, \mathbf{y}, \mathbf{z}] \,.$$

Clearly if the rows are linearly independent, $\Delta \neq 0$ and $R = 3$. Suppose the rows are linearly dependent then $\Delta = 0$. If two of the rows are linearly independent (those represented by the vectors **x** and **y** for example), then **x** and **y** are not parallel and the determinants $\begin{vmatrix} x_1 & x_3 \\ y_1 & y_3 \end{vmatrix}$, $\begin{vmatrix} x_1 & x_2 \\ y_1 & y_2 \end{vmatrix}$ cannot both be zero. Hence R = 2.

If, however, no two rows are linearly independent the vectors **x**, **y**, **z** are all parallel (or zero). Hence all the (2 × 2) submatrices are singular and R = 1.

This proves the theorem for (3 × 3) matrices.† It is also true for matrices of any dimensions whether square or not.

Examples 1.7

(1) The vectors (1, 2, 3), (2, 4, 6) are linearly dependent because they are parallel. In general the vectors **a** and k**a** are linearly dependent.

(2) The vectors **a** = (1, 2, 0), **b** = (–6, 4, 0), **c** = (2, –1, 0) are linearly dependent because they are coplanar. This is evident without the need to evaluate [**a**, **b**, **c**]. If we write down the determinant for [**a**, **b**, **c**] namely

$$\begin{vmatrix} 1 & 2 & 0 \\ -6 & 4 & 0 \\ 2 & -1 & 0 \end{vmatrix},$$

this too is clearly zero because one column is zero.

(3) Given any three linearly indpendent vectors $\mathbf{a}_1, \mathbf{a}_2, \mathbf{a}_3$, in R^3 an orthogonal triad of vectors $\mathbf{e}_1, \mathbf{e}_2, \mathbf{e}_3$, may be formed by writing:

$$\mathbf{e}_1 = \mathbf{a}_1 ,$$

$$\mathbf{e}_2 = \mathbf{a}_2 - \lambda_1 \mathbf{e}_1 ,$$

$$\mathbf{e}_3 = \mathbf{a}_3 - \mu_1 \mathbf{e}_1 - \mu_2 \mathbf{e}_2 .$$

where $\lambda_1 = \dfrac{\mathbf{a}_2.\mathbf{e}_1}{\mathbf{e}_1^2}$, $\mu_1 = \dfrac{\mathbf{a}_3.\mathbf{e}_1}{\mathbf{e}_1^2}$, $\mu_2 = \dfrac{\mathbf{a}_3.\mathbf{e}_2}{\mathbf{e}_2^2}$.

This process, which may be extended to deal with n linearly independent vectors in R^n, is known as the **Gram-Schmidt orthogonalization process**. Note that the orthogonal vectors $\mathbf{e}_1, \mathbf{e}_2, \mathbf{e}_3$, are not necessarily unit vectors.

Problems 1.7

(1) Prove that if **x**, **y**, **z**, **w** are any four vectors,

$$(\mathbf{x} \times \mathbf{y}).(\mathbf{z} \times \mathbf{w}) = (\mathbf{y}.\mathbf{w})(\mathbf{x}.\mathbf{z}) - (\mathbf{y}.\mathbf{z})(\mathbf{x}.\mathbf{w}).$$

†For a proof of the more general theorem the reader is referred to *Advanced Engineering Mathematics* by Erwin Kreyszig, published by John Wiley.

Hence show that the plane of **x** and **y** is perpendicular to the plane of **z** and **w** if and only if

$$\begin{vmatrix} \mathbf{x}.\mathbf{z} & \mathbf{x}.\mathbf{w} \\ \mathbf{y}.\mathbf{z} & \mathbf{y}.\mathbf{w} \end{vmatrix} = 0 \ .$$

(2) Prove that $(\mathbf{x} \times \mathbf{y}) \times (\mathbf{z} \times \mathbf{w}) = [\mathbf{x}, \mathbf{y}, \mathbf{w}]\mathbf{z} - [\mathbf{x}, \mathbf{y}, \mathbf{z}]\mathbf{w}$, and deduce that $(\mathbf{x} \times \mathbf{y}) \times (\mathbf{y} \times \mathbf{z}) = [\mathbf{x}, \mathbf{y}, \mathbf{z}]\mathbf{y}$. Hence prove that $[\mathbf{x} \times \mathbf{y}, \mathbf{y} \times \mathbf{z}, \mathbf{z} \times \mathbf{x}] = [\mathbf{x}, \mathbf{y}, \mathbf{z}]^2$.
(3) Prove that the condition (1.105) for the linear dependence of three vectors $\mathbf{x}_1, \mathbf{x}_2, \mathbf{x}_3$ is equivalent to the condition (1.106) that the triple scalar product shall vanish.
(4) Given that the three vectors $\mathbf{x}_1, \mathbf{x}_2, \mathbf{x}_3$ are linearly independent, show how to express any fourth vector $\mathbf{x}_4$, in terms of $\mathbf{x}_1, \mathbf{x}_2, \mathbf{x}_3$.
(5) Determine whether the following sets of vectors are linearly dependent: (i) (1, 0, 2), (2, 0, 4); (ii) (1, 1, 2), (3, −2, 4), (−3, 7, −2); (iii) (1, 1, 2), (1, 2, 3), (−1, −1, 1); (iv) (1, 3, 4), (1, 0, −1), (2, 1, 3), (−1, 1, −1). For each set of vectors which is linearly dependent, find a set of numbers $\alpha_1, \alpha_2 \ldots$ such that $\alpha_1\mathbf{x}_1 + \alpha_2\mathbf{x}_2 + \ldots = \mathbf{0}$.

1.8 VECTOR FIELDS

1.8.1 Definition

Recalling the definition of a scalar field given in section 1.3.2, we define a vector field as follows. Suppose a vector **v** is defined at every point P within a domain D by the relation $\mathbf{v} = \mathbf{v}(P)$. If, for every point P, there is a unique value **v**(P), then $\mathbf{v} = \mathbf{v}(P)$ defines a vector function of position called a **vector field**. The gravitational field of the Earth, the electromagnetic field of a charged particle in motion and the velocity distribution throughout a fluid are examples of vector fields.

1.8.2 Differentiation of Scalar Fields

The reader will recall the definition of the derivative of a function ϕ of a single variable x:

$$\phi'(x) = \lim_{\delta x \to 0} \frac{\phi(x + \delta x) - \phi(x)}{\delta x} \ . \tag{1.109}$$

This definition may be recast as follows: If there exists a function ϕ' such that

$$\phi(x + \delta x) - \phi(x) = \delta x \phi'(x) + \delta x \eta \ , \tag{1.110}$$

where $\eta \to 0$ as $\delta x \to 0$, then ϕ is said to be differentiable at x, and $\phi'(x)$ is called the **derivative** of ϕ at x. The function ϕ' defined by (1.110) is identical with that defined by (1.109). It is the second form of the definition, however, which proves to be the more readily extended to deal with derivatives of scalar and vector fields.

The scalar function of position $\phi = \phi(\mathrm{P})$, known as a **scalar field**, was defined in section 1.3.2. Let us agree to write $\phi(\mathbf{r})$ in place of $\phi(\mathrm{P})$, where $\mathbf{r} = (x_1, x_2, x_3)$ is the position vector of P. The rate of change of ϕ with position will in general depend on direction. It is therefore reasonable to expect that any spatial derivative of ϕ will be vectorial in character. With this point in mind we may frame our definition of derivative as follows: if there exists a vector function of position, $\nabla\phi$, independent of $\mathbf{h}$, such that

$$\phi(\mathbf{r}+\mathbf{h}) - \phi(\mathbf{r}) = \mathbf{h}.\nabla\phi + \mathbf{h}.\boldsymbol{\eta} \ , \tag{1.111}$$

where $|\boldsymbol{\eta}| \to 0$ as $|\mathbf{h}| \to 0$, then ϕ is said to be differentiable at $\mathbf{r}$ and the derivative $\nabla\phi$ is called the **gradient** of ϕ at $\mathbf{r}$.

If $\nabla\phi$ exists at all points of the domain D, it defines a vector field on D, an explicit expression for which may be obtained as follows: Let $\mathbf{h} = h\mathbf{e}_i$.

Then $$\phi(\mathbf{r}+h\mathbf{e}_i) - \phi(\mathbf{r}) = h(\nabla\phi)_i + h\eta_i \ , \tag{1.112}$$

where $$\eta_i \to 0 \text{ as } h \to 0 \ ,$$

therefore $$(\nabla\phi)_i = \lim_{h\to 0} \frac{\phi(\mathbf{r} + h\mathbf{e}_i) - \phi(\mathbf{r})}{h} = \frac{\partial\phi}{\partial x_i} \ . \tag{1.113}$$

Hence $$\nabla\phi = \sum_{i=1}^{3} \frac{\partial\phi}{\partial x_i}\mathbf{e}_i \ . \tag{1.114}$$

Examples 1.8

(1) The function ϕ given by

$$\phi(x_1, x_2, x_3) = x_1^2 + 2x_2 + x_1x_2$$

defines a scalar field. The derivative $\nabla\phi$ given by

$$\nabla\phi = (2x_1 + x_2, x_1 + 2, 0)$$

defines a vector field.

(2) Let ϕ be a spherically symmetric function of position,

i.e. $$\phi = \phi(r), \text{ where } r^2 = x_1^2 + x_2^2 + x_3^2 \ .$$

then $$\nabla\phi = \sum_{i=1}^{3} \frac{\partial\phi}{\partial x_i}\mathbf{e}_i = \sum_{i=1}^{3} \frac{\partial\phi}{\partial r}\frac{\partial r}{\partial x_i}\mathbf{e}_i = \sum_{i=1}^{3} \phi'(r)\frac{x_i}{r}\mathbf{e}_i \ ,$$

i.e. $$\nabla\phi = \frac{1}{r}\phi'(r)\mathbf{r}.$$

(3) If $\phi = \phi(\mathbf{r})$ the total differential $d\phi$ is given by:

$$d\phi = \frac{\partial\phi}{\partial x_1} dx_1 + \frac{\partial\phi}{\partial x_2} dx_2 + \frac{\partial\phi}{\partial x_3} dx_3 ,$$
$$= \nabla\phi.d\mathbf{r} .$$

(4) The **directional derivative** of ϕ in the direction of the unit vector $\mathbf{n}$ is $\mathbf{n}.\nabla\phi$, i.e. the component of $\nabla\phi$ in the direction of $\mathbf{n}$. The maximum directional derivative occurs therefore when $\mathbf{n}$ is in the direction of $\nabla\phi$. The magnitude of this maximum is $|\nabla\phi|$.

Problems 1.8

(1) Let $\phi = \phi(\mathbf{r})$, $\psi = \psi(\mathbf{r})$ define two scalar fields. Show that
(i) $\nabla(\phi + \psi) = \nabla\phi + \nabla\psi$; (ii) $\nabla(\phi\psi) = \phi\nabla\psi + \psi\nabla\phi$.
(2) Find $\nabla\phi$ where $\phi =$ (i) $x_1^2 + 2x_2$; (ii) $2r^4 + 3r^3 + 4r^2 + r$; (iii) $1/r$; (iv) r^n; (v) $\ln r$; (vi) $f(r)$; (vii) x_1/r.
(3) Show that the vector $\nabla\phi$ is normal to the surface $\phi =$ const. Hence find the unit vector normal to the ellipsoid

$$x^2 + y^2 + 4(z^2 - 1) = 0$$

at the point $(2/\sqrt{3}, 2/\sqrt{3}, 1/\sqrt{3})$.
(4) Find the directional derivative of $\phi = x^2 + y^2 + 4(z^2 - 1)$ in the direction of the vector (1, 1, 1), at the point $(2/\sqrt{3}, 2/\sqrt{3}, 1/\sqrt{3})$.
(5) Find the angle between the surfaces $x^2 + y^2 + 4(z^2 - 1) = 0$ and $x^2 + y^2 - 8z^2 = 0$ at the point $(2/\sqrt{3}, 2/\sqrt{3}, 1/\sqrt{3})$.

1.9 THE SUMMATION CONVENTION

1.9.1 Suffices

By labelling the coordinate axes x_1, x_2, x_3 rather than x, y, z, and the unit vectors along the axes $\mathbf{e}_1, \mathbf{e}_2, \mathbf{e}_3$ rather than $\mathbf{i}, \mathbf{j}, \mathbf{k}$ we have from time to time been able to make use of the summation symbol Σ to shorten our expressions [see for example (1.75), (1.77), (1.114)]. In subsequent chapters we shall make use of quantities with multiple subscripts and there will be considerable scope for a shorthand notation in which to write expressions of increasing length and complexity. The convention known as the **summation convention** may be illustrated by an example. We have seen (1.75) that the vector $\mathbf{x} = (x_1, x_2, x_3)$ may be written

$$\mathbf{x} = \sum_{i=1}^{3} x_i \mathbf{e}_i \qquad (1.115)$$

Let us agree to omit the summation sign and write simply

$$\mathbf{x} = x_i \mathbf{e}_i , \qquad (1.116)$$

adopting the convention that where a suffix occurs twice a summation is implied. By the same convention we may write the expression for $\nabla\phi$ in (1.114) as

$$\nabla\phi = \frac{\partial\phi}{\partial x_i}\mathbf{e}_i \ . \tag{1.117}$$

We may now write down the summation convention formally as follows.

(1) Any suffix occurring only once in a product is called a **free suffix** and does not imply any summation.

(2) Any suffix occurring exactly twice in a product is called a **repeated suffix** and implies a summation over the values 1, 2, 3.

(3) No suffix may occur more than twice in a product.

Notice that the summation convention is framed in terms of products rather than expressions. Thus relations of the type

$$\mathbf{a} + \mathbf{b} = a_i\mathbf{e}_i + b_i\mathbf{e}_i \tag{1.118}$$

are perfectly permissible and imply a summation over the suffix i which occurs twice in each product.

1.9.2 The Kronecker Delta

It is useful to define the following symbol:

$$\delta_{ik} = \begin{cases} 1 & i = k \\ 0 & i \neq k \ . \end{cases}$$

δ_{ik} is called the **Kronecker Delta**. Thus

$$\delta_{11} = \delta_{22} = \delta_{33} = 1, \delta_{23} = \delta_{32} = \delta_{31} = \delta_{13} = \delta_{12} = \delta_{21} = 0.$$

Alternatively the symbol is known as the **substitution operator**, for employing the summation convention established in section 1.9.1

we have: $$\delta_{ik}a_k = \delta_{i1}a_1 + \delta_{i2}a_2 + \delta_{i3}a_3 \ . \tag{1.119}$$

Now i must have one of the values 1, 2 or 3. If $i = 1, \delta_{i1} = 1$ and $\delta_{i2} = \delta_{i3} = 0$, hence $\delta_{1k}a_k = a_1$. Similarly if $i = 2, \delta_{2k}a_k = a_2$. Thus in all cases:

$$\delta_{ik}a_k = a_i \ , \tag{1.120}$$

thus δ_{ik} operating on a_k has *substituted* the index i for the index k.

Examples 1.9

(1) An alternative term for repeated suffix is **dummy suffix**. This is because it may be replaced by any other symbol *not already in use*.

Thus $$a_ib_i = a_kb_k = a_1b_1 + a_2b_2 + a_3b_3 \ .$$

(2) $$\frac{\partial x_i}{\partial x_k} = \begin{cases} 1 & i = k \\ 0 & i \neq k \end{cases}$$

where x_1, x_2, x_3 are independent variables, i.e. $\partial x_i/\partial x_k = \delta_{ik}$.

(3) $\delta_{ii} = \delta_{11} + \delta_{22} + \delta_{33} = 3$.

(4) $\mathbf{e}_i.\mathbf{e}_k = \delta_{ik}$.

(5) $r^2 = \mathbf{r}^2 = x_i x_i$, therefore $2r\dfrac{\partial r}{\partial x_k} = 2x_i\dfrac{\partial x_i}{\partial x_k} = 2x_i\delta_{ik} = 2x_k$,

i.e. $\partial r/\partial x_k = x_k/r$.

Problems 1.9

(1) Simplify (i) $a_1\mathbf{e}_1 + a_2\mathbf{e}_2 + a_3\mathbf{e}_3$; (ii) $a_1 b_1 + a_2 b_2 + a_3 b_3$.
(2) Show that (i) $a_i\mathbf{e}_i \,.\, \mathbf{e}_j = a_j$; (ii) $(\mathbf{a} \,.\, \mathbf{e}_i)\mathbf{e}_i = \mathbf{a}$.
(3) Expand, if possible, the expressions (i) $a_i + b_i$; (ii) $a_k c_k$; (iii) $a_k(b_k + c_k)$.
(4) Evaluate (i) $\delta_{ik}\delta_{ik}$; (ii) $\delta_{ik}\delta_{im}$.
(5) Re-work Problem 1.8 No. 2, (ii) - (vii) making use of the suffix notation and the properties of δ_{ik}.
(6) Two perpendicular lines have direction cosines $(\alpha_1, \alpha_2, \alpha_3)$, $(\beta_1, \beta_2, \beta_3)$ respectively. Show that (i) $\alpha_i\alpha_i = 1$; (ii) $\alpha_i\beta_i = 0$.
(7) In section 1.2.4 it was stated that the (i,k) element of the product matrix BA was the scalar product of the ith row of B and the kth column of A. If $C = BA$ show that $C_{ik} = B_{ip}A_{pk}$.

1.10 TRANSFORMATIONS OF VECTORS

1.10.1 The Invariance of Vectors – Transformation Properties

In section 1.3.1 stress was laid on the fact that scalars are invariant with respect to coordinate transformations. The same is true of vectors. A velocity, for example, is unaltered in speed and direction by a change of basis, so also are force, acceleration or any other vector quantity. We remarked further in section 1.3.1 that the scalar is the simplest example of a Cartesian tensor - a tensor of order zero. The vector, which shares with the scalar the property of invariance, is a **Cartesian tensor of order one.**

We now come to an important question. Is the invariance property of vectors equally obvious when the vector is considered in the abstract? The answer to this question depends on the representation employed. It illustrates an important difference between the two representations so far considered. If we choose to represent a vector by a directed line segment its invariant property is self evident. If, however, we choose our original representation, the ordered triple, the property of invariance is no longer obvious, because a transformation will in general cause the components of the vector to alter. In the sections which follow we shall examine how these components are affected by a particular type of transformation, namely a rotation about the origin.

1.10.2 Rotations

Consider a Cartesian coordinate system whose axes are parallel respectively to the orthogonal triad of unit vectors $\mathbf{e}_1, \mathbf{e}_2, \mathbf{e}_3$. (Fig. 1.14). The axes are rotated about the origin O until they are parallel to the vectors of a second orthogonal triad $\mathbf{e}_1^*, \mathbf{e}_2^*, \mathbf{e}_3^*$.

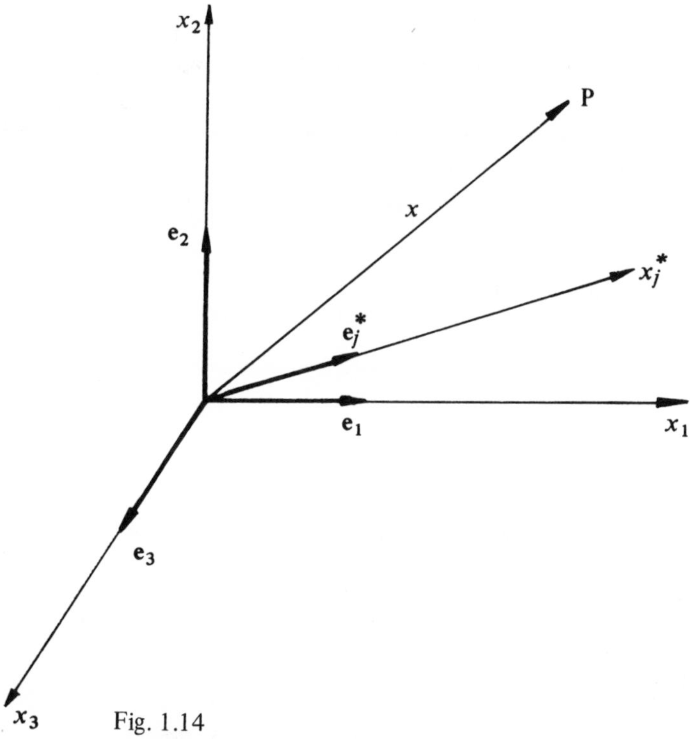

Fig. 1.14

Let $\mathbf{x}$ be the position vector of a point P fixed in space, and let the components of $\mathbf{x}$ be (x_1, x_2, x_3) with respect to the original axes and (x_1^*, x_2^*, x_3^*) with respect to the new axes. We wish to express the components (x_1^*, x_2^*, x_3^*) in terms of the components (x_1, x_2, x_3), and the vectors $\mathbf{e}_1^*, \mathbf{e}_2^*, \mathbf{e}_3^*$, in terms of the vectors $\mathbf{e}_1, \mathbf{e}_2, \mathbf{e}_3$.

Let $$\theta_{ji} = \mathbf{e}_j^* . \mathbf{e}_i \ . \tag{1.121}$$

$\theta_{j1}, \theta_{j2}, \theta_{j3}$ are then the direction cosines (see section 1.6.5) of the vector $\mathbf{e}_j^*$ with respect to the original axes.

Now $$\mathbf{e}_j^* = (\mathbf{e}_j^* . \mathbf{e}_i)\mathbf{e}_i \ , \tag{1.122}$$

[see Problems 1.9 No. 2 (ii)]

therefore $$\mathbf{e}_j^* = \theta_{ji}\mathbf{e}_i \ . \tag{1.123}$$

To obtain an expression for x_j^* we note that

$$x_j^* = \mathbf{x}.\mathbf{e}_j^* , \tag{1.124}$$

$$= (x_i \, \mathbf{e}_i).\mathbf{e}_j^* , \tag{1.125}$$

$$= \theta_{ji} x_i . \tag{1.126}$$

We have obtained the **transformation laws** for unit vectors which rotate (1.123) and for components of vectors as the axes rotate (1.126). These transformation laws are fundamental to the study of Cartesian tensors, so much so that it is perfectly valid to define a tensor as something which obeys such a law. Here, however, alternative definitions are prefered for tensors of order zero, one, two etc., and the transformation laws will follow as natural consequences of these definitions.

1.10.3 The Rotation Matrix

The nine quantities θ_{ji}, defined by (1.121), may be arranged in an array:

$$\Theta = \begin{bmatrix} \theta_{11} & \theta_{12} & \theta_{13} \\ \theta_{21} & \theta_{22} & \theta_{23} \\ \theta_{31} & \theta_{32} & \theta_{33} \end{bmatrix} . \tag{1.127}$$

The matrix Θ is called the **rotation matrix**; it completely determines the rotation. The elements of Θ are not, however, all independent as we shall see. Let Θ' be the transpose of Θ. We shall first prove that

$$\Theta\Theta' = \begin{bmatrix} 1 & 0 & 0 \\ 0 & 1 & 0 \\ 0 & 0 & 1 \end{bmatrix} . \tag{1.128}$$

We have, from the theory of matrix multiplication (see Problems 1.9 No. 7):

$$(\Theta\Theta')_{jl} = \theta_{ji}\theta'_{il} , \tag{1.129}$$

$$= \theta_{ji}\theta_{li} ,$$

$$= (\mathbf{e}_j^* .\mathbf{e}_i) \, (\mathbf{e}_l^* .\mathbf{e}_i) ,$$

$$= (\mathbf{e}_j^*)_i \, (\mathbf{e}_l^*)_i ,$$

$$= \mathbf{e}_j^* .\mathbf{e}_l^* ,$$

$$= \delta_{jl} . \tag{1.130}$$

This proves the desired result. Now, since $\Theta\Theta'$ is the unit matrix, its determinant is equal to unity. Hence, since det Θ = det Θ', it follows that

$$\det \Theta = \pm 1 \ . \tag{1.131}$$

We have seen that the elements of any rotation matrix must satisfy (1.131). If (1.131) is not satisfied the matrix cannot be a rotation matrix.

It is of interest to examine the significance of the sign in (1.131). We shall see (Problem 1.10, No. 2) that a rotation in the sense that we have described gives rise to a matrix with positive determinant. With such a rotation a right-handed system (i.e. one in which $\mathbf{e}_1 \times \mathbf{e}_2 = \mathbf{e}_3$) remains right-handed after rotation. A rotation which preserves right-handedness is called a **proper rotation**. A rotation which consists of a proper rotation together with a reflection gives rise to a rotation matrix with negative determinant. Such a rotation is called an **improper rotation**. Under an improper rotation a right-handed system is changed into a left-handed system (i.e. one in which $\mathbf{e}_1 \times \mathbf{e}_2 = -\mathbf{e}_3$).

Examples 1.10

(1) Interpret the rotation represented by the matrix

$$\begin{bmatrix} \cos\alpha & \sin\alpha & 0 \\ -\sin\alpha & \cos\alpha & 0 \\ 0 & 0 & 1 \end{bmatrix} .$$

We have for our rotation:

$$\begin{bmatrix} x_1^* \\ x_2^* \\ x_3^* \end{bmatrix} = \begin{bmatrix} \cos\alpha & \sin\alpha & 0 \\ -\sin\alpha & \cos\alpha & 0 \\ 0 & 0 & 1 \end{bmatrix} \begin{bmatrix} x_1 \\ x_2 \\ x_3 \end{bmatrix} ,$$

i.e.

$$x_1^* = x_1 \cos\alpha + x_2 \sin\alpha \ ,$$

$$x_2^* = -x_1 \sin\alpha + x_2 \cos\alpha \ ,$$

$$x_3^* = x_3 \ .$$

The matrix represents a rotation of angle α about the x_3-axis in the positive sense.

(2) Use the transformation law for vectors (1.126) to show that, for any two vectors $\mathbf{x}$, $\mathbf{y}$, the scalar product $\mathbf{x}.\mathbf{y}$ remains invariant under rotations.

We have $\quad x_i{}^*y_i{}^* = \theta_{ij}x_j\theta_{il}y_l \ ,$

$\qquad = \delta_{jl}x_jy_l \ ,\quad$ [see Problems 1.10, No. 1 (i)],

$\qquad = x_jy_j \ .$

This is the desired result.

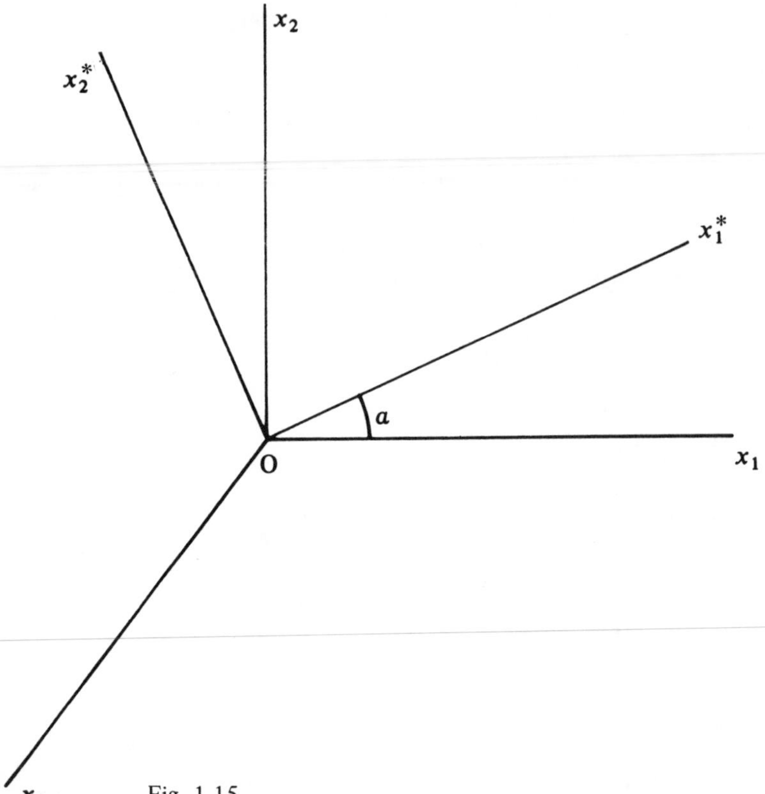

Fig. 1.15

Problems 1.10

(1) Prove that (i) $\theta_{ij}\theta_{il} = \delta_{jl}$; (ii) $\mathbf{e}_i = \theta_{ji}\mathbf{e}_j^*$; (iii) $x_i = \theta_{ji}x_j^*$.

(2) Use the results (1.94) and (1.121) to prove that if Θ is a rotation matrix, $\det\Theta = +1$.

(3) Interpret the rotations (proper or improper) represented by the following matrices:

$$\text{(i)} \begin{bmatrix} 0 & 1 & 0 \\ 1 & 0 & 0 \\ 0 & 0 & 1 \end{bmatrix} \text{(ii)} \begin{bmatrix} 1 & 0 & 0 \\ 0 & 1 & 0 \\ 0 & 0 & -1 \end{bmatrix} \text{(iii)} \begin{bmatrix} 0 & 1 & 0 \\ 0 & 0 & 1 \\ 1 & 0 & 0 \end{bmatrix}$$

$$\text{(iv)} \begin{bmatrix} 1 & 0 & 0 \\ 0 & \cos\alpha & \sin\alpha \\ 0 & -\sin\alpha & \cos\alpha \end{bmatrix} \text{(v)} \begin{bmatrix} \cos\alpha & \sin\alpha & 0 \\ -\sin\alpha\cos\beta & \cos\alpha\cos\beta & \sin\beta \\ \sin\alpha\sin\beta & -\cos\alpha\sin\beta & \cos\beta \end{bmatrix}.$$

(4) The vector **a** has components (a_1, a_2, a_3) in a given basis. A new coordinate system is formed by rotating the original system by an angle α about the x_1-axis in the positive sense, followed by a rotation of β about the new x_2-axis. Find the rotation matrix for the combined rotation. Hence determine the components of **a** in the new system when $\alpha = \beta = \pi/6$.

1.11 RATE OF CHANGE OF VECTOR IN ROTATING FRAME OF REFERENCE

Suppose the vector **v** is a function of the time t; we may define the rate of change with respect to t in the usual way.

Thus $$\frac{\mathrm{d}\mathbf{v}}{\mathrm{d}t} = \lim_{\delta t \to 0} \frac{\mathbf{v}(t+\delta t) - \mathbf{v}(t)}{\delta t} \quad . \tag{1.132}$$

The usual rules for the derivatives of products hold.

Thus $$\frac{\mathrm{d}}{\mathrm{d}t}(\mathbf{u}.\mathbf{v}) = \mathbf{u}.\dot{\mathbf{v}} + \dot{\mathbf{u}}.\mathbf{v} \quad , \tag{1.133}$$

where $\dot{\mathbf{v}}$ represents $\mathrm{d}\mathbf{v}/\mathrm{d}t$.

Also $$\frac{\mathrm{d}}{\mathrm{d}t}(\mathbf{u} \times \mathbf{v}) = \mathbf{u} \times \dot{\mathbf{v}} + \dot{\mathbf{u}} \times \mathbf{v} \quad . \tag{1.134}$$

These rules may be proved from first principles. A case of particular interest arises when **v** is a vector of constant magnitude, i.e. v^2 is a constant. Then, from (1.133) with $\mathbf{u} = \mathbf{v}$ we have:

$$\mathbf{v}.\dot{\mathbf{v}} = 0 \quad , \tag{1.135}$$

i.e. for a vector **v** of constant magnitude, $\mathrm{d}\mathbf{v}/\mathrm{d}t$ is perpendicular to **v**.

Consider a vector **a**, of fixed length a, rotating with angular speed ω about the x_3-axis (Fig. 1.16). At a given instant t we may represent **a** by:

$$\mathbf{a} = a[\cos(\omega t)\mathbf{e}_1 + \sin(\omega t)\mathbf{e}_2] \cos\phi + a \sin\phi\, \mathbf{e}_3 \quad ,$$

therefore $$\mathrm{d}\mathbf{a}/\mathrm{d}t = a\omega \cos\phi[-\sin(\omega t)\mathbf{e}_1 + \cos(\omega t)\mathbf{e}_2] \quad ,$$

i.e. $$\mathrm{d}\mathbf{a}/\mathrm{d}t = \boldsymbol{\omega} \times \mathbf{a} \quad , \tag{1.136}$$

where $\boldsymbol{\omega}$ is the vector $\omega\mathbf{e}_3$ called the **angular velocity** vector. The result (1.136) is universally true for a vector **a** of constant magnitude a rotating with angular velocity $\boldsymbol{\omega}$.

Now suppose $\mathbf{a} = a_i\mathbf{e}_i^*$ is specified with respect to a frame of reference F* rotating with angular velocity $\boldsymbol{\omega}$ with respect to a fixed frame F. Let $\partial\mathbf{a}/\partial t$ be

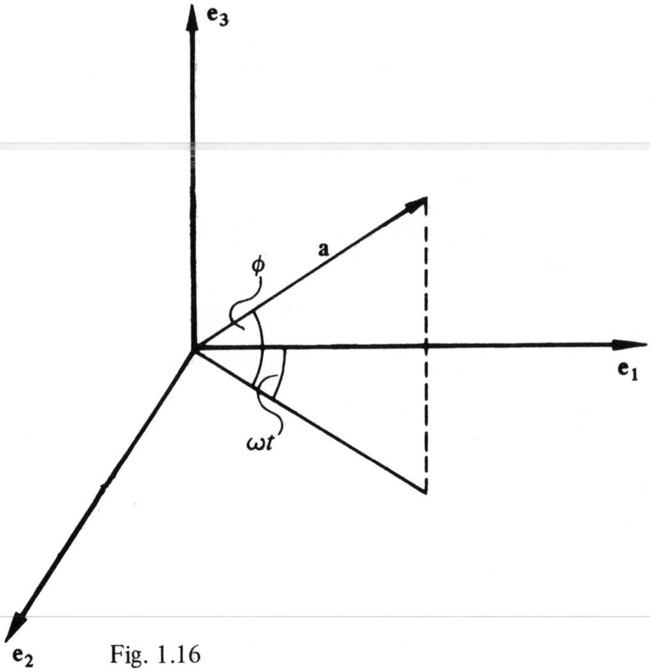

Fig. 1.16

the apparent rate of change of **a** with respect to the rotating frame F*,

i.e. $$\frac{\partial \mathbf{a}}{\partial t} = \frac{\mathrm{d}a_i}{\mathrm{d}t}\mathbf{e}_i^* \; . \qquad (1.137)$$

Let $\mathrm{d}\mathbf{a}/\mathrm{d}t$ be the rate of change of **a** with respect to the fixed frame F.

Then $$\frac{\mathrm{d}\mathbf{a}}{\mathrm{d}t} = \frac{\mathrm{d}}{\mathrm{d}t}(a_i\mathbf{e}_i^*) \; ,$$

$$= \frac{\mathrm{d}a_i}{\mathrm{d}t}\mathbf{e}_i^* + \frac{\mathrm{d}\mathbf{e}_i^*}{\mathrm{d}t} \; ,$$

$$= \partial\mathbf{a}/\partial t + a_i\boldsymbol{\omega} \times \mathbf{e}_i^* \; ,$$

$$= \partial\mathbf{a}/\partial t + \boldsymbol{\omega} \times \mathbf{a} \; . \qquad (1.138)$$

We may of course apply the formula

$$\frac{\mathrm{d}}{\mathrm{d}t} = \frac{\partial}{\partial \mathrm{t}} + \boldsymbol{\omega} \times$$

to the vector $d\mathbf{a}/dt$ to obtain the second derivative:

$$\frac{d^2\mathbf{a}}{dt^2} = \left(\frac{\partial}{\partial t} + \boldsymbol{\omega}\times\right)\left(\frac{\partial \mathbf{a}}{\partial t} + \boldsymbol{\omega}\times \mathbf{a}\right),$$

$$= \frac{\partial^2 \mathbf{a}}{\partial t^2} + \frac{\partial \boldsymbol{\omega}}{\partial t} \times \mathbf{a} + 2\boldsymbol{\omega} \times \frac{\partial \mathbf{a}}{\partial t} + \boldsymbol{\omega} \times (\boldsymbol{\omega} \times \mathbf{a})\,. \tag{1.139}$$

By putting $\mathbf{a} = \boldsymbol{\omega}$ in (1.138) we see that there is no distinction between $d\boldsymbol{\omega}/dt$ and $\partial\boldsymbol{\omega}/\partial t$. Hence we may write:

$$\frac{d^2\mathbf{a}}{dt^2} = \frac{\partial^2 \mathbf{a}}{\partial t^2} + \dot{\boldsymbol{\omega}} \times \mathbf{a} + 2\boldsymbol{\omega} \times \frac{\partial \mathbf{a}}{\partial t} + \boldsymbol{\omega} \times (\boldsymbol{\omega} \times \mathbf{a})\,. \tag{1.140}$$

In the special case where $\boldsymbol{\omega}$ is constant we have:

$$\frac{d^2\mathbf{a}}{dt^2} = \frac{\partial^2 \mathbf{a}}{\partial t^2} + 2\boldsymbol{\omega}\times\frac{\partial \mathbf{a}}{\partial t} + \boldsymbol{\omega}\times (\boldsymbol{\omega} \times \mathbf{a})\,. \tag{1.141}$$

If in addition, $|\boldsymbol{\omega}|$ is small,

$$\frac{d^2\mathbf{a}}{dt^2} \approx \frac{\partial^2 \mathbf{a}}{\partial t^2} + 2\boldsymbol{\omega}\times\frac{\partial \mathbf{a}}{\partial t}\,. \tag{1.142}$$

CHAPTER 2

Second-Order Cartesian Tensors

2.1 INTRODUCTION

2.1.1 Linear Vector Functions

In Chapter 5 it will be shown that the angular momentum, **H**, of a rigid body rotating with angular velocity $\boldsymbol{\omega}$, may be written as a vector-valued function $\mathbf{I}.\boldsymbol{\omega}$ of the vector $\boldsymbol{\omega}$, which satisfies the relation:

$$\mathbf{I}.(\lambda_1 \boldsymbol{\omega}_1 + \lambda_2 \boldsymbol{\omega}_2) = \lambda_1 \mathbf{I}.\boldsymbol{\omega}_1 + \lambda_2 \mathbf{I}.\boldsymbol{\omega}_2 \ , \tag{2.1}$$

for any pair of numbers λ_1, λ_2. Such a function is known as a **linear vector function of a vector**. To every vector $\boldsymbol{\omega}$ it assigns a vector $\mathbf{I}.\boldsymbol{\omega}$.

The reader will observe that we have assigned an extended meaning to the dot symbol hitherto reserved for the scalar product of two vectors. We shall see later (section 2.7.1) that these are both particular cases of a process known as **contraction**.

We shall see in the sections which follow that linear vector functions of vectors define **second-order tensors**. It is to the study and application of second-order tensors that the remainder of this book is chiefly devoted. In section 2.1.2 we shall introduce a basic type of second-order tensor called a dyad, and examine its properties in section 2.2.1. By the end of section 2.2.1 it should be clear to the reader why a linear vector function defines a *second*-order tensor, a term which is consistent with the terms zero- and first-order tensors for scalars and vectors. In the remainder of this book it may be assumed that all tensors are second-order tensors unless it is specifically stated otherwise.

2.1.2 Dyads

Let $\mathbf{a}, \mathbf{b}, \mathbf{c}$ be vectors where $\mathbf{a}$ and $\mathbf{b}$ are fixed but $\mathbf{c}$ is allowed to vary. Consider the vector $\mathbf{T.c}$ defined by:

$$\mathbf{T.c} = \mathbf{a}(\mathbf{b.c}) \ . \tag{2.2}$$

$\mathbf{T.c}$ is clearly linear in $\mathbf{c}$ since

$$\mathbf{a}[\mathbf{b}.(\lambda_1 \mathbf{c}_1 + \lambda_2 \mathbf{c}_2)] = \lambda_1 \mathbf{a}(\mathbf{b.c}_1) + \lambda_2 \mathbf{a}(\mathbf{b.c}_2) \ . \tag{2.3}$$

Hence **T.c** is a linear function of the vector **c**. The quantity **T** is said to be a second-order tensor.

Let us agree to introduce a new symbol $\otimes$.
We write:

$$(\mathbf{a} \otimes \mathbf{b}).\mathbf{c} = \mathbf{a}(\mathbf{b}.\mathbf{c}) \; . \tag{2.4}$$

The right-hand side of (2.4) is a vector, and so therefore is the left-hand side.

The quantity $(\mathbf{a} \otimes \mathbf{b}).\mathbf{c}$ is a linear vector function of the vector **c**. Hence $\mathbf{a} \otimes \mathbf{b}$ is a tensor.

The tensor $\mathbf{a} \otimes \mathbf{b}$ is called a **dyad** or **tensor product of two vectors**. Equation (2.4) defines the dyad in terms of its operation on a vector **c**. The equation is true for all vectors **c**, but the dyad cannot be defined except in terms of its operation on a vector. Equation (2.4) also defines the operation indicated by the dot symbol.

Examples 2.1

(1) $\mathbf{F}(\mathbf{v}) = \mathbf{v}$ defines a linear vector function **F**, since **v** is a vector linear in **v**.
(2) $F(\mathbf{v}) = \mathbf{a}.\mathbf{v}$ does *not* define a linear vector function because **a.v** is not a vector.
(3) $\mathbf{F}(\mathbf{v}) = \mathbf{a} + \mathbf{v}$, where **a** is a constant vector, does *not* define a linear vector function because $\mathbf{a} + \mathbf{v}$ is not linear in **v** according to the definition

$$\mathbf{F}(\lambda_1 \mathbf{v}_1 + \lambda_2 \mathbf{v}_2) = \lambda_1 \mathbf{F}(\mathbf{v}_1) + \lambda_2 \mathbf{F}(\mathbf{v}_2) \; .$$

Problems 2.1

(1) For each of the following definitions of a function **F** or F of a vector **v**, state whether the function is a linear vector function of **v**. Give reasons for your answers.

(i) $\mathbf{F}(\mathbf{v}) = \alpha \mathbf{V}$ (α a scalar);
(ii) $\mathbf{F}(\mathbf{v}) = \mathbf{a} \times \mathbf{v}$ (**a** a vector);
(iii) $\mathbf{F}(\mathbf{v}) = \mathbf{v} + \mathbf{a}|\mathbf{v}|$;
(iv) $\mathbf{F}(\mathbf{v}) = \mathbf{a} \times (\mathbf{b} \times \mathbf{v}) + (\mathbf{a} \times \mathbf{v}) \times \mathbf{v}$ (**b** a vector);
(v) $\mathbf{F}(\mathbf{v}) = \mathbf{a} \times (\mathbf{b} \times \mathbf{v}) + \mathbf{a} \times (\mathbf{v} \times \mathbf{v})$;
(vi) $F(\mathbf{v}) = \mathbf{a}.(\mathbf{b} \times \mathbf{v})$;
(vii) $\mathbf{F}(\mathbf{v}) = (\mathbf{a}.\mathbf{v})\mathbf{b} + 3\mathbf{v}$;
(viii) $\mathbf{F}(\mathbf{v}) = \mathbf{a} \times \mathbf{v} + \mathbf{v} \times \mathbf{a}$;
(ix) $F(\mathbf{v}) = \exp(\mathbf{a}.\mathbf{v})$;
(x) $\mathbf{F}(\mathbf{v}) = \mathbf{a} \times \mathbf{b}$;
(xi) $F(\mathbf{v}) = |\mathbf{a} \times \mathbf{v}|$.

(2) Calculate (i) $(\mathbf{e}_1 \otimes \mathbf{e}_2).\mathbf{e}_3$; (ii) $(\mathbf{e}_1 \otimes \mathbf{e}_2).\mathbf{e}_2$; (iii) $(\mathbf{e}_1 \otimes \mathbf{e}_2).\mathbf{e}_1$; (iv) $(\mathbf{e}_2 \otimes \mathbf{e}_1).\mathbf{e}_3$; (v) $(\mathbf{e}_2 \otimes \mathbf{e}_1).\mathbf{e}_2$.
(3) Let $\mathbf{a} = (3, 4, 1)$; $\mathbf{b} = (1, 0, 1)$; $\mathbf{c} = (2, 1, 0)$. Calculate (i) $(\mathbf{a} \otimes \mathbf{b}).\mathbf{c}$; (ii) $(\mathbf{b} \otimes \mathbf{c}).\mathbf{a}$; (iii) $(\mathbf{c} \otimes \mathbf{a}).\mathbf{b}$; (iv) $(\mathbf{b} \otimes \mathbf{a}).\mathbf{c}$.
(4) Given $\mathbf{a} \times \mathbf{b} = \mathbf{c} \times \mathbf{d}$, show that $\mathbf{a} \otimes \mathbf{b} - \mathbf{b} \otimes \mathbf{a} = \mathbf{c} \otimes \mathbf{d} - \mathbf{d} \otimes \mathbf{c}$.

2.2 PROPERTIES OF SECOND ORDER TENSORS

2.2.1 Properties of Dyads

It follows from the definition of a dyad (2.4) that the sum of two dyads and the product of a dyad and a scalar are given by:

$$(\lambda_1 \mathbf{a}_1 \otimes \mathbf{b}_1 + \lambda_2 \mathbf{a}_2 \otimes \mathbf{b}_2).\mathbf{c} = \lambda_1 \mathbf{a}_1 (\mathbf{b}_1.\mathbf{c}) + \lambda_2 \mathbf{a}_2 (\mathbf{b}_2.\mathbf{c}) \ . \qquad (2.5)$$

We showed in section 2.1.2 that a dyad is a second-order tensor. We shall now show that any second-order tensor $\mathbf{T}$ may be written uniquely as the sum of 9 dyads.

Now $\mathbf{T}.\mathbf{v}$ is a vector and, in particular, $\mathbf{T}.\mathbf{e}_1$ is a vector and, as such, may be expressed as a linear combination of the base vectors $\mathbf{e}_1$, $\mathbf{e}_2$ and $\mathbf{e}_3$ (see section 1.7.4). Hence, for some set of numbers λ, μ, ν:

$$\mathbf{T}.\mathbf{e}_1 = \lambda \mathbf{e}_1 + \mu \mathbf{e}_2 + \nu \mathbf{e}_3 \ , \qquad (2.6)$$

or, by re-naming the numbers λ, μ, ν:

$$\mathbf{T}.\mathbf{e}_1 = T_{11}\mathbf{e}_1 + T_{21}\mathbf{e}_2 + T_{31}\mathbf{e}_3 \ ,$$

i.e. $$\mathbf{T}.\mathbf{e}_i = T_{ki}\,\mathbf{e}_k \ . \qquad (2.7)$$

We now suppose that $\mathbf{T}$ may be written as the sum of 9 dyads:

thus $$\mathbf{T} = T_{ki}\,\mathbf{e}_k \otimes \mathbf{e}_i \ , \qquad (2.8)$$

then $$\mathbf{T}.\mathbf{e}_m = T_{ki}\,(\mathbf{e}_k \otimes \mathbf{e}_i).\mathbf{e}_m \ ,$$

$$= T_{ki}\,\mathbf{e}_k\,(\mathbf{e}_i.\mathbf{e}_m) = T_{km}\mathbf{e}_k \ .$$

Hence $$\mathbf{T}.\mathbf{e}_i = T_{ki}\mathbf{e}_k \ . \qquad (2.9)$$

Thus the supposition that $\mathbf{T}$ may be written as the sum of 9 dyads is consistent with (2.7). To show that we may always do this and that the expansion (2.8) is unique, it is necessary to derive explicit expressions for the numbers T_{ki}.

From (2.7) $$T_{ki}\mathbf{e}_k.\mathbf{e}_m = \mathbf{e}_m.(\mathbf{T}.\mathbf{e}_i) \ , \qquad (2.10)$$

i.e. $$T_{mi} = \mathbf{e}_m.(\mathbf{T}.\mathbf{e}_i) \ . \qquad (2.11)$$

The numbers T_{mi} are called the **components** of the tensor $\mathbf{T}$ in the basis $\mathbf{e}_i$. They may be written conveniently in matrix form:

$$\mathbf{T} = \begin{bmatrix} T_{11} & T_{12} & T_{13} \\ T_{21} & T_{22} & T_{23} \\ T_{31} & T_{32} & T_{33} \end{bmatrix} . \qquad (2.12)$$

The matrix (2.12) is a representation of the tensor $\mathbf{T}$ in a given basis or with respect to a given Cartesian coordinate system. Here T_{ik} is the element in the ith row and the kth column. Rows and columns are not, in general, interchangeable.

It is now apparent why a linear vector function of a vector defines a *second-order* tensor. In R^3 the scalar has one ($= 3^0$) component, the vector has three ($= 3^1$) components, and the second-order tensor has 9 ($= 3^2$) components. The order of the tensor is equal to the number of sufficies needed for each component.

Before leaving this section the reader is invited to look again at the technique used for extracting an expression for the individual components T_{ki} from expression (2.8). The steps represented by equations (2.8), (2.10), (2.11) should be thoroughly understood as the technique will be required again.

2.2.2 The Components of a Dyad

In the previous section an expression (2.11) was obtained for the components of a tensor. It is interesting to apply this expression to the dyad $\mathbf{a} \otimes \mathbf{b}$. We have, from (2.11):

$$(\mathbf{a} \otimes \mathbf{b})_{ik} = \mathbf{e}_i.[(\mathbf{a} \otimes \mathbf{b}).\mathbf{e}_k] \ , \tag{2.13}$$

$$= \mathbf{e}_i.[\mathbf{a} b_k] = a_i b_k \ . \tag{2.14}$$

In matrix notation, corresponding to (2.12), this result becomes:

$$\mathbf{a} \otimes \mathbf{b} = \begin{bmatrix} a_1 b_1 & a_1 b_2 & a_1 b_3 \\ a_2 b_1 & a_2 b_2 & a_2 b_3 \\ a_3 b_1 & a_3 b_2 & a_3 b_3 \end{bmatrix} . \tag{2.15}$$

2.2.3 The Components of the Dot Product of a Tensor and a Vector

We remarked in section 2.2.1 that $\mathbf{T}.\mathbf{v}$ is a vector. It is natural to enquire what are its components in a given basis, in terms of the components of the tensor $\mathbf{T}$. We have, from (2.9):

$$\mathbf{T}.\mathbf{v} = T_{ki}(\mathbf{e}_k \otimes \mathbf{e}_i).\mathbf{v} \ , \tag{2.16}$$

$$= T_{ki}\mathbf{e}_k v_i \ .$$

Hence $$(\mathbf{T}.\mathbf{v})_k = T_{ki} v_i \ . \tag{2.17}$$

In practice an operation such as that represented by (2.17) is probably most conveniently carried out with the help of matrix multiplication. Thus we have:

$$\mathbf{T}.\mathbf{v} = \begin{bmatrix} T_{11} & T_{12} & T_{13} \\ T_{21} & T_{22} & T_{23} \\ T_{31} & T_{32} & T_{33} \end{bmatrix} \begin{bmatrix} v_1 \\ v_2 \\ v_3 \end{bmatrix} = \begin{bmatrix} T_{11} v_1 + T_{12} v_2 + T_{13} v_3 \\ T_{21} v_1 + T_{22} v_2 + T_{23} v_3 \\ T_{31} v_1 + T_{32} v_2 + T_{33} v_3 \end{bmatrix} . \tag{2.18}$$

2.2.4 Properties of Second-Order Tensors

Two tensors are said to be equal if their dot products with all possible vectors are equal, i.e. $\mathbf{S} = \mathbf{T}$ if and only if $\mathbf{S}.\mathbf{a} = \mathbf{T}.\mathbf{a}, \forall \mathbf{a}$.

The **sum** of two tensors **S** and **T** is given by $\mathbf{S} + \mathbf{T}$, where

$$(\mathbf{S} + \mathbf{T}).\mathbf{a} = \mathbf{S}.\mathbf{a} + \mathbf{T}.\mathbf{a}, \forall \mathbf{a} \ . \tag{2.19}$$

The **zero tensor** is given by **S**, where

$$\mathbf{S}.\mathbf{a} = \mathbf{0}, \forall \mathbf{a} \ . \tag{2.20}$$

The **unit tensor** is given by **1**, where

$$\mathbf{1}.\mathbf{a} = \mathbf{a}, \forall \mathbf{a} \ . \tag{2.21}$$

We now prove two uniqueness theorems.

Theorem 2.2.4 (i) A second-order tensor is defined uniquely by giving its value on three linearly independent vectors $\mathbf{a}_1, \mathbf{a}_2, \mathbf{a}_3$.

Proof. Let $\mathbf{T}.\mathbf{a}_i = \mathbf{p}_i$ and let **v** be any vector which for some set of numbers λ_i, is given by $\mathbf{v} = \lambda_i \mathbf{a}_i$.

Then
$$\mathbf{T}.\mathbf{v} = \mathbf{T}.(\lambda_i \mathbf{a}_i) \ ,$$
$$= \lambda_i \mathbf{p}_i \ .$$

T.**v** is known for any vector **v**. Thus **T** is defined uniquely.

Theorem 2.2.4 (ii) A second-order tensor is defined uniquely by giving its nine components with respect to a Cartesian basis.

Proof. Let $\mathbf{T} = T_{ik}\mathbf{e}_i \otimes \mathbf{e}_k$, and let **v** be any vector.

Then
$$\mathbf{T}.\mathbf{v} = T_{ik}(\mathbf{e}_i \otimes \mathbf{e}_k).\mathbf{v} \ ,$$
$$= T_{ik}\mathbf{e}_i v_k \ .$$

T.**v** is known for any vector **v**. Thus **T** is defined uniquely.

Examples 2.2

(1) In matrix notation $\mathbf{e}_1 \otimes \mathbf{e}_2 = \begin{bmatrix} 0 & 1 & 0 \\ 0 & 0 & 0 \\ 0 & 0 & 0 \end{bmatrix}$.

(2) Suppose $\mathbf{a} = (1, 3, 4)$, $\mathbf{b} = (2, -1, -2)$, $\mathbf{c} = (2, 1, 0)$, then

$$\mathbf{a} \otimes \mathbf{b} = \begin{bmatrix} 2 & -1 & -2 \\ 6 & -3 & -6 \\ 8 & -4 & -8 \end{bmatrix} , \quad (\mathbf{a} \otimes \mathbf{b}).\mathbf{c} = (3, 9, 12) = \mathbf{a}(\mathbf{b}.\mathbf{c}) \ .$$

(3) Not every tensor may be expressed as a single dyad. To see this we need only examine a single counter-example. Consider the unit tensor:

$$\mathbf{1} = \mathbf{e}_1 \otimes \mathbf{e}_1 + \mathbf{e}_2 \otimes \mathbf{e}_2 + \mathbf{e}_3 \otimes \mathbf{e}_3 \ .$$

Suppose $\mathbf{1}$ may be expressed as a single dyad $\mathbf{a} \otimes \mathbf{b}$.

Then $\mathbf{1}.\mathbf{e}_1 = \mathbf{a}(\mathbf{b}.\mathbf{e}_1) = \mathbf{a}b_1 = \mathbf{e}_1$,

$$\mathbf{1}.\mathbf{e}_2 = \mathbf{a}(\mathbf{b}.\mathbf{e}_2) = \mathbf{a}b_2 = \mathbf{e}_2 \ .$$

This is a contradiction because $\mathbf{a}$ cannot be parallel to both $\mathbf{e}_1$ and $\mathbf{e}_2$.

Problems 2.2

(1) Write down the following dyads in matrix notation: (i) $\mathbf{e}_1 \otimes \mathbf{e}_1$; (ii) $\mathbf{e}_1 \otimes \mathbf{e}_3$; (iii) $\mathbf{e}_2 \otimes (\mathbf{e}_2 + \mathbf{e}_3)$.

(2) Use the definition of section 2.2.4 to prove that two tensors are equal if and only if their components with respect to a given Cartesian basis are equal.

(3) Write down the components of the unit tensor and show that they are independent of Cartesian basis. Write down the matrix representation of the tensor.

(4) Prove that the components of the zero tensor are all zero with respect to any Cartesian basis.

[In Problems 5-8 let $\mathbf{a}_1 = (3, 4, 1)$; $\mathbf{a}_2 = (1, 0, 1)$; $\mathbf{a}_3 = (2, 1, 0)$; $\mathbf{c}_1 = (3, 0, -6)$; $\mathbf{c}_2 = (-9, 0, -6)$; $\mathbf{c}_3 = (3, 0, -3)$; $\mathbf{d}_1 = (15, -12, -1)$; $\mathbf{d}_2 = (0, 0, 6)$; $\mathbf{d}_3 = (6, -6, 2)$; $\mathbf{v} = (3, 4, 5)$.]

(5) Find the components of the dyad $\mathbf{a}_1 \otimes \mathbf{a}_2$, and verify that

$$(\mathbf{a}_1 \otimes \mathbf{a}_2).\mathbf{a}_3 = \mathbf{a}_1(\mathbf{a}_2.\mathbf{a}_3) \ .$$

(6) $\mathbf{T}$ is a tensor such that $\mathbf{T}.\mathbf{e}_i = \mathbf{a}_i, i = 1, 2, 3$. Find the components of the vector $\mathbf{T}.\mathbf{v}$ and of the tensor $\mathbf{T}$.

(7) $\mathbf{T}$ is a tensor such that $\mathbf{T}.\mathbf{a}_i = \mathbf{c}_i, i = 1, 2, 3$. Find the components of the vector $\mathbf{T}.\mathbf{v}$ and of the tensor $\mathbf{T}$.

(8) $\mathbf{T}$ is a tensor such that $\mathbf{T}.\mathbf{a}_i = \mathbf{d}_i, i = 1, 2, 3$. Find the components of the vector $\mathbf{T}.\mathbf{v}$ and of the tensor $\mathbf{T}$.

(9) $\mathbf{T}$ is a tensor such that $\mathbf{T}.\mathbf{b}_i = \mathbf{f}_i, i = 1, 2, 3$, where the $\mathbf{b}_i$ and $\mathbf{f}_i$ are any vectors. Obtain an explicit expression for the components T_{jl} of $\mathbf{T}$, and show that, in general

$$\mathbf{T} = \frac{\mathbf{f}_1 \otimes (\mathbf{b}_2 \times \mathbf{b}_3) + \mathbf{f}_2 \otimes (\mathbf{b}_3 \times \mathbf{b}_1) + \mathbf{f}_3 \otimes (\mathbf{b}_1 \times \mathbf{b}_2)}{[\mathbf{b}_1, \mathbf{b}_2, \mathbf{b}_3]} \ .$$

Explain what happens when $[\mathbf{b}_1, \mathbf{b}_2, \mathbf{b}_3] = 0$.

2.3 FURTHER PROPERTIES OF SECOND ORDER TENSORS

2.3.1 Pre- and Post-Multiplication of a Tensor and a Vector

The dot product of a tensor and a vector was defined in section 2.2.3 and its components given by (2.17). The product in which the vector follows the tensor is known as **post-multiplication**. We now define a new product, namely **pre-multiplication**, by the relation:

$$(\mathbf{a}.\mathbf{T}).\mathbf{c} = \mathbf{a}.(\mathbf{T}.\mathbf{c}), \forall \mathbf{c} \ . \tag{2.22}$$

a.T is a vector whose components may be found by letting $\mathbf{c} = \mathbf{e}_i$.

Thus
$$(\mathbf{a}.\mathbf{T})_i = \mathbf{a}.(T_{ki}\mathbf{e}_k) = a_k T_{ki} \ . \tag{2.23}$$

By virtue of the truth of (2.22) we may omit the brackets and write either product as **a.T.c**. It is not true however that **a.T.c.** and **c.T.a** have the same value in general. The verification of this fact is left as an exercise for the reader.

We notice that in (2.17) and (2.23) the repeated suffices are adjacent. This point will be brought out again when dealing with the subject of **contraction** of which (2.17) and (2.23) are particular cases.

Whereas the matrix equivalent of (2.17) gives a column vector, the matrix equivalent of (2.23) gives a row vector.

Thus
$$(a_1, a_2, a_3)\begin{bmatrix} T_{11} & T_{12} & T_{13} \\ T_{21} & T_{22} & T_{23} \\ T_{31} & T_{32} & T_{33} \end{bmatrix}$$

$$= (a_1T_{11} + a_2T_{21} + a_3T_{31},\ a_1T_{12} + a_2T_{22} + a_3T_{32},\ a_1T_{13} + a_2T_{23} + a_3T_{33}) \ .$$

It should be emphasized that the terms 'row vector' and 'column vector' only have distinct meanings in the language of matrices. They both represent the same vector.

2.3.2 The Transpose Tensor

Suppose the tensor **T** is expressed by the sum of dyads:

$$\mathbf{T} = \mathbf{a} \otimes \mathbf{b} + \mathbf{c} \otimes \mathbf{d} + \ldots \ . \tag{2.24}$$

The tensor $\mathbf{T}'$, expressed by the sum of dyads:

$$\mathbf{T}' = \mathbf{b} \otimes \mathbf{a} + \mathbf{d} \otimes \mathbf{c} + \ldots \ , \tag{2.25}$$

is called the **transpose** of **T**. To show that this definition is independent of the particular sum of dyads chosen to represent **T**, we choose any vector **v** and form the dot product:

$$\mathbf{T}'.\mathbf{v} = (\mathbf{b} \otimes \mathbf{a} + \mathbf{d} \otimes \mathbf{c} + ..).\mathbf{v} \ ,$$

$$= \mathbf{b}(\mathbf{a}.\mathbf{v}) + \mathbf{d}(\mathbf{c}.\mathbf{v}) + \ldots \ .$$

Thus $$\mathbf{T}'.\mathbf{v} = \mathbf{v}.\mathbf{T} \ . \tag{2.26}$$

Equation (2.26), true for all vectors **v**, provides an alternative definition for the transpose tensor (see Problems 2.3, No. 3).

The Cartesian components of $\mathbf{T}'$ may be found by writing $\mathbf{a} = a_i\mathbf{e}_i$, etc.

Thus $$\mathbf{T}' = (b_i a_k + d_i c_k + \ldots)\mathbf{e}_i \otimes \mathbf{e}_k \ , \tag{2.27}$$

therefore $$T'_{ik} = b_i a_k + d_i c_k + \ldots \tag{2.28}$$

$$= T_{ki} \ . \tag{2.29}$$

The components of $\mathbf{T}'$ in any basis are therefore determined by the components of **T**.

Examples 2.3

(1) The transpose of the tensor **T** which has the matrix representation:

$$\mathbf{T} = \begin{bmatrix} T_{11} & T_{12} & T_{13} \\ T_{21} & T_{22} & T_{23} \\ T_{31} & T_{32} & T_{33} \end{bmatrix},$$

with respect to a given basis, has the matrix representation:

$$\mathbf{T}' = \begin{bmatrix} T_{11} & T_{21} & T_{31} \\ T_{12} & T_{22} & T_{32} \\ T_{13} & T_{23} & T_{33} \end{bmatrix},$$

with respect to the same basis.

(2) The transpose of the transpose of a tensor is the tensor itself, i.e. $(\mathbf{T}')' \equiv \mathbf{T}$.

Problems 2.3

(1) Let $\mathbf{a} = (3, 4, 1)$; $\mathbf{b} = (1, 0, 1)$; $\mathbf{c} = (2, 1, 0)$;

$$\mathbf{T} = \begin{bmatrix} 3 & 1 & 1 \\ 2 & 4 & -2 \\ 1 & -1 & 0 \end{bmatrix}; \quad \mathbf{S} = \begin{bmatrix} 1 & 1 & -1 \\ 2 & 2 & 3 \\ 4 & 0 & 2 \end{bmatrix}.$$

Write down (i) **T.a**; (ii) **a.T**; (iii) **T.b**; (iv) **a.(T.b)**; (v) **(a.T).b**; (vi) **b.(T.a)**; (vii) **a.S.b**; (viii) **b.S.a**.

(2) **S** and **T** are two tensors such that $S_{ik} = T_{ki}$ in any frame of reference. Prove that $\mathbf{S} = \mathbf{T}'$.

(3) **S** and **T** are two tensors such that **S.v** = **v.T**, for all vectors **v**. Prove that **S** = **T**′.

(4) **T** is a tensor such that **a.T** = **T.a** for all vectors **a**. Show that, with respect to any Cartesian basis, $T_{ik} = T_{ki}$.

(5) **a** and **b** are vectors such that **T.a** = **T.b** for all tensors **T**. What can be deduced about the vectors **a** and **b**?

(6) Prove that the tensor **1** defined by the relation:

$$\mathbf{a.1} = \mathbf{a}, \text{ for all vectors } \mathbf{a},$$

is the same in all respects as the tensor **1** defined in section 2.2.4.

(7) **a**, **b**, **c** are any three vectors and **S** is any tensor. Prove that:

(i) $\mathbf{a}.(\mathbf{b} \otimes \mathbf{c}) = (\mathbf{a}.\mathbf{b})\mathbf{c}$; (ii) $\mathbf{a.S.b} = \mathbf{b.S}'.\mathbf{a}$.

(8) Prove that for any two tensors **S** and **T**

(i) $(\mathbf{S} + \mathbf{T})' = \mathbf{S}' + \mathbf{T}'$; (ii) $(\lambda\mathbf{T})' = \lambda\mathbf{T}'$.

(9) Demonstrate the truth of the results proved in questions (7) and (8) for the vectors and tensors given in question (1).

2.4 SYMMETRIC AND ANTISYMMETRIC TENSORS

A tensor which is equal to its transpose is said to be **symmetric**, i.e. **T** = **T**′. Great emphasis will be laid on symmetric tensors here because most tensors which arise naturally in physics are symmetric. For example the inertia tensor, the stress tensor, the strain tensor and the rate of strain tensor are all symmetric. It will be shown that symmetric tensors have a number of mathematical properties which are not shared by tensors in general, the most remarkable of these being that by a suitable choice of Cartesian axes, the off-diagonal terms can be made to vanish.

Since **T** = **T**′, we have, from (2.29):

$$T_{ki} = T_{ik} \quad . \tag{2.30}$$

Equation (2.30) is true in any frame of reference. The matrix representation of a symmetric tensor therefore takes the form:

$$\mathbf{T} = \begin{bmatrix} T_{11} & T_{12} & T_{13} \\ T_{12} & T_{22} & T_{23} \\ T_{13} & T_{23} & T_{33} \end{bmatrix} . \tag{2.31}$$

Hence the symmetric tensor has only six independent components rather than nine.

A tensor which is equal to minus its transpose is said to be **antisymmetric** or

skew-symmetric. Antisymmetric tensors also occur in physics and tend to be associated with rotation. The spin tensor, introduced in Chapter 6, is an example of an antisymmetric tensor.

Arguments similar to those employed for symmetric tensors lead to the relations

$$\mathbf{T} = -\mathbf{T}' \ , \tag{2.32}$$

$$T_{ki} = -T_{ik} \ , \tag{2.33}$$

$$\mathbf{T} = \begin{bmatrix} 0 & T_{12} & T_{13} \\ -T_{12} & 0 & T_{23} \\ -T_{13} & -T_{23} & 0 \end{bmatrix} , \tag{2.34}$$

for any antisymmetric tensor $\mathbf{T}$. It will be noted that the components on the leading diagonal are identically zero and that there are but three independent components.

The classification of tensors into symmetric tensors, antisymmetric tensors and others is of considerable mathematical significance as the following theorem will show. It can also be of some physical significance as we shall see when we examine the tensor gradient of the velocity field in Chapter 6.

Theorem 2.4.1 Every second-order tensor may be expressed *uniquely* as the sum of a symmetric and an antisymmetric tensor.

Proof. Suppose it is possible to do this, then any tensor $\mathbf{T}$ may be written:

$$\mathbf{T} = \mathbf{S} + \mathbf{A} \ , \tag{2.35}$$

where $\mathbf{S}$ is a symmetric and $\mathbf{A}$ an antisymmetric tensor, with the properties $\mathbf{S} = \mathbf{S}'$, $\mathbf{A} = -\mathbf{A}'$. Taking the transpose of (2.35) we have:

$$\mathbf{T}' = \mathbf{S} - \mathbf{A} \ . \tag{2.36}$$

Hence, from (2.35) and (2.36):

$$\left.\begin{aligned} \mathbf{S} &= \tfrac{1}{2}(\mathbf{T} + \mathbf{T}') \ , \\ \mathbf{A} &= \tfrac{1}{2}(\mathbf{T} - \mathbf{T}') \ . \end{aligned}\right\} \tag{2.37}$$

Therefore $\mathbf{T} \equiv \tfrac{1}{2}(\mathbf{T} + \mathbf{T}') + \tfrac{1}{2}(\mathbf{T} - \mathbf{T}')$. (2.38)

Equation (2.38) shows that the tensor may be split in the desired way and (2.37) shows that the split is unique. It is therefore meaningful to talk of the **symmetric part** of a tensor $\mathbf{T}$, i.e. $\tfrac{1}{2}(\mathbf{T} + \mathbf{T}')$, and the **antisymmetric part**, i.e. $\tfrac{1}{2}(\mathbf{T} - \mathbf{T}')$. All tensors may be split in this way if the zero tensor is regarded as being both symmetric and antisymmetric. Thus the symmetric part of a symmetric tensor is the tensor itself, and the antisymmetric part is the zero tensor. Similarly the antisymmetric part of an antisymmetric tensor is the tensor itself and the symmetric part is the zero tensor.

Examples 2.4

(1) The tensors $\mathbf{1}$, $\mathbf{e}_1 \otimes \mathbf{e}_2 + \mathbf{e}_2 \otimes \mathbf{e}_1$, $\mathbf{a} \otimes \mathbf{b} + \mathbf{b} \otimes \mathbf{a}$, are all symmetric as may be readily verified by writing down their matrix representations. Similarly the tensors $\mathbf{e}_1 \otimes \mathbf{e}_2 - \mathbf{e}_2 \otimes \mathbf{e}_1$, $\mathbf{a} \otimes \mathbf{b} - \mathbf{b} \otimes \mathbf{a}$, are antisymmetric.

(2) The tensor $\mathbf{a} \otimes \mathbf{b}$ is in general neither symmetric nor antisymmetric. The symmetric part is $\frac{1}{2}(\mathbf{a} \otimes \mathbf{b} + \mathbf{b} \otimes \mathbf{a})$ and the antisymmetric part is $\frac{1}{2}(\mathbf{a} \otimes \mathbf{b} - \mathbf{b} \otimes \mathbf{a})$.

Problems 2.4

(1) State whether the following tensors are symmetric, antisymmetric or neither:

(i) $\mathbf{e}_1 \otimes \mathbf{e}_2$; (ii) $\mathbf{e}_1 \otimes \mathbf{e}_1$; (iii) $\mathbf{e}_2 \otimes \mathbf{e}_3 - \mathbf{e}_3 \otimes \mathbf{e}_2$;

(iv) $\begin{bmatrix} 1 & 3 & 1 \\ 3 & 0 & -2 \\ 1 & -2 & -2 \end{bmatrix}$; (v) $\begin{bmatrix} 3 & 2 & -7 \\ -2 & 1 & 4 \\ 7 & -4 & 5 \end{bmatrix}$; (vi) $\begin{bmatrix} 3 & 2 & -7 \\ 2 & 1 & -4 \\ -7 & -4 & 5 \end{bmatrix}$; (vii) $\begin{bmatrix} 0 & 2 & -2 \\ 2 & 0 & -3 \\ -2 & -3 & 0 \end{bmatrix}$;

(viii) $\mathbf{a} \otimes (\mathbf{b} + \mathbf{c}) - (\mathbf{b} + \mathbf{c}) \otimes \mathbf{a}$; (ix) $\mathbf{a} \otimes (\mathbf{b} + \mathbf{c}) + \mathbf{b} \otimes (\mathbf{a} + \mathbf{c})$;
(x) $\mathbf{e}_1 \otimes \mathbf{e}_2 + \mathbf{e}_2 \otimes \mathbf{e}_3 + \mathbf{e}_3 \otimes \mathbf{e}_1$.

(2) Split all the tensors of question (1) into their symmetric and antisymmetric parts.

(3) The vectors $\mathbf{a}$ and $\mathbf{b}$ are given by $\mathbf{a} = (1, 3, 4)$; $\mathbf{b} = (2, -1, -2)$. Find the symmetric and antisymmetric parts of the dyad $\mathbf{a} \otimes \mathbf{b}$, and express them in matrix form.

(4) $\mathbf{S}$ is a symmetric tensor and $\mathbf{A}$ is an antisymmetric tensor. Prove that

(i) $A_{ii} = 0$; (ii) $S_{ik}A_{ki} = 0$.

2.5 TRANSFORMATION PROPERTIES OF TENSORS

2.5.1 The Invariance of Tensors – Transformation Law

We have seen already (sections 1.3.1, 1.10.1) that scalars and vectors are invariant with respect to coordinate transformations. It is absolutely fundamental that second - (and higher-) order tensors share this property of invariance. This result is obvious for physical reasons as soon as we begin to discuss particular tensors, but more particularly it is obvious from our definition of section 2.1.1, which is entirely basis-free.

In sections 1.10.1, 1.10.2 we pursued the matter of the invariance of vectors and noted that, while the vectors themselves were invariant, their components with respect to a particular coordinate system were not. In fact they varied according to equation (1.126) under a rotation specified by the rotation matrix. We now face the task of formulating a law, corresponding to (1.126), for the transformation of second-order tensor components. This too is of unusual importance because many authors (Jeffreys, Milne, etc.) take this law as their starting point, that is to say they define a second-order Cartesian tensor as a quantity whose components transform according to the law which we are about to prove (2.43), when the basis suffers a rotation Θ about the origin.

We shall use the notation of section 1.10.2 in which the Cartesian axes were rotated from a triad $(\mathbf{e}_1, \mathbf{e}_2, \mathbf{e}_3)$ to a new triad $(\mathbf{e}_1^*, \mathbf{e}_2^*, \mathbf{e}_3^*)$. We consider a typical tensor $\mathbf{T}$ whose components with respect to the original basis are T_{ik}, and with respect to the new basis are T_{ik}^*. As before, we define the components θ_{ji} of the rotating tensor by the relation:

$$\theta_{ji} = \mathbf{e}_j^*.\mathbf{e}_i \; . \tag{2.39}$$

Proceeding as in section 1.10.2 we have, corresponding to (1.124):

$$T_{jl}^* = \mathbf{e}_j^*.\mathbf{T}.\mathbf{e}_l^* \; , \tag{2.40}$$

$$= \mathbf{e}_j^*.(T_{ik}\mathbf{e}_i \otimes \mathbf{e}_k).\mathbf{e}_l^* \; , \tag{2.41}$$

$$= T_{ik}\theta_{ji}\theta_{lk} \; , \tag{2.42}$$

$$= \theta_{ji}\theta_{lk}T_{ik} \; . \tag{2.43}$$

Equation (2.43) expresses the transformation law for the components of Cartesian tensors. It corresponds to equation (1.126), the transformation law for the components of vectors. As mentioned above, the law (2.43) provides an alternative starting point for the theory of Cartesian tensors. In other words a tensor may be defined as a quantity which can be represented by an array whose components obey the transformation law (2.43). We have obtained the transformation law from our definition of a tensor contained in (2.4) and (2.8). In order to establish the precise equivalence of the two definitions it is necessary to derive the linear function definition from the transformation law. This we shall now do.

Let us suppose that $\mathbf{T}$ is a quantity whose components obey (2.43).

Let $\mathbf{c} = \mathbf{T}.\mathbf{a}$, where $\mathbf{a}$ is a vector. (2.44)

Our object is to show that $\mathbf{c}$ is also a vector. In component form (2.44) becomes:

$$c_i = T_{ik}a_k.$$

We now rotate the axes to obtain:

$$\begin{aligned} c_i^* &= T_{ik}^* a_k^* , \\ &= \theta_{ij}\theta_{kl}T_{jl}\theta_{kn}a_n , \\ &= \theta_{ij}\delta_{ln}T_{jl}a_n , \\ &= \theta_{ij}T_{jn}a_n , \\ &= \theta_{ij}c_j . \end{aligned} \tag{2.45}$$

Equation (2.45) shows that **c** is a quantity whose components obey the transformation law for vectors. Thus **c** is a vector. **T.c** is therefore a linear vector-valued function of a vector, in other words **T** is a second-order tensor according to our original definition.

2.5.2 The Practical Transformation of Tensor Components

In practice the matrix equivalent of the transformation law (2.43) is easier to use than (2.43) itself. Thus we have:

$$\mathbf{T}^* = \Theta \mathbf{T} \Theta' , \tag{2.46}$$

which, when written out in full becomes:

$$\begin{bmatrix} T_{11}^* & T_{12}^* & T_{13}^* \\ T_{21}^* & T_{22}^* & T_{23}^* \\ T_{31}^* & T_{32}^* & T_{33}^* \end{bmatrix} =$$

$$\begin{bmatrix} \theta_{11} & \theta_{12} & \theta_{13} \\ \theta_{21} & \theta_{22} & \theta_{23} \\ \theta_{31} & \theta_{32} & \theta_{33} \end{bmatrix} \begin{bmatrix} T_{11} & T_{12} & T_{13} \\ T_{21} & T_{22} & T_{23} \\ T_{31} & T_{32} & T_{33} \end{bmatrix} \begin{bmatrix} \theta_{11} & \theta_{21} & \theta_{31} \\ \theta_{12} & \theta_{22} & \theta_{32} \\ \theta_{13} & \theta_{23} & \theta_{33} \end{bmatrix} . \tag{2.47}$$

Θ is the rotation matrix discussed in section 1.10.3.

Example 2.5

(1) A tensor **T** has components $\begin{bmatrix} 1 & 0 & 0 \\ 0 & 0 & 1 \\ 0 & 1 & 0 \end{bmatrix}$ with respect to a certain set of coordinate axes. The axes are rotated by an angle θ in the positive sense about the x_3-axis (Fig. 1.15). Write down the components of the tensor with respect to the new axes.

The rotation matrix is given by:

$$\Theta = \begin{bmatrix} \cos\theta & \sin\theta & 0 \\ -\sin\theta & \cos\theta & 0 \\ 0 & 0 & 1 \end{bmatrix} .$$

The transformed tensor is therefore given by:

$$\mathbf{T}^* = \begin{bmatrix} \cos\theta & \sin\theta & 0 \\ -\sin\theta & \cos\theta & 0 \\ 0 & 0 & 1 \end{bmatrix} \begin{bmatrix} 1 & 0 & 0 \\ 0 & 0 & 1 \\ 0 & 1 & 0 \end{bmatrix} \begin{bmatrix} \cos\theta & -\sin\theta & 0 \\ \sin\theta & \cos\theta & 0 \\ 0 & 0 & 1 \end{bmatrix} ,$$

$$= \begin{bmatrix} \cos^2\theta & -\cos\theta\sin\theta & \sin\theta \\ -\cos\theta\sin\theta & -\sin^2\theta & \cos\theta \\ \sin\theta & \cos\theta & 0 \end{bmatrix} .$$

Problems 2.5

(1) A tensor **T** has components $\begin{bmatrix} 1 & 0 & 0 \\ 0 & 1 & 0 \\ 1 & 0 & 0 \end{bmatrix}$ with respect to a certain set of coordinate axes. The axes are rotated by an angle θ in the positive sense about the x_1-axis. Write down the components of the tensor with respect to the new axes, and evaluate these components when $\theta = \pi/2, \pi, 3\pi/2$, and 2π.

(2) Find the components of the dyad $\mathbf{a} \otimes \mathbf{b}$ where $\mathbf{a} = (2, 1, 1), \mathbf{b} = (-3, -2, 4)$, with respect to a certain set of coordinate axes. Rotate the axes by $\pi/4$ about the x_2-axis in the negative sense and find the new components of the dyad. Find also the new components of the vectors **a** and **b** and hence the components of $\mathbf{a} \otimes \mathbf{b}$. Compare the two results.

(3) A quadric surface has an equation of the form:

$$ax^2 + by^2 + cz^2 + 2fyz + 2gzx + 2hxy = 1 .$$

Show that, on a change of coordinate system, the coefficients transform like the components of a symmetric second-order tensor. Hence show that the equation may be written in the form:

$$\mathbf{x}.\mathbf{T}.\mathbf{x} = 1 .$$

(4) Prove that the determinant of a tensor is unchanged by a rotation.

(5) A set of Cartesian axes is rotated about the origin to coincide with the unit vectors:

$$\mathbf{e}_1^* = (\sin\theta\cos\phi, \sin\theta\sin\phi, \cos\theta) ,$$

$$\mathbf{e}_2^* = (\cos\theta\cos\phi, \cos\theta\sin\phi, -\sin\theta) ,$$

$$\mathbf{e}_3^* = (-\sin\phi, \cos\phi, 0) .$$

Write down the rotation matrix corresponding to this rotation and transform the components of the tensor

$$\mathbf{T} = \begin{bmatrix} 2 & 1 & 0 \\ 1 & -3 & -1 \\ 0 & -1 & 2 \end{bmatrix}$$

when (i) $\theta = \phi = \pi/4$; (ii) $\theta = \pi/4, \phi = \pi/3$.

(6) In Problems 2.2, No. 3 we proved that the components of the unit tensor were scalars, independent of Cartesian basis. Now prove the same result using the transformation law (2.43). Hint: the result of equation (1.130) may be useful.

2.6 THE SCALAR INVARIANTS OF A SECOND-ORDER TENSOR

The invariance properties of tensors of various orders have been stressed at appropriate points in this book (see sections 1.3.1, 1.10.1, 2.5.1). It has also been stressed that the individual components of tensors are not, in general, invariant, the only exception to this rule which we have encountered so far being the components of the unit tensor. There are, however, a number of scalar invariants associated with every second-order tensor, i.e. numbers which depend on the tensor itself, and not on the matrix representing it, or its individual components. One such invariant, namely the determinant of the matrix, has already been found (Problems 2.5, No. 4). The invariance of the determinant may be used to establish two further invariants.

Let $\mathbf{T} = \mathbf{S} - \lambda\mathbf{1}$, where $\mathbf{S}$ is any second-order tensor, $\mathbf{1}$ is the unit tensor and λ is a number. The determinant of $\mathbf{T}$ is unchanged by a rotation. Written out in full, this determinant is:

$$\det \mathbf{T} = \begin{vmatrix} S_{11} - \lambda & S_{12} & S_{13} \\ S_{21} & S_{22} - \lambda & S_{23} \\ S_{31} & S_{32} & S_{33} - \lambda \end{vmatrix} . \tag{2.48}$$

The determinant in (2.48) may be expanded in powers of λ to give:

$$\det \mathbf{T} = -\lambda^3 + \lambda^2(S_{11} + S_{22} + S_{33}) - \lambda \begin{vmatrix} S_{22} & S_{23} \\ S_{32} & S_{33} \end{vmatrix}$$

$$-\lambda \begin{vmatrix} S_{33} & S_{31} \\ S_{13} & S_{11} \end{vmatrix} - \lambda \begin{vmatrix} S_{11} & S_{12} \\ S_{21} & S_{22} \end{vmatrix} + \begin{vmatrix} S_{11} & S_{12} & S_{13} \\ S_{21} & S_{22} & S_{23} \\ S_{31} & S_{32} & S_{33} \end{vmatrix} . \tag{2.49}$$

Since det **T** must be invariant with respect to λ, the coefficients of the powers of λ must themselves be invariants.

The coefficient of λ^2 is the first invariant:

$$I_1 = S_{ii} . \tag{2.50}$$

S_{ii} is known as the **trace** of the tensor **S**, and is written Tr(**S**).

It is the sum of the terms on the leading diagonal. Its invariance may, of course, also be proved by a direct appeal to the transformation law for tensor components [see Problems 2.6 No. 1(i)]. This result is especially significant when the trace has a physical interpretation as in the case of the rate of strain tensor for example.

The second invariant, I_2, is minus the coefficient of λ, that is:

$$I_2 = S_{22}S_{33} - S_{23}S_{32} + S_{33}S_{11} - S_{13}S_{31} + S_{11}S_{22} - S_{12}S_{21} . \tag{2.51}$$

This result may be written more compactly using the suffix notation (see Examples 2.6, No. 2).

The third invariant, I_3, is the determinant of **S**:

$$I_3 = \det \mathbf{S} = \begin{vmatrix} S_{11} & S_{12} & S_{13} \\ S_{21} & S_{22} & S_{23} \\ S_{31} & S_{32} & S_{33} \end{vmatrix} . \tag{2.52}$$

Expanding (2.52) we obtain:

$$I_3 = S_{11}(S_{22}S_{33} - S_{23}S_{32}) + S_{12}(S_{23}S_{31} - S_{21}S_{33}) + S_{13}(S_{21}S_{32} - S_{22}S_{31}),$$

$$= S_{11}S_{22}S_{33} + S_{12}S_{23}S_{31} + S_{13}S_{21}S_{32} - S_{11}S_{23}S_{32} - S_{22}S_{13}S_{31} - S_{33}S_{12}S_{21} .$$

To write this result in suffix notation we must wait until third-order tensors have been studied (see Problems 3.2, No. 2).

The three scalar invariants have an important role to play as the theory of Cartesian tensors is developed, particularly with regard to the theory of elasticity and the representation theorems for isotropic tensors.

Strictly speaking one should speak of the trace and determinant of the matrix representing the tensor rather than the trace and determinant of the tensor. However, once the invariance of these quantities has been established the actual matrix used to represent the tensor is of no consequence and we can speak of the trace and determinant of the tensor without ambiguity.

Examples 2.6

(1) $I_2 = A_{ii}$, where A_{ik} is the cofactor of S_{ik} [see (2.51), (2.52)].

(2) $I_2 = \frac{1}{2}(S_{ii}S_{kk} - S_{ik}S_{ki})$, for $S_{ik}S_{ki} = S_{11}^2 + S_{22}^2 + S_{33}^2 + 2(S_{23}S_{32} + S_{31}S_{13} + S_{12}S_{21})$, and $S_{ii}S_{kk} = S_{11}^2 + S_{22}^2 + S_{33}^2 + 2(S_{22}S_{33} + S_{33}S_{11} + S_{11}S_{22})$.

(3) A fourth scalar invariant is given by:

$$I_4 = S_{ik}S_{ik} ,$$

$$\text{for } S_{jl}^*S_{jl}^* = \theta_{ji}\theta_{lk}\theta_{jm}\theta_{lp}S_{ik}S_{mp},$$

$$= \delta_{im}\delta_{kp}S_{ik}S_{mp}, = S_{ik}S_{ik} .$$

I_4 is independent of I_1, I_2 and I_3 except when **S** is symmetric.

Problems 2.6

(1) Use the transformation laws for vectors and tensors to show that the following quantities are scalar invariants:

(i) S_{ii}; (ii) $S_{ik}S_{ki}$; (iii) **u.S.v**.

(2) Prove that the relation **S.v** = λ**v** is invariant under rotations.

(3) Prove that symmetry is an invariant property of a tensor.

2.7 INNER PRODUCTS

2.7.1 The Inner Product of Two Tensors - Contraction

The dot product of two vectors was defined at an early stage (section 1.5.2). With the introduction of the summation convention in section 1.9.1 we were able to write the dot product of two vectors **a** and **b**, as:

$$\mathbf{a.b} = a_i b_i . \tag{2.53}$$

In section 2.2.3 of the present chapter the dot product of a tensor and a vector was given explicitly as:

$$(\mathbf{T.v})_k = T_{ki}v_i . \tag{2.54}$$

In each case the repeated suffices were adjacent. Such adjacent suffices are alternatively termed **inner suffices** and the dot product is alternatively known as the **inner product**. The next obvious use of the dot symbol is for an inner product of two tensors. We shall define the product in a basis-free way:

Definition. Let **S** and **T** be two tensors and **v** any vector, then the inner product of **S** and **T** is given by:

$$(\mathbf{S}.\mathbf{T}).\mathbf{v} = \mathbf{S}.(\mathbf{T}.\mathbf{v}) \qquad \forall \mathbf{v} \; . \tag{2.55}$$

Equation (2.55) defines the inner product of two tensors in terms of its inner product with an arbitrary vector **v**. The right-hand side of (2.55) is a vector linear in **v** therefore the same is true of the left-hand side. Thus the inner product **S.T** is itself a second-order tensor. The components of **S.T** are found as follows:

$$\begin{aligned}(\mathbf{S}.\mathbf{T})_{ik} &= \mathbf{e}_i.(\mathbf{S}.\mathbf{T}).\mathbf{e}_k \; , \\ &= \mathbf{e}_i.[\mathbf{S}.(\mathbf{T}.\mathbf{e}_k)] \; , \\ &= \mathbf{e}_i.[\mathbf{S}.(T_{mk}\mathbf{e}_m)] \; , \\ &= \mathbf{e}_i.(S_{pm}\mathbf{e}_p T_{mk}) \; , \\ &= S_{im}T_{mk} \; . \end{aligned} \tag{2.56}$$

We observe again that it is the inner suffices which are repeated.

The three inner products which are given explicitly in equations (2.53), (2.54) and (2.56) all represent particular cases of a process known as **contraction**. This is a process by which two of the indices of the tensors present are put equal and the implied summation carried out. The number of free suffices is therefore reduced by two, with a corresponding reduction in the sum of the orders. The orders of the tensors on the right-hand sides of (2.53), (2.54) and (2.56) are given respectively by:

$$\left.\begin{aligned} 1+1-2 &= 0 \\ 2+1-2 &= 1 \\ 2+2-2 &= 2 \end{aligned}\right\} . \tag{2.57}$$

2.7.2 The Double Inner Product of Two Tensors

The third equation of (2.57) indicates that there is further scope for reduction in the sum of the orders of two tensors. This reduction may be effected by equating the two free indices in (2.56), but again we prefer a basis-free definition. The symbol used is two dots.

Definition. The double inner product of two tensors **S** and **T** is given by

$$\mathbf{S}:\mathbf{T} = \mathrm{Tr}(\mathbf{S}.\mathbf{T}) \; . \tag{2.58}$$

Hence $\quad \mathbf{S}:\mathbf{T} = S_{im}T_{mi} \; . \quad$ (2.59)

We have already seen (Problems 2.6, No. 1) that the trace of any tensor is a scalar invariant. The double dot therefore represents a double contraction in which the sum of the orders has been reduced by four.

Example 2.7

(1) The operation of finding the trace of the tensor **T** is an example of contraction, for the indices of the components T_{ik} are put equal.

Problems 2.7

(1) Show that the components of **S.T** as given by (2.56), obey the transformation law for second-order tensors.
(2) Prove that for any tensor **T** and any vectors **a** and **b**:

$$\mathbf{T}:\mathbf{a} \otimes \mathbf{b} = \mathbf{b}.\mathbf{T}.\mathbf{a} \ .$$

(3) Prove that the trace of any tensor **T** is given by:

$$\mathrm{Tr}(\mathbf{T}) = \mathbf{T}:\mathbf{1} \ .$$

(4) Show that (**A.B**):(**C.D**), where **A, B, C, D** are second-order tensors, is a scalar invariant, and write it down in suffix notation.
(5) Use the transformation law for tensors to show that the double inner product **S**:**T** is a scalar invariant.

2.8 VECTORS AND TENSORS WITH COMPLEX COMPONENTS

We propose now to extend the definitions of scalar and vector products to include vectors with complex components.

The **complex scalar product** of two vectors **x** and **y** is defined to be:

$$\mathbf{x}.\mathbf{y} = \bar{x}_i y_i \ , \tag{2.60}$$

where $\bar{x}_i$ is the complex conjugate of x_i. This definition reduces to the definition (1.69) when all the components are real. The same is true of all subsequent definitions in this section.

The complex scalar product does not possess the commutative property. In fact

$$\mathbf{y}.\mathbf{x} = \bar{y}_i x_i = \overline{\mathbf{x}.\mathbf{y}} \ . \tag{2.61}$$

When **x** = **y** we obtain the result:

$$\mathbf{x}.\mathbf{x} = x_i \bar{x}_i \ , \tag{2.62}$$

i.e. **x.x** is real. This is the square of the **modulus** of **x** defined by

$$|\mathbf{x}| = \sqrt{(\mathbf{x}.\mathbf{x})} \ . \tag{2.63}$$

The **complex vector product** of **x** and **y** is defined to be:

$$\mathbf{x} \times \mathbf{y} = \begin{vmatrix} \mathbf{e}_1 & \mathbf{e}_2 & \mathbf{e}_3 \\ \bar{x}_1 & \bar{x}_2 & \bar{x}_3 \\ \bar{y}_1 & \bar{y}_2 & \bar{y}_3 \end{vmatrix} \ . \tag{2.64}$$

If $\mathbf{x}$ is a vector whose elements are x_i, the vector whose corresponding elements are $\bar{x}_i$ is called the **conjugate vector** $\bar{\mathbf{x}}$. In a similar way we may define the conjugate tensor $\bar{\mathbf{T}}$.

Thus, from (2.61) and (2.64) we have:

$$\bar{\mathbf{x}}.\bar{\mathbf{y}} = \overline{\mathbf{x}.\mathbf{y}} \;, \quad \bar{\mathbf{x}} \times \bar{\mathbf{y}} = \overline{\mathbf{x} \times \mathbf{y}} \;. \tag{2.65}$$

The two inner products of a tensor and a vector, both of which may have complex elements, are defined by:

$$\begin{aligned} \mathbf{T}.\mathbf{x} &= T_{ki}x_i\mathbf{e}_k \;, \\ \mathbf{x}.\mathbf{T} &= x_i\bar{T}_{ik}\mathbf{e}_k \;. \end{aligned} \tag{2.66}$$

We define now the **Hermitian Transpose** of $\mathbf{T}$, denoted by $\mathbf{T}^{\mathrm{H}}$, as the transpose of $\bar{\mathbf{T}}$.

Thus $$\mathbf{T}^{\mathrm{H}} = \bar{\mathbf{T}}' \;. \tag{2.67}$$

From (2.66) we have:

$$(\mathbf{x}.\mathbf{T}).\mathbf{y} = \bar{x}_i T_{ik} y_k = \mathbf{x}.(\mathbf{T}.\mathbf{y}) \;. \tag{2.68}$$

Hence we may continue to write $\mathbf{x}.\mathbf{T}.\mathbf{y}$ without ambiguity.

Example 2.8

(1) Prove that for any tensor $\mathbf{T}$ and any vectors $\mathbf{x}$ and $\mathbf{y}$:

$$\mathbf{x}.\mathbf{T}^{\mathrm{H}}.\mathbf{y} = \overline{\mathbf{y}.\mathbf{T}.\mathbf{x}} \;.$$

We have $$\begin{aligned} \mathbf{x}.\mathbf{T}^{\mathrm{H}}.\mathbf{y} &= \bar{x}_i T^{\mathrm{H}}_{ik} y_k \;, \\ &= \bar{x}_i \bar{T}_{ki} y_k = y_k \bar{T}_{ki} \bar{x}_i \;, \\ &= \overline{\mathbf{y}.\mathbf{T}.\mathbf{x}} \;. \end{aligned}$$

Problems 2.8

(1) Let $\mathbf{x} = (1, 2, i), \mathbf{y} = (1, -i, 1), \mathbf{T} = \begin{bmatrix} 1 & -i & 3 \\ 2 & i & 2 \\ 1 & 0 & -1 \end{bmatrix}$.

Find $\mathbf{x}.\mathbf{y}$; $\mathbf{y}.\mathbf{x}$; $\mathbf{x} \times \mathbf{y}$; $|\mathbf{x}|$; $\mathbf{x}.(\mathbf{x} \times \mathbf{y})$; $(\mathbf{x} \times \mathbf{y}).\mathbf{y}$; $\mathbf{y}.(\mathbf{T}.\mathbf{x})$; $(\mathbf{y}.\mathbf{T}).\mathbf{x}$; $\mathbf{x}.\mathbf{T}^{\mathrm{H}}.\mathbf{y}$.

Show that the result of Example 2.8 No. 1 is satisfied.

(2) $\mathbf{x}, \mathbf{y}, \mathbf{z}$ are vectors with components (x_1, x_2, x_3) etc., which may be complex. Using the definitions given in (2.60) and (2.64) deduce an expression for the triple scalar product $(\mathbf{x} \times \mathbf{y}).\mathbf{z}$, and show that $(\mathbf{x} \times \mathbf{y}).\mathbf{y} = 0$. Prove that:

$\mathbf{x}.(\mathbf{y} \times \mathbf{z}) = \overline{(\mathbf{x} \times \mathbf{y}).\mathbf{z}}$. Find $(\mathbf{x} \times \mathbf{y}).\mathbf{z}$ and $\mathbf{x}.(\mathbf{y} \times \mathbf{z})$ when $\mathbf{x} = (1, 2, i)$, $\mathbf{y} = (1, -i, 1), \mathbf{z} = (1, i, i)$.

(3) Prove that:

(i) $i(\mathbf{x} \times \mathbf{y}) = -(i\mathbf{x}) \times \mathbf{y}$;

(ii) $(\mathbf{x} \times \mathbf{y}) \times \mathbf{z} = (\mathbf{z}.\mathbf{x})\mathbf{y} - (\mathbf{z}.\mathbf{y})\mathbf{x}$;

(iii) $\mathbf{x} \times (\mathbf{y} \times \mathbf{z}) = (\mathbf{x}.\mathbf{z})\mathbf{y} - (\mathbf{x}.\mathbf{y})\mathbf{z}$.

2.9 THE EIGENVALUE PROBLEM

2.9.1 Statement of the Problem

The eigenvector problem is that of finding values of λ for which the linear equation

$$\mathbf{T}.\mathbf{v} = \lambda\mathbf{v} \tag{2.69}$$

has non-trivial solutions for $\mathbf{v}$. Values of λ for which solutions of (2.69) exist are known as **eigenvalues**, **principal values**, **characteristic values** or **latent roots** of the tensor $\mathbf{T}$. The vectors $\mathbf{v}$ which are solutions of (2.69) are called **eigenvectors** (or **latent vectors** etc.). We note that eigenvectors are, by definition, non-zero.

We have seen already (Problems 2.6, No. 2) that equation (2.69) remains invariant under rotations. It therefore yields the same solution irrespective of basis. In component form (2.69) becomes:

$$T_{ik}v_k = \lambda v_i = \lambda\delta_{ik}v_k ,$$

or

$$(T_{ik} - \lambda\delta_{ik})v_k = 0 . \tag{2.70}$$

Equation (2.70) represents the three homogeneous equations:

$$\left.\begin{aligned} (T_{11} - \lambda)v_1 + T_{12}v_2 + T_{13}v_3 &= 0 , \\ T_{21}v_1 + (T_{22} - \lambda)v_2 + T_{23}v_3 &= 0 , \\ T_{31}v_1 + T_{32}v_2 + (T_{33} - \lambda)v_3 &= 0 , \end{aligned}\right\} \tag{2.71}$$

in the unknowns v_1, v_2 and v_3. In matrix form (2.71) becomes:

$$\begin{bmatrix} T_{11} - \lambda & T_{12} & T_{13} \\ T_{21} & T_{22} - \lambda & T_{23} \\ T_{31} & T_{32} & T_{33} - \lambda \end{bmatrix} \begin{bmatrix} v_1 \\ v_2 \\ v_3 \end{bmatrix} = 0 . \tag{2.72}$$

It is known from the theory of linear homogeneous simultaneous equations that non-trivial solutions of (2.71) exist if and only if

$$\Delta = \det(T_{ik} - \lambda\delta_{ik}) = 0 , \tag{2.73}$$

i.e.
$$\begin{vmatrix} T_{11}-\lambda & T_{12} & T_{13} \\ T_{21} & T_{22}-\lambda & T_{23} \\ T_{31} & T_{32} & T_{33}-\lambda \end{vmatrix} = 0 \ , \tag{2.74}$$

or
$$|\,\mathbf{T}-\lambda\mathbf{1}\,| = 0 \ . \tag{2.75}$$

The equation expressed as (2.73), (2.74) or (2.75) is known as the **determinantal equation**, **characteristic equation** or **secular equation** for the tensor **T**. It is a cubic in λ whose three roots are the eigenvalues of **T**. The eigenvalues may be distinct, or two or three may be repeated. We will suppose for the moment that they are distinct, in which case there are three linearly independent eigenvectors, one corresponding to each eigenvalue, as the theorem below shows.

Reference to section 2.6. shows that the characteristic equation written out in full becomes:

$$\lambda^3 - I_1\lambda^2 + I_2\lambda - I_3 = 0 \ , \tag{2.76}$$

where I_1, I_2 and I_3 are the scalar invariants defined in that section.

Theorem 2.9.1 (i) Eigenvectors corresponding to distinct eigenvalues are linearly independent.

Proof. Let **T** be a tensor and let $\mathbf{v}_1, \mathbf{v}_2$ be eigenvectors corresponding to distinct eigenvalues λ_1, λ_2.

Then
$$\left.\begin{aligned} \mathbf{T}.\mathbf{v}_1 &= \lambda_1\mathbf{v}_1 \ , \\ \mathbf{T}.\mathbf{v}_2 &= \lambda_2\mathbf{v}_2 \ . \end{aligned}\right\} \tag{2.77}$$

Suppose $\mathbf{v}_1, \mathbf{v}_2$ are linearly dependent, i.e. for some pair of numbers α, β (both non-zero since $\mathbf{v}_1$ and $\mathbf{v}_2$, being eigenvectors, are non-zero):

$$\alpha\mathbf{v}_1 = \beta\mathbf{v}_2 \ , \tag{2.78}$$

then
$$\mathbf{T}.(\alpha\mathbf{v}_1) = \mathbf{T}.(\beta\mathbf{v}_2) \ ,$$

or
$$\alpha\mathbf{T}.\mathbf{v}_1 = \beta\lambda_2\mathbf{v}_2 \ . \tag{2.79}$$

But
$$\alpha\mathbf{T}.\mathbf{v}_1 = \alpha\lambda_1\mathbf{v}_1 \ . \tag{2.80}$$

Now $\lambda_1 \neq \lambda_2$, hence (2.79) and (2.80) are contradictory. We conclude therefore that the initial assumption was false, and $\mathbf{v}_1, \mathbf{v}_2$ are linearly independent.

The converse of Theorem 2.9.1(i) is expressed by Theorem 2.9.1(ii) below.

Theorem 2.9.1(ii). A tensor with three distinct eigenvalues has only one linearly independent eigenvector corresponding to each eigenvalue.

Proof. Let $\mathbf{T}$ be the tensor and let $\mathbf{v}_1, \mathbf{v}_2, \mathbf{v}_3$ be linearly independent eigenvectors corresponding to the distinct eigenvalues $\lambda_1, \lambda_2, \lambda_3$.

Then $\mathbf{T}.\mathbf{v}_i = \lambda_i \mathbf{v}_i$, $i = 1, 2, 3$, where no summation is implied over the index i.

Let $\mathbf{T}.\mathbf{u} = \lambda_1 \mathbf{u}$, and assume that $\mathbf{u}$ and $\mathbf{v}_1$ are linearly independent. Now, since $\mathbf{v}_1, \mathbf{v}_2, \mathbf{v}_3$ are linearly independent, we may find numbers $\alpha_1, \alpha_2, \alpha_3$, such that

$$\mathbf{u} = \alpha_i \mathbf{v}_i \ .$$

Therefore $\mathbf{T}.(\alpha_i \mathbf{v}_i) = \lambda_1 (\alpha_i \mathbf{v}_i)$,

i.e. $\mathbf{T}.(\alpha_2 \mathbf{v}_2 + \alpha_3 \mathbf{v}_3) = \lambda_1 (\alpha_2 \mathbf{v}_2 + \alpha_3 \mathbf{v}_3)$.

But $\mathbf{T}.(\alpha_2 \mathbf{v}_2) = \alpha_2 \lambda_2 \mathbf{v}_2$,

and $\mathbf{T}.(\alpha_3 \mathbf{v}_3) = \alpha_3 \lambda_3 \mathbf{v}_3$.

Hence $\alpha_2 (\lambda_2 - \lambda_1) \mathbf{v}_2 + \alpha_3 (\lambda_3 - \lambda_1) \mathbf{v}_3 = \mathbf{0}$.

Now, since $\mathbf{u}$ and $\mathbf{v}_1$ are assumed to be linearly independent, α_2, α_3 cannot *both* be zero. Hence

$$\lambda_2 = \lambda_1 \text{ and/or } \lambda_3 = \lambda_1 \ .$$

Neither of these results is possible. Hence we conclude that the initial assumption was false, and $\mathbf{u}$, $\mathbf{v}_1$ cannot be linearly independent.

2.9.2 Eigenvectors of the Hermitian Transpose Tensor

Some interesting results may be deduced concerning the eigenvalues and eigenvectors of the Hermitian transpose tensor. These are set out in the following theorems.

Theorem 2.9.2(i) The eigenvalues of the Hermitian transpose tensor $\mathbf{T}^{\mathrm{H}}$ are the complex conjugates of the eigenvalues of $\mathbf{T}$.

Proof. The eigenvalues of $\mathbf{T}$ are given by equation (2.75),

i.e $$|\mathbf{T} - \lambda \mathbf{1}| = 0 \ , \tag{2.81}$$

while the eigenvalues of $\mathbf{T}^{\mathrm{H}}$ are given by:

$$|\mathbf{T}^{\mathrm{H}} - \lambda \mathbf{1}| = 0 \ . \tag{2.82}$$

Now the determinant of a matrix is equal to that of its transpose, hence we may replace $(\mathbf{T}^{\mathrm{H}} - \lambda \mathbf{1})$ in (2.82) by its transpose $(\bar{\mathbf{T}} - \lambda \mathbf{1})$,

i.e $$|\bar{\mathbf{T}} - \lambda \mathbf{1}| = 0 \ . \tag{2.83}$$

Clearly the solutions for λ of (2.83) will be the complex conjugates of the solutions of (2.81). This proves the theorem.

Theorem 2.9.2(ii) If $\mathbf{T}$ is a tensor with distinct eigenvalues λ, μ, ν, and corresponding eigenvectors **u**, **v**, **w**, then the eigenvectors of **u**′, **v**′, **w**′ of $\mathbf{T}^H$ corresponding to the eigenvalues $\bar{\lambda}, \bar{\mu}, \bar{\nu}$ satisfy the relations:

$$\mathbf{v}.\mathbf{w}' = \mathbf{v}'.\mathbf{w} = \mathbf{w}.\mathbf{u}' = \mathbf{w}'.\mathbf{u} = \mathbf{u}.\mathbf{v}' = \mathbf{u}'.\mathbf{v} = 0 \ . \qquad (2.84)$$

Proof. We have

$$\mathbf{T}.\mathbf{u} = \lambda\mathbf{u}, \quad \mathbf{T}.\mathbf{v} = \mu\mathbf{v}, \quad \mathbf{T}.\mathbf{w} = \nu\mathbf{w},$$

$$\mathbf{T}^H.\mathbf{u}' = \bar{\lambda}\mathbf{u}', \quad \mathbf{T}^H.\mathbf{v}' = \bar{\mu}\mathbf{v}', \quad \mathbf{T}^H.\mathbf{w}' = \bar{\nu}\mathbf{w}' \ .$$

Therefore $\mathbf{w}'.\mathbf{T}.\mathbf{v} = \mu\mathbf{w}'.\mathbf{v}$, (2.85)

and $\mathbf{v}.\mathbf{T}^H.\mathbf{w}' = \bar{\nu}\mathbf{v}.\mathbf{w}'$. (2.86)

Taking the complex conjugate of (2.86) and using the results of (2.61) and Example 2.8 No. 1, we obtain:

$$\mathbf{w}'.\mathbf{T}.\mathbf{v} = \nu\mathbf{w}'.\mathbf{v} \ . \qquad (2.87)$$

From (2.85) and (2.87), remembering $\mu \neq \nu$, we obtain: $\mathbf{w}'.\mathbf{v} = 0$. The remaining results of (2.84) are obtained in a similar way.

Example 2.9

(1) Find all the eigenvalues and eigenvectors of the tensor

$$\mathbf{T} = \begin{bmatrix} 2 & 0 & 3i \\ 0 & 1 & 2 \\ 0 & 0 & -1 \end{bmatrix},$$

and of the Hermitian transpose tensor $\mathbf{T}^H$. Verify that Theorem 2.9.2(ii) is satisfied.

The characteristic equation is:

$$\begin{vmatrix} 2-\lambda & 0 & 3i \\ 0 & 1-\lambda & 2 \\ 0 & 0 & -1-\lambda \end{vmatrix} = 0 \ ,$$

i.e. $(\lambda - 2)(\lambda - 1)(\lambda + 1) = 0$.

From equation (2.72) the eigenvectors are $\mathbf{u} = (1, 0, 0)$, $\mathbf{v} = (0, 1, 0)$, $\mathbf{w} = (i, 1, -1)$. The Hermitian transpose tensor is:

$$\mathbf{T}^H = \begin{bmatrix} 2 & 0 & 0 \\ 0 & 1 & 0 \\ -3i & 2 & -1 \end{bmatrix},$$

with the same set of eigenvalues (since they are real) and eigenvectors $\mathbf{u}' = (i, 0, 1), \mathbf{v}' = (0, 1, 1), \mathbf{w}' = (0, 0, 1)$. Clearly equation (2.84) of Theorem 2.9.2(ii) is satisfied.

Problems 2.9

(1) Find all the eigenvalues and eigenvectors of the given tensors, and of their Hermitian transposes. Verify in each case the truth of Theorem 2.9.2(ii).

$$\text{(i)} \begin{bmatrix} 1 & 1 & 1 \\ 1 & -1 & i \\ 0 & 0 & 2 \end{bmatrix}; \quad \text{(ii)} \begin{bmatrix} 1 & 0 & 0 \\ -1 & 2 & 3i \\ -2 & 3i & 2 \end{bmatrix}; \quad \text{(iii)} \begin{bmatrix} 1 & 0 & 0 \\ 1+i & 0 & 0 \\ 1-i & 2 & 2 \end{bmatrix};$$

$$\text{(iv)} \begin{bmatrix} 1+i & -2 & i \\ \frac{1}{8}+i & 0 & 0 \\ 0 & 2 & 1 \end{bmatrix}.$$

(2) Find all the eigenvalues and eigenvectors of the given tensors with real components, and of their transposes. Verify in each case the truth of Theorem 2.9.2(ii).

$$\text{(i)} \begin{bmatrix} 2 & 0 & 3 \\ 0 & 1 & 2 \\ 0 & 0 & -1 \end{bmatrix}; \quad \text{(ii)} \begin{bmatrix} 3 & 4 & 2 \\ 1 & 0 & -2 \\ -1 & 3 & 3 \end{bmatrix}; \quad \text{(iii)} \begin{bmatrix} 2 & 4 & -1 \\ 6 & -7 & 10 \\ 3 & -4 & 6 \end{bmatrix}.$$

(3) Verify that the matrix

$$\Theta = \begin{bmatrix} \tfrac{1}{2} & 1/\sqrt{2} & \tfrac{1}{2} \\ -1/\sqrt{2} & 0 & 1/\sqrt{2} \\ \tfrac{1}{2} & -1/\sqrt{2} & \tfrac{1}{2} \end{bmatrix}$$

is a rotation matrix. Find all the eigenvalues and the eigenvector corresponding to the real eigenvalue. What is the geometrical significance of this result?

(4) The matrix in Example 2.9 No. 1 was **upper triangular**, i.e. all the elements below the leading diagonal were zero. Prove that the eigenvalues of any such matrix are the elements on the leading diagonal. Find all the eigenvalues and eigenvectors of the **lower triangular** matrix:

$$\begin{bmatrix} 2 & 0 & 0 \\ 0 & 1 & 0 \\ 3 & 2 & -1 \end{bmatrix}.$$

(5) Show that the real antisymmetric tensor (2.34) has one zero eigenvalue and two conjugate pure imaginary eigenvalues. Find the eigenvector corresponding to the zero eigenvalue.

(6) If two tensors **S** and **T** have the same eigenvectors, show that $\mathbf{S.T} = \mathbf{T.S}$.

(7) If a tensor **T** satisfies $\sum_{n=0}^{k} a_n \mathbf{T}^n = \mathbf{0}$ where $\mathbf{T}^2 = \mathbf{T.T}, \mathbf{T}^3 = \mathbf{T.T}^2$, etc., show that the eigenvalues of **T** satisfy $\sum_{n=0}^{k} a_n \lambda^n = 0$.

(8) Determine the values of λ such that $\mathbf{T.X} = \lambda\mathbf{X}$, where $\mathbf{T} = \begin{bmatrix} 4 & 1 \\ -1 & 2 \end{bmatrix}$ and **X** is a (2×2) tensor. Find the most general form of **X**.

(9) Show that the characteristic equation of the tensor **T** in question No. 8 is $\lambda^2 - 6\lambda + 9 = 0$. Show that **T** satisfies the equation:

$$\mathbf{T}^2 - 6\mathbf{T} + 9\mathbf{1} = \mathbf{0} .$$

(10) Show that the eigenvalues of the tensor $\mathbf{T}^{-1}$, represented by the inverse matrix, are the reciprocals of the eigenvalues of **T**, and that the eigenvectors of $\mathbf{T}^{-1}$ coincide with those of **T**.

(11) Let the eigenvalues of **T** be $\lambda_1, \lambda_2, \lambda_3$. Show that the eigenvalues of $\mathbf{T}^n$, where n is a positive integer, are $\lambda_1{}^n, \lambda_2{}^n, \lambda_3{}^n$ and that $\mathbf{T}^n$ has the same eigenvectors as **T**.

(12) Let the eigenvalues of a tensor **T** be $\lambda_1, \lambda_2, \lambda_3$, and let the scalar invariants defined in section 2.6. be I_1, I_2, I_3. Prove that:

$$I_1 = \lambda_1 + \lambda_2 + \lambda_3 ,$$

$$I_2 = \lambda_2\lambda_3 + \lambda_3\lambda_1 + \lambda_1\lambda_2 ,$$

$$I_3 = \lambda_1\lambda_2\lambda_3 .$$

2.10 THE EIGENVECTOR PROBLEM: DEGENERATE CASES

2.10.1 Double Root of the Characteristic Equation

Theorems 2.9.1(i) and 2.9.1(ii) show between them that, in general, the characteristic equation (2.74) will have three distinct roots for λ, each leading to a single eigenvector which is linearly independent from the other two. If, however, equation (2.74) has a double root λ for a tensor **T**, λ is said to be an eigenvalue of **multiplicity** 2. We shall show that, corresponding to such an eigenvalue, there may be one or two linearly independent eigenvectors, depending on the rank of the matrix $(\mathbf{T} - \lambda\mathbf{1})$. A tensor with only one eigenvector corresponding to the double eigenvalue, i.e. with less than three eigenvectors altogether, is said to be **defective**.

We first prove a useful theorem;

Theorem 2.10.1(i) The number of linearly independent solutions of equation (2.71) is equal to $(3 - R)$ where R is the rank of the matrix $(\mathbf{T} - \lambda\mathbf{1})$.

Proof. From Theorem 1.7.4, R is the number of linearly independent scalar equations in (2.71). We may therefore eliminate R unknowns from (2.71) leaving $(3 - R)$ unknowns. There are thus $(3 - R)$ linearly independent vector solutions of (2.71).

We are now in a position to prove:

Theorem 2.10.1(ii) The number of linearly independent eigenvectors corresponding to an eigenvalue of multiplicity 2 is at most 2.

Proof. Let R be the rank of the matrix $(\mathbf{T} - \lambda\mathbf{1})$. Suppose $R = 0$. Then $(\mathbf{T} - \lambda\mathbf{1})$ is the null matrix. Hence T is a multiple of the unit tensor and λ is an eigenvalue of multiplicity 3. Hence $R > 0$ or $(3 - R) < 3$. Thus we have proved that there are less than 3 linearly independent eigenvectors corresponding to the double eigenvalue λ.

Theorem 2.10.1(ii) is a special case of the general result† that the number of linearly independent eigenvectors cannot exceed the multiplicity of that eigenvalue. Theorem 2.9.1(ii) is also a special case of this result. It may be re-stated 'The number of linearly independent eigenvectors corresponding to an eigenvalue of multiplicity 1 is exactly 1.'

Theorems 2.10.1(iii) and 2.10.1(iv) show how the eigenvectors of a tensor with a double eigenvalue are related to the eigenvectors of the Hermitian transpose tensor.

Theorem 2.10.1(iii) If $\mathbf{u}, \mathbf{v}, \mathbf{w}$ are linearly independent eigenvectors of $\mathbf{T}$ corresponding to one distinct and one double eigenvalue of $\mathbf{T}$, then there exists a set of linearly independent eigenvectors $\mathbf{u}', \mathbf{v}', \mathbf{w}'$ of the Hermitian transpose tensor $\mathbf{T}^{\mathrm{H}}$ such that

$$\mathbf{v}.\mathbf{w}' = \mathbf{v}'.\mathbf{w} = \mathbf{w}.\mathbf{u}' = \mathbf{w}'.\mathbf{u} = \mathbf{u}.\mathbf{v}' = \mathbf{u}'.\mathbf{v} = 0 \ . \tag{2.88}$$

Proof. We have

$$\begin{aligned} &\mathbf{T}.\mathbf{u} = \lambda\mathbf{u}, \qquad \mathbf{T}.\mathbf{v} = \mu\mathbf{v}, \qquad \mathbf{T}.\mathbf{w} = \mu\mathbf{w} \ , \\ &\mathbf{T}^{\mathrm{H}}.\mathbf{u}' = \bar{\lambda}\mathbf{u}', \qquad \mathbf{T}^{\mathrm{H}}.\mathbf{v}' = \bar{\mu}\mathbf{v}', \qquad \mathbf{T}^{\mathrm{H}}.\mathbf{w}' = \bar{\mu}\mathbf{w}' \ , \end{aligned} \tag{2.89}$$

where μ is the double eigenvalue of $\mathbf{T}$. Proceeding as in Theorem 2.9.2(ii) we may obtain the final four results of (2.88).

To obtain the two remaining results we proceed as follows: From (2.89) we observe that if $\mathbf{v}'$ and $\mathbf{w}'$ are eigenvectors of $\mathbf{T}^{\mathrm{H}}$ corresponding to the eigenvalue of $\bar{\mu}$, then any linear combination of $\mathbf{v}'$ and $\mathbf{w}'$ is also an eigenvector of $\mathbf{T}^{\mathrm{H}}$ corresponding to $\bar{\mu}$. Hence we may choose any vector $\mathbf{v}'$ which is orthogonal

†The reader requiring a proof of the general theorem is referred to *Applied Linear Algebra* by Ben Noble, published by Prentice Hall Inc., 1969, page 351 (Theorem 11.6).

to **u**, and any vector **v** which is orthogonal to **u**′. The vectors **w** and **w**′ are then found from the relations:

$$\mathbf{w} = \mathbf{u}' \times \mathbf{v}', \qquad \mathbf{w}' = \mathbf{u} \times \mathbf{v} ,$$

and the first two results of (2.88) are obtained. This proves the theorem.

Theorem 2.10.1(iv) If **u** and **v** are the only linearly independent eigenvectors of the defective tensor **T**, corresponding respectively to one distinct and one double eigenvalue, and **u**′, **v**′ are the corresponding eigenvectors of the Hermitian transpose tensor $\mathbf{T}^{\mathrm{H}}$, then:

$$\mathbf{u}'.\mathbf{v} = \mathbf{u}.\mathbf{v}' = \mathbf{v}.\mathbf{v}' = 0 . \tag{2.90}$$

Proof. We have:

$$\left.\begin{aligned} \mathbf{T}.\mathbf{u} &= \lambda\mathbf{u}, \qquad & \mathbf{T}.\mathbf{v} &= \mu\mathbf{v} , \\ \mathbf{T}^{\mathrm{H}}.\mathbf{u}' &= \bar{\lambda}\mathbf{u}', \qquad & \mathbf{T}^{\mathrm{H}}.\mathbf{v}' &= \bar{\mu}\mathbf{v}' , \end{aligned}\right\} \tag{2.91}$$

where μ is the repeated eigenvalue and **v** is the only eigenvector corresponding to μ. The first two results of (2.90) are readily obtained as in Theorem 2.9.2(ii). We have only to prove that **v** and **v**′ are orthogonal, for which we need to use the condition that μ is a double root of the determinantal equation (2.74). We do this by differentiating (2.74) with respect to λ and equating to zero with $\lambda = \mu$. This yields the result:

$$\begin{vmatrix} T_{11}-\mu & T_{12} \\ T_{21} & T_{22}-\mu \end{vmatrix} + \begin{vmatrix} T_{22}-\mu & T_{23} \\ T_{32} & T_{33}-\mu \end{vmatrix} + \begin{vmatrix} T_{33}-\mu & T_{31} \\ T_{13} & T_{11}-\mu \end{vmatrix} = 0 . \tag{2.92}$$

We now define a tensor $\mathbf{A} = (\mathbf{T} - \mu\mathbf{1})$ with components

$$a_{ik} = T_{ik} - \mu\delta_{ik} . \tag{2.93}$$

Equation (2.70) with $\lambda = \mu$ becomes:

$$a_{ik}v_k = 0 , \tag{2.94}$$

while the equation for the components of **v**′ is

$$\bar{a}_{ki}v'_k = 0 , \tag{2.95}$$

or

$$a_{ki}\bar{v}'_k = 0 . \tag{2.96}$$

Now the expansion of the determinant of **A** in terms of cofactors or the expansion by alien cofactors is

$$a_{ik}A_{jk} = 0 , \qquad \text{for all } i, j \tag{2.97}$$

where A_{jk} is the cofactor of a_{jk} and the matrix of cofactors is called the **adjugate matrix** adj **A**. Hence, comparing (2.94) and (2.97), since **v** is the only vector satisfying (2.94),

$$v_k = A_{jk} \ \forall j \ . \tag{2.98}$$

Similarly $$\bar{v}_k{}' = A_{kj} \ \forall j \ . \tag{2.99}$$

The double root condition (2.92) yields:

$$A_{ii} = 0 \ . \tag{2.100}$$

Now, since, according to (2.98), the elements of any row of the adjugate matrix are proportional to the components of the eigenvector **v**, the vectors forming the rows of the adjugate matrix are all parallel. This fact may be expressed by the relation

$$A_{ij}A_{kl} = A_{il}A_{kj} \ , \tag{2.101}$$

therefore $$A_{ik}A_{kj} = A_{ij}A_{kk} \ .$$

Hence, from (2.100)

$$A_{ik}A_{kj} = 0 \ , \tag{2.102}$$

therefore $$v_k\bar{v}_k{}' = 0 \ , \tag{2.103}$$

and **v**, **v**′ are orthogonal.

2.10.2 Treble Root of the Characteristic Equation

We consider now the case in which the characteristic equation (2.75) has a treble root. Here the rank of $(\mathbf{T} - \lambda\mathbf{1})$ may be 0, 1 or 2, but by far the most interesting results are found when $R = 1$.

The case of $R = 0$ has already been mentioned incidentally in Theorem 2.10.1(iii). **T** is necessarily a multiple of the unit tensor **1** and all vectors satisfy equation (2.69). Thus any vector is an eigenvector.

When $R = 1$ the treble eigenvalue leads to two linearly independent eigenvectors which are the subject of Theorem 2.10.2(i). Finally Theorem 2.10.2(ii) deals with the case when $R = 2$.

Theorem 2.10.2(i) If **u** and **v** are the only linearly independent eigenvectors of a defective tensor **T** with a triple eigenvalue λ, the components of **u** are given by the elements of any non-zero column of the matrix $(\mathbf{T} - \lambda\mathbf{1})$. Similarly, if **u**′ and **v**′ are the eigenvectors of the Hermitian transpose tensor $\mathbf{T}^{\mathrm{H}}$, the components of **u**′ are given by the elements of any non-zero column of the matrix $(\mathbf{T}^{\mathrm{H}} - \lambda\mathbf{1})$. Further $\mathbf{u}.\mathbf{u}' = 0$ and $\mathbf{v} = \mathbf{v}' = \mathbf{u} \times \mathbf{u}'$.

Proof. Proceeding as in Theorem 2.10.1(iv) we let

$$a_{ik} = T_{ik} - \lambda\delta_{ik} \ . \tag{2.104}$$

Hence $$a_{ik}u_k = 0 \tag{2.105}$$

and $$a_{ki}\bar{u}_k{}' = 0 \qquad \text{or} \qquad \bar{a}_{ki}u'_k = 0 \ . \tag{2.106}$$

Now, if **T** is to yield two linearly independent eigenvectors from the triple eigenvalue λ, the rank of $(\mathbf{T} - \lambda\mathbf{1})$ must be 1. Hence the vectors whose components are the rows or columns of $(\mathbf{T} - \lambda\mathbf{1})$ are all parallel or zero,

i.e. $$a_{ij}a_{kl} = a_{il}a_{kj} \ . \tag{2.107}$$

The double root condition (2.92) is automatically satisfied but, since λ is a triple eigenvalue, we may differentiate (2.92) with respect to μ and equate to zero to obtain the result:

$$a_{ii} = 0 \ , \tag{2.108}$$

which, using (2.107) yields:

$$a_{ik}a_{kj} = 0 \ . \tag{2.109}$$

A comparison of (2.109) with (2.105) and (2.106) shows that any non-zero vector whose components are a column of $(\mathbf{T} - \lambda\mathbf{1})$ is an eigenvector of **T**, while any non-zero vector whose components are a column of $(\mathbf{T}^{\mathrm{H}} - \lambda\mathbf{1})$ is an eigenvector of $\mathbf{T}^{\mathrm{H}}$. We conclude also from (2.105), (2.106) and (2.109) that

$$\mathbf{u}.\mathbf{u}' = 0 \ . \tag{2.110}$$

Finally, since (2.110) is essentially equivalent to the eigenvector equations (2.105) and (2.106), we conclude that any vector which is orthogonal to $\mathbf{u}'$ is an eigenvector of **T**, while any vector orthogonal to **u** is an eigenvector of $\mathbf{T}^{\mathrm{H}}$. Hence the vector given by

$$\mathbf{v} = \mathbf{v}' = \mathbf{u} \times \mathbf{u}' \tag{2.111}$$

is an eigenvector of **T** and of $\mathbf{T}^{\mathrm{H}}$.

Theorem 2.10.2(ii) If **u** is the only eigenvector of a defective tensor **T** with a triple eigenvalue λ, and $\mathbf{u}'$ is the eigenvector of the Hermitian transpose tensor $\mathbf{T}^{\mathrm{H}}$, then $\mathbf{u}.\mathbf{u}' = 0$.

Proof. The proof of this result, which is essentially part of Theorem 2.10.2(i), is left to the reader.

Examples 2.10

(1) Find all the eigenvalues and eigenvectors of the tensor **T** and its Hermitian

transpose, where

$$\mathbf{T} = \begin{bmatrix} 2i & 0 & 0 \\ 3 & 1 & 1 \\ 2 & 0 & 2i \end{bmatrix} .$$

The characteristic equation is:

$$\begin{vmatrix} 2i-\lambda & 0 & 0 \\ 3 & 1-\lambda & 1 \\ 2 & 0 & 2i-\lambda \end{vmatrix} = 0 ,$$

or $$(\lambda - 1)(\lambda - 2i)^2 = 0 .$$

When $\lambda = 1, \mathbf{u} = \mathbf{e}_2, \mathbf{u}' = (17-4i, 5+10i, 5)$. When $\lambda = 2i$ the rank of $(\mathbf{T} - \lambda\mathbf{1})$ is 2 and $\mathbf{v} = (0, 1, 2i-1)$. When $\bar{\lambda} = -2i$ we obtain $\mathbf{v}' = \mathbf{e}_1$. We observe that equation (2.90) of Theorem 2.10.1 (iv) is satisfied.

(2) Find the eigenvalues and eigenvectors of the matrix $\mathbf{T}$ and its Hermitian transpose, where

$$\mathbf{T} = \begin{bmatrix} 12 & 12 & 18i \\ 3 & 12 & 9i \\ 4i & 8i & -6 \end{bmatrix} .$$

The characteristic equation is:

$$\begin{vmatrix} 12-\lambda & 12 & 18i \\ 3 & 12-\lambda & 9i \\ 4i & 8i & -6-\lambda \end{vmatrix} = 0 ,$$

or $$(\lambda - 6)^3 = 0 .$$

When $\lambda = 6$ the rank of $(\mathbf{T} - \lambda\mathbf{1})$ is 1. Hence by Theorem 2.10.2(i) there are exactly two linearly independent eigenvectors of $\mathbf{T}$, of which one is given by $\mathbf{u} = (6, 3, 4i)$. There are also two linearly independent eigenvectors of $\mathbf{T}^H$, of which one is given by $\mathbf{u}' = (6, 12, 18i)$ or $\mathbf{u}' = (1, 2, 3i)$. From (2.111) the eigenvector which is common to $\mathbf{T}$ and $\mathbf{T}^H$ is given by
$\mathbf{v} = \mathbf{v}' = \mathbf{u} \times \mathbf{u}' = (1, -14, 9i)$.

(3) Find the eigenvalues and linearly independent eigenvectors of the following matrices.

$$(i)\begin{bmatrix}1&0&0\\0&1&0\\0&0&1\end{bmatrix};\ (ii)\begin{bmatrix}3&0&1\\0&3&0\\0&0&3\end{bmatrix};\ (iii)\begin{bmatrix}4&2&2\\0&4&2\\0&0&4\end{bmatrix}.$$

(i) We recognize this as the unit tensor. The characteristic equation becomes $(\lambda - 1)^3 = 0$, leading to the triple eigenvalue $\lambda = 1$. The rank of $(\mathbf{T} - \lambda\mathbf{1})$ is zero hence there are three linearly independent eigenvectors in R^3, e.g. $\mathbf{e}_1$, $\mathbf{e}_2$, $\mathbf{e}_3$.

(ii) $\lambda = 3$ is a triple eigenvalue and the rank of $(\mathbf{T} - \lambda\mathbf{1})$ is 1. Hence, by Theorem 2.10.2 (i), $\mathbf{e}_1$ is an eigenvector of $\mathbf{T}$ and $\mathbf{e}_3$ is an eigenvector of $\mathbf{T}'$. Also $\mathbf{e}_1 \times \mathbf{e}_3 = -\mathbf{e}_2$ is an eigenvector of $\mathbf{T}$ and $\mathbf{T}'$.

(iii) $\lambda = 4$ is a triple eigenvalue and the rank of $(\mathbf{T} - \lambda\mathbf{1})$ is 2. Hence by Theorem 2.10.2(ii) there is only one eigenvector $\mathbf{u}$ of $\mathbf{T}$, and one eigenvector $\mathbf{u}'$ of $\mathbf{T}'$, such that $\mathbf{u}.\mathbf{u}' = 0$. Equations (2.71) yield $\mathbf{u} = \mathbf{e}_1$, $\mathbf{u}' = \mathbf{e}_3$.

Problem 2.10

(1) Find all the eigenvalues, and eigenvectors corresponding to distinct eigenvalues, of the following tensors and their Hermitian transposes. For repeated eigenvalues λ, determine the rank of the matrix of $(\mathbf{T} - \lambda\mathbf{1})$ and find the appropriate number of linearly independent eigenvectors.

$$(i)\begin{bmatrix}3&2&0\\1&0&-1\\-1&5&4\end{bmatrix};\ (ii)\begin{bmatrix}2&0&0\\3&i&i\\2&0&2\end{bmatrix};\ (iii)\begin{bmatrix}0&0&1\\0&-1&0\\2&2i&1\end{bmatrix};$$

$$(iv)\begin{bmatrix}2&2&-2\\2&-1&-1\\-6&-3&1\end{bmatrix};\ (v)\begin{bmatrix}3&2&1\\2&3&1\\-8&-8&-3\end{bmatrix};\ (vi)\begin{bmatrix}3&1&0\\-1&4&3\\1&-2&-1\end{bmatrix};$$

$$(vii)\begin{bmatrix}3&-1&2\\-33&0&-1\\-19&-2&3\end{bmatrix}.$$

2.11 HERMITIAN TENSORS

2.11.1 The Eigenvectors of Hermitian Tensors

A **Hermitian tensor** is a tensor which is equal to its Hermitian transpose. Thus, if

T is Hermitian:

$$\mathbf{T}^{\mathrm{H}} = \mathbf{T} \ , \tag{2.112}$$

$$\bar{T}_{ki} = T_{ik} \ . \tag{2.113}$$

Hermitian tensors have important properties, two of which are summarized in Theorems 2.11.1(i) and 2.11.1(ii).

Theorem 2.11.1(i) The eigenvalues of a Hermitian tensor are real.

Lemma. We first prove that if **T** is Hermitian then for any vector **u**, the scalar **u.T.u** is real.
We have, from (2.68):

$$\mathbf{u.T.u} = \bar{u}_i u_k T_{ik} \ , \tag{2.114}$$

therefore $\overline{\mathbf{u.T.u}} = u_i \bar{u}_k \bar{T}_{ik}$,

$$= u_i \bar{u}_k T_{ki} \ ,$$

$$= \mathbf{u.T.u} \ . \tag{2.115}$$

This proves the lemma.

Proof. Suppose λ is an eigenvalue of a Hermitian tensor **T**, with corresponding eigenvector **u**,

i.e $\quad \mathbf{T.u} = \lambda\mathbf{u}$,

therefore $\quad \mathbf{u.T.u} = \lambda\mathbf{u.u}$. $\quad$ (2.116)

We have shown that **u.T.u** is real and we know already that **u.u** is real (see (2.62)). Hence λ is also real. This proves the theorem.

Theorem 2.11.1(ii). The eigenvectors corresponding to distinct eigenvalues of a Hermitian tensor are orthogonal.

Proof. Suppose **u** and **v** are eigenvectors corresponding to distinct eigenvalues λ, μ of the Hermitian tensor **T**.

i.e. $\quad \mathbf{T.u} = \lambda\mathbf{u}, \qquad \mathbf{T.v} = \mu\mathbf{v}$,

then $\quad \lambda\mathbf{u.v} = (\mathbf{T.u}).\mathbf{v} = \bar{T}_{ik}\bar{u}_k v_i$, $\quad$ (2.117)

and $\quad \mu\mathbf{u.v} = \mathbf{u.T.v} = \bar{u}_i T_{ik} v_k = \bar{u}_k \bar{T}_{ik} v_i$, $\quad$ (2.118)

or $\quad (\lambda - \mu)\mathbf{u.v} = 0$.

But $\lambda \neq \mu$, hence $\mathbf{u.v} = 0$. This proves the theorem.

2.11.2 Hermitian Tensors - The Degenerate Cases

Suppose the characteristic equation (2.75) has a double root λ. Theorems

2.10.1(iii) and 2.10.1(iv) dealt respectively with the cases in which tensors with an eigenvalue of multiplicity 2 had two or one eigenvectors associated with that eigenvalue. We shall now prove that a Hermitian tensor with a double eigenvalue always has two linearly independent eigenvectors associated with it, i.e. a Hermitian tensor is not defective.

Theorem 2.11.2(i) If λ and μ are eigenvalues of multiplicity 1 and 2 respectively of the Hermitian tensor **T**, and **u** is the eigenvector corresponding to the eigenvalue λ, then there are two linearly independent eigenvectors, **v** and **w**, corresponding to the eigenvalue μ such that

$$\mathbf{v}.\mathbf{w} = \mathbf{w}.\mathbf{u} = \mathbf{u}.\mathbf{v} = 0 \ . \tag{2.119}$$

Proof. We have:

$$\mathbf{T}.\mathbf{u} = \lambda\mathbf{u}, \qquad \mathbf{T}.\mathbf{v} = \mu\mathbf{v} \ .$$

Suppose **v** is the only eigenvector corresponding to the eigenvalue μ. Then, since the eigenvectors of **T** are equal to those of the Hermitian transpose $\mathbf{T}^{\mathrm{H}}$, we have, from Theorem 2.10.1(iv):

$$\mathbf{v}.\mathbf{v} = 0 \ . \tag{2.120}$$

The result (2.120) is clearly impossible. There must therefore be two linearly independent eigenvectors corresponding to the eigenvalue μ. The result (2.119) follows immediately from Theorem 2.10.1(iii), equation (2.88).

We now consider the case in which the characteristic equation (2.75) has a treble root, λ. Theorems 2.10.2(i) and 2.10.2(ii) dealt respectively with the cases in which the tensor **T** had two or one eigenvectors. We propose to show now that neither of these cases is possible if **T** is Hermitian.

Theorem 2.11.2(ii) The only Hermitian tensor with a treble eigenvalue is a multiple of the unit tensor.

Proof. Suppose the Hermitian tensor **T** has a treble root λ and two linearly independent eigenvectors. From Theorem 2.10.2(i), remembering that $\mathbf{T}^{\mathrm{H}}$ has the same eigenvectors as **T** (since $\mathbf{T}^{\mathrm{H}} = \mathbf{T}$), we have:

$$\mathbf{v} = \mathbf{u} \times \mathbf{u}, \qquad \mathbf{u}.\mathbf{u} = 0 \ . \tag{2.121}$$

Neither of the equations in (2.121) is possible. Hence **T** cannot have exactly two eigenvectors.

Suppose now that **T** has only one eigenvector. Then, by Theorem 2.10.2(ii):

$$\mathbf{u}.\mathbf{u} = 0 \ .$$

Again the result is impossible and we conclude that **T** must have three linearly independent eigenvectors. As we have already seen, [Theorem 2.10.1(ii)], three linearly independent eigenvectors arising from a single eigenvalue implies that the rank of $(\mathbf{T} - \lambda\mathbf{1})$ is zero. Hence $\mathbf{T} = \lambda\mathbf{1}$. This proves the theorem.

Example 2.11

(1) Find all the eigenvalues and eigenvectors of the tensor

$$\begin{bmatrix} 2 & 4i & -2i \\ -4i & 2 & 2 \\ 2i & 2 & 5 \end{bmatrix},$$

and show that Theorems 2.11.1(i) and 2.11.2(i) are satisfied.

The characteristic equation is:

$$\begin{vmatrix} 2-\lambda & 4i & -2i \\ -4i & 2-\lambda & 2 \\ 2i & 2 & 5-\lambda \end{vmatrix} = 0 \ ,$$

or

$$\begin{vmatrix} 2-\lambda & 4 & -2 \\ 4 & 2-\lambda & 2 \\ -2 & 2 & 5-\lambda \end{vmatrix} = 0 \ ,$$

i.e. $$(\lambda-6)^2\,(\lambda+3) = 0 \ .$$

When $\lambda = -3, \mathbf{v} = \mathbf{v}_1 = (2i, -2, 1)$;
when $\lambda = 6, \mathbf{v} = (\alpha, \beta, 2i\alpha + 2\beta)$.
Let $\mathbf{v}_2 = (1, 0, 2i)$, (i.e. $\alpha = 1, \beta = 0$)

then $$\mathbf{v}_3 = \begin{vmatrix} \mathbf{e}_1 & \mathbf{e}_2 & \mathbf{e}_3 \\ -2i & -2 & 1 \\ 1 & 0 & -2i \end{vmatrix} = (4i, 5, 2), \text{ (i.e. } \alpha = 4i, \beta = 5) \ .$$

Clearly the eigenvalues are real and Theorem 2.11.1(i) is satisfied. Also, using definition (2.60), we see that $\mathbf{v}_2 . \mathbf{v}_3 = \mathbf{v}_3 . \mathbf{v}_1 = \mathbf{v}_1 . \mathbf{v}_2 = 0$, and Theorem 2.11.2(i) is also satisfied.

Problem 2.11

(1) Find the eigenvalues and eigenvectors of the following tensors, and show that Theorems 2.11.1(i) and 2.11.1(ii) are satisfied.

$$\text{(i)} \begin{bmatrix} 1 & -i & 1 \\ i & 0 & -i \\ 1 & i & 1 \end{bmatrix}; \ \text{(ii)} \begin{bmatrix} 1 & 1+i & 0 \\ 1-i & 1 & 0 \\ 0 & 0 & 0 \end{bmatrix}; \ \text{(iii)} \begin{bmatrix} 1 & 0 & i \\ 0 & 1 & 0 \\ -i & 0 & 1 \end{bmatrix}.$$

2.12 REAL SYMMETRIC TENSORS

2.12.1 The Eigenvectors of Real Symmetric Tensors

The special case of a Hermitian tensor with real components is a real symmetric tensor. The results proved for Hermitian tensors are therefore true for real symmetric tensors. They may be summarized as follows

1. A real symmetric tensor has three real eigenvalues of which two or three may be repeated.

2(a). A real symmetric tensor with distinct eigenvalues has a unique set of three mutually orthogonal eigenvectors.

2(b). A real symmetric tensor with one distinct and one repeated eigenvalue has a unique eigenvector corresponding to the distinct eigenvalue. Two eigenvectors may be found corresponding to the repeated eigenvalue such that the three eigenvectors form a mutually orthogonal set.

2(c). The only real symmetric tensor with a triple eigenvalue is a multiple of the unit tensor. Any vector is an eigenvector of such a tensor.

2.12.2 Diagonalization of a Symmetric Tensor

Suppose we refer a symmetric tensor to a basis consisting of its three eigenvectors. The eigenvectors are now $\mathbf{e}_1$, $\mathbf{e}_2$, $\mathbf{e}_3$ and the eigenvector equation (2.69) becomes:

$$\left.\begin{aligned} \mathbf{T}.\mathbf{e}_1 &= T_{i1}\mathbf{e}_i = \lambda\mathbf{e}_1 \ , \\ \mathbf{T}.\mathbf{e}_2 &= T_{i2}\mathbf{e}_i = \mu\mathbf{e}_2 \ , \\ \mathbf{T}.\mathbf{e}_3 &= T_{i3}\mathbf{e}_i = \nu\mathbf{e}_3 \ . \end{aligned}\right\} \qquad (2.122)$$

The solution of the three vector equations (2.122) is:

$$\mathbf{T} = \begin{bmatrix} \lambda & 0 & 0 \\ 0 & \mu & 0 \\ 0 & 0 & \nu \end{bmatrix} . \qquad (2.123)$$

The tensor $\mathbf{T}$ in (2.123), all of whose non-zero elements are on the leading diagonal, is known as a **diagonal tensor**. The process of referring a symmetric tensor to a basis consisting of its eigenvectors is known as **diagonalizing** the tensor.

Examples 2.12

(1) Find all the eigenvalues and eigenvectors of the real symmetric tensor:

$$\begin{bmatrix} 1 & 2 & 1 \\ 2 & 3 & 0 \\ 1 & 0 & 0 \end{bmatrix}.$$

Write down the tensor referred to a basis consisting of its eigenvectors.
The characteristic equation is:

$$\begin{vmatrix} 1-\lambda & 2 & 1 \\ 2 & 3-\lambda & 0 \\ 1 & 0 & -\lambda \end{vmatrix} = 0 ,$$

i.e. $(\lambda + 1)(\lambda^2 - 5\lambda + 3) = 0.$

Hence $\lambda = -1, \mu = ½(5 + \sqrt{13}), \nu = ½(5 - \sqrt{13}).$

Therefore
$$\begin{aligned} \mathbf{u} &= (2, -1, -2) , \\ \mathbf{v} &= (5 + \sqrt{13}, 6 + 2\sqrt{13}, 2) , \\ \mathbf{w} &= (5 - \sqrt{13}, 6 - 2\sqrt{13}, 2) . \end{aligned}$$

We note that $\mathbf{v}.\mathbf{w} = \mathbf{w}.\mathbf{u} = \mathbf{u}.\mathbf{v} = 0.$

Referred to a basis consisting of its eigenvectors, the tensor becomes:

$$\begin{bmatrix} -1 & 0 & 0 \\ 0 & ½(5 + \sqrt{13}) & 0 \\ 0 & 0 & ½(5 - \sqrt{13}) \end{bmatrix}.$$

(2) Find all the eigenvalues and eigenvectors of the real symmetric tensor:

$$\begin{bmatrix} 2 & 4 & -2 \\ 4 & 2 & 2 \\ -2 & 2 & 5 \end{bmatrix}.$$

The characteristic equation is:

$$\begin{vmatrix} 2-\lambda & 4 & -2 \\ 4 & 2-\lambda & 2 \\ -2 & 2 & 5-\lambda \end{vmatrix} = 0 ,$$

i.e. $$(\lambda + 3)(\lambda - 6)^2 = 0 .$$

Hence $\lambda = -3$, $\mu = 6$ (repeated). The eigenvector corresponding to the distinct eigenvalue $\lambda = -3$ is:

$$\mathbf{u} = (2, -2, 1) .$$

We know from Theorem 2.11.2(i) that there are two linearly independent eigenvectors corresponding to the eigenvalue $\mu = 6$. By putting $\lambda = 6$ in (2.71) we find that these eigenvectors have the general form $(\alpha, \beta, -2\alpha + 2\beta)$. This is the most general form for a vector perpendicular to **u**. We may arbitrarily choose $\alpha = \beta = 1$ to obtain:

$$\mathbf{v} = (1, 1, 0).$$

Then $$\mathbf{w} = \mathbf{u} \times \mathbf{v} = (1, -1, -4) .$$

Problems 2.12

(1) Find all the eigenvalues and eigenvectors of the real symmetric tensors

(i) $\begin{bmatrix} 4 & 0 & 1 \\ 0 & -2 & -1 \\ 1 & -1 & 1 \end{bmatrix}$; (ii) $\begin{bmatrix} 3 & 1 & 0 \\ 1 & 3 & 0 \\ 0 & 0 & 2 \end{bmatrix}$; (iii) $\begin{bmatrix} 7 & -1 & -2 \\ -1 & 7 & 2 \\ -2 & 2 & 4 \end{bmatrix}$;

(iv) $\begin{bmatrix} 3 & -1 & -\sqrt{2} \\ -1 & 3 & \sqrt{2} \\ -\sqrt{2} & \sqrt{2} & 4 \end{bmatrix}$.

(2) If $\lambda_1, \lambda_2, \lambda_3$ are the eigenvalues of the symmetric tensor **T**, show that the scalar invariants of section 2.6.1 are given by:

$$\text{(i) } I_1 = \text{Tr}(\mathbf{T}) = \lambda_1 + \lambda_2 + \lambda_3 ,$$

$$\text{(ii) } I_2 = \tfrac{1}{2}(T_{ii}T_{kk} - T_{ik}T_{ki}) = \lambda_2\lambda_3 + \lambda_3\lambda_1 + \lambda_1\lambda_2 ,$$

$$\text{(iii) } I_3 = \det \mathbf{T} = \lambda_1\lambda_2\lambda_3 .$$

2.13 The Cayley – Hamilton Theorem

Problems 2.9, No. 9 is a special case of the **Cayley – Hamilton theorem** which states that: 'A square matrix satisfies its own characteristic equation'. We shall give here a simple proof, valid for non-defective matrices, and in particular for non-defective second-order Cartesian tensors.

Proof. Let the tensor **T** have three linearly independent eigenvectors $\mathbf{v}_1, \mathbf{v}_2, \mathbf{v}_3$, and characteristic equation:

$$\sum_{n=0}^{3} a_n \lambda^n = 0 \ .$$

Then $$\sum_{n=0}^{3} a_n \lambda^n \mathbf{v}_i = \mathbf{0} \ .$$

Now $\mathbf{T}.\mathbf{v}_i = \lambda_i \mathbf{v}_i$ (not summed over i) ,

and $\mathbf{T}^2.\mathbf{v}_i = \lambda_i^2 \mathbf{v}_i$ etc.

Therefore $$\sum_{n=0}^{3} a_n \mathbf{T}^n.\mathbf{v}_i = \mathbf{0} \ , \tag{2.124}$$

where $\mathbf{T}^0$ is to be interpreted as the unit tensor $\mathbf{1}$. Now (2.124) is true for three linearly independent vectors $\mathbf{v}_1, \mathbf{v}_2, \mathbf{v}_3$.

Therefore $$\sum_{n=0}^{3} a_n \mathbf{T}^n = \mathbf{0} \ .$$

The characteristic equation (2.76) is:

$$\lambda^3 - I_1\lambda^2 + I_2\lambda - I_3 = 0 \ .$$

Hence $$\mathbf{T}^3 - I_1\mathbf{T}^2 + I_2\mathbf{T} - I_3\mathbf{1} = \mathbf{0} \ , \tag{2.125}$$

where I_1, I_2, I_3 are the scalar invariants defined in section 2.6.1. Thus we have proved the Cayley - Hamilton theorem for Cartesian tensors and (3×3) matrices. The method of proof given here may, of course, be extended to matrices of any dimension.

A modified form of (2.125) may be obtained for the degenerate cases. Suppose the eigenvalues are $\lambda_1, \lambda_2, \lambda_2$, then in place of the cubic characteristic equation, we have:

$$\lambda^2 - (\lambda_1 + \lambda_2)\lambda + \lambda_1\lambda_2 = 0 \ .$$

Consequently we obtain in place of (2.125):

$$\mathbf{T}^2 - (\lambda_1 + \lambda_2)\mathbf{T} + \lambda_1\lambda_2\mathbf{1} = \mathbf{0} \ . \tag{2.126}$$

Finally, if $\lambda = \lambda_1$ is a treble root of the characteristic equation, we may write the equation as:

$$\lambda - \lambda_1 = 0 \ ,$$

leading to $$\mathbf{T} - \lambda_1\mathbf{1} = \mathbf{0} \ . \tag{2.127}$$

Thus we have again obtained the result first mentioned in sections 2.10.1 and 2.10.2, that the only non-defective tensor with triple eigenvalue λ is a multiple of the unit tensor.

Example 2.13

(1) Find the characteristic equation of the tensor

$$\mathbf{T} = \begin{bmatrix} 1 & 1 & 0 \\ 0 & 1 & 1 \\ 0 & 0 & 0 \end{bmatrix}, \text{ and deduce that } \mathbf{T}^3 - 2\mathbf{T}^2 + \mathbf{T} = \mathbf{0}.$$

Check your result numerically.

The characteristic equation is:

$$\begin{vmatrix} 1-\lambda & 1 & 0 \\ 0 & 1-\lambda & 1 \\ 0 & 0 & -\lambda \end{vmatrix} = 0 \ ,$$

i.e. $$\lambda^3 - 2\lambda^2 + \lambda = 0 \ .$$

Hence, by the Cayley – Hamilton theorem:

$$\mathbf{T}^3 - 2\mathbf{T}^2 + \mathbf{T} = \mathbf{0} \ ,$$

or $$\mathbf{T}.(\mathbf{T} - \mathbf{1})^2 = \mathbf{0} \ .$$

Now $$(\mathbf{T} - \mathbf{1})^2 = \begin{bmatrix} 0 & 1 & 0 \\ 0 & 0 & 1 \\ 0 & 0 & -1 \end{bmatrix} \begin{bmatrix} 0 & 1 & 0 \\ 0 & 0 & 1 \\ 0 & 0 & -1 \end{bmatrix} = \begin{bmatrix} 0 & 0 & 1 \\ 0 & 0 & -1 \\ 0 & 0 & 1 \end{bmatrix}$$

Therefore $$\mathbf{T}.(\mathbf{T} - \mathbf{1})^2 = \begin{bmatrix} 1 & 1 & 0 \\ 0 & 1 & 1 \\ 0 & 0 & 0 \end{bmatrix} \begin{bmatrix} 0 & 0 & 1 \\ 0 & 0 & -1 \\ 0 & 0 & 1 \end{bmatrix} = \begin{bmatrix} 0 & 0 & 0 \\ 0 & 0 & 0 \\ 0 & 0 & 0 \end{bmatrix} .$$

Problems 2.13

(1) Find the characteristic equations of the given tensors and use the Cayley - Hamilton theorem to write them down in the form of (2.125). Check your results numerically.

$$\text{(i)} \begin{bmatrix} 0 & -1 & 0 \\ 2 & 0 & 0 \\ 0 & 0 & -3 \end{bmatrix}; \text{ (ii)} \begin{bmatrix} 1 & 3 & -\sqrt{2} \\ 3 & 1 & \sqrt{2} \\ -\sqrt{2} & \sqrt{2} & -2 \end{bmatrix}; \text{ (iii)} \begin{bmatrix} -4 & 0 & \sqrt{2} \\ 0 & -4 & \sqrt{2} \\ \sqrt{2} & \sqrt{2} & -4 \end{bmatrix} .$$

(2) Show that, for any tensor $\mathbf{T}$, the tensor $\mathbf{T}^{-1}$ represented by the inverse matrix is given by:

$$\mathbf{T}^{-1} = (1/I_3)\,(\mathbf{T}^2 - I_1\mathbf{T} + I_2\mathbf{1}) \ ,$$

where I_1, I_2, I_3 are the scalar invariants defined in section 2.6.1. Find the inverse of the tensor **T** given by:

$$\mathbf{T} = \begin{bmatrix} 0 & 0 & -2 \\ 0 & 1 & 0 \\ 2 & 3 & 1 \end{bmatrix} .$$

2.14 THE REPRESENTATION THEOREM FOR SYMMETRIC TENSORS

Suppose **A** and **B** are symmetric Cartesian tensors such that **B** may be written as a function of **A**. If the functional relationship remains the same in any Cartesian coordinate system, **B** is said to be an **isotropic** function of **A**. An example of such an isotropic function is

$$\mathbf{B} = \alpha \mathbf{A}^n , \tag{2.128}$$

where α is some function of the scalar invariants I_1, I_2, I_3 of **A**, which clearly will not be affected by a rotation. To show that (2.128) is an isotropic function consider a rotation characterized by the rotation matrix Θ. We have:

$$\begin{aligned} \Theta \mathbf{B} \Theta' &= \alpha \Theta \mathbf{A}^n \Theta' , \\ &= \alpha \Theta \mathbf{A} \Theta' \Theta \mathbf{A} \Theta' \ldots . \Theta \mathbf{A} \Theta' , \end{aligned}$$

since $\Theta'\Theta$ is a unit matrix.

Hence $$\Theta \mathbf{B} \Theta' = \alpha (\Theta \mathbf{A} \Theta')^n . \tag{2.129}$$

In fact, as we shall see, (2.128) represents the only type of isotropic function, so that the most general isotropic function is a linear combination of powers of **A**. However, we may employ the Cayley - Hamilton theorem to show that **B** may be written as a linear combination of **1**, **A** and $\mathbf{A}^2$. This result is the subject of the following theorem.

Theorem 2.14.1(i) Let **A** be a symmetric tensor with invariants I_1, I_2, I_3, and let **B** be a symmetric tensor such that **B** is an isotropic function of **A**. Then scalar functions α, β, γ, of I_1, I_2, I_3 exist such that

$$\mathbf{B} = \alpha \mathbf{1} + \beta \mathbf{A} + \gamma \mathbf{A}^2 . \tag{2.130}$$

Further, if two eigenvalues of **A** coincide, then α and β may be chosen such that $\gamma = 0$, and if all three eigenvalues of **A** coincide then α may be chosen such that $\beta = \gamma = 0$.

Proof. We establish first that if **B** is an isotropic function of **A**, then **A** and **B** have the same set of eigenvectors. Let us choose axes such that **A** is in diagonal form, i.e.

$$\mathbf{A} = \begin{bmatrix} \lambda_1 & 0 & 0 \\ 0 & \lambda_2 & 0 \\ 0 & 0 & \lambda_3 \end{bmatrix}, \tag{2.131}$$

where $\lambda_1, \lambda_2, \lambda_3$ are the eigenvalues of **A** which we shall, for the time being, assume to be distinct.

Consider a rotation represented by the matrix

$$\Theta = \begin{bmatrix} (-1)^p & 0 & 0 \\ 0 & (-1)^q & 0 \\ 0 & 0 & (-1)^r \end{bmatrix},$$

where $p + q + r$ is even. The matrix **B** is transformed to

$$\begin{bmatrix} b_{11} & (-1)^r b_{12} & (-1)^q b_{13} \\ (-1)^r b_{12} & b_{22} & (-1)^p b_{23} \\ (-1)^q b_{13} & (-1)^p b_{23} & b_{33} \end{bmatrix},$$

whereas **A** remains unchanged.

Now, if **A** remains unchanged, **B** must also remain unchanged. Hence the off-diagonal terms of **B** must be zero,

i.e.

$$\mathbf{B} = \begin{bmatrix} \mu_1 & 0 & 0 \\ 0 & \mu_2 & 0 \\ 0 & 0 & \mu_3 \end{bmatrix}, \tag{2.132}$$

where μ_1, μ_2, μ_3 are the eigenvalues of **B**. Thus we have shown that when **A** is in diagonal form **B** is also in diagonal form. Therefore they have the same eigenvectors.

We may now replace the functional relationship between the tensors **A** and **B** by a relationship between their eigenvalues. Thus, for some set of functions f_1, f_2, f_3:

$$\mu_i = f_i(\lambda_1, \lambda_2, \lambda_3) \ .$$

Now consider a rotation which interchanges λ_2 and λ_3 in **A**.

Then $\mu_1 = f_1(\lambda_1, \lambda_3, \lambda_2)$,

i.e. f_1 is symmetric in λ_2 and λ_3. Hence there exists a function F such that

$$\mu_1 = F[\lambda_1, \phi(\lambda_1, \lambda_2, \lambda_3)] \ ,$$

where ϕ is symmetric in $\lambda_1, \lambda_2, \lambda_3$. Now, any symmetric function of the eigenvalues may be written in terms of the invariants I_1, I_2, I_3.† Hence, there exists a function ψ such that

$$\mu_1 = \psi(\lambda_1; I_1, I_2, I_3) \ .$$

We consider next a rotation which interchanges λ_1, λ_2 in **A** and μ_1, μ_2 in **B**.

Then $\quad \mu_2 = f_1(\lambda_2, \lambda_1, \lambda_3) = \psi(\lambda_2; I_1, I_2, I_3) \ .$

Similarly $\quad \mu_3 = \psi(\lambda_3; I_1, I_2, I_3) \ .$

We now expand ψ in a Taylor series to obtain:

$$\psi(\lambda; I_1, I_2, I_3) = \psi(0) + \lambda\psi'(0) + \frac{\lambda^2}{2!}\psi''(0) + \ldots. ,$$

where $\psi(0)$ has been written for $\psi(0; I_1, I_2, I_3)$.

Therefore $\quad \mathbf{B} = \psi(0)\mathbf{1} + \psi'(0)\mathbf{A} + \tfrac{1}{2}\psi''(0)\mathbf{A}^2 + \ldots.$

But, by the Cayley - Hamilton theorem:

$$\mathbf{A}^3 = I_1\mathbf{A}^2 - I_2\mathbf{A} + I_3\mathbf{1} \ ,$$

and $\quad \mathbf{A}^4 = I_1\mathbf{A}^3 - I_2\mathbf{A}^2 + I_3\mathbf{A}$, etc.

Therefore $\quad \mathbf{B} = \alpha\mathbf{1} + \beta\mathbf{A} + \gamma\mathbf{A}^2 \ , \qquad (2.133)$

where $\quad \alpha = \alpha(I_1, I_2, I_3)$ etc.

If λ is a double root of the characteristic equation we may deduce, with the aid of (2.126), that if **B** is an isotropic function of **A**, then **B** may be written as:

$$\mathbf{B} = \alpha\mathbf{1} + \beta\mathbf{A} \ . \qquad (2.134)$$

Finally, if λ is a treble root of the characteristic equation, the only isotropic relation is:

$$\mathbf{B} = \alpha\mathbf{1} \ . \qquad (2.135)$$

2.15 TENSOR FIELDS

2.15.1 Definition

We defined scalar and vector fields in sections 1.3.2 and 1.8.1 respectively. The gradient of a scalar field was then defined as a vector field in section 1.8.2. We shall now define a second-order tensor field, and the gradient of a vector field as an example of such a field.

†For a proof of this result see, for example, *Vector Analysis and Cartesian Tensors* (second edition) by D. E. Bourne and P. C. Kendall, §.9.3.

Suppose a second-order tensor **T** is defined at every point P within a domain $\mathscr{D}$ by the relation $\mathbf{T} = \mathbf{T}(P)$. If for every point P there is a unique value $\mathbf{T}(P)$, then $\mathbf{T} = \mathbf{T}(P)$ defines a second-order tensor function of position called a **second-order tensor field**. The distributions of stress and strain in an elastic medium are examples of second-order tensor fields. It is also possible to regard the inertia tensor as an example of a tensor field.

2.15.2 Differentiation of Vector Fields

We are now in a position to define the derivative of a vector field following the method of section 1.8.2. The definition is as follows.

Let **v** be a vector field. If there exists a second-order tensor $\nabla\mathbf{v}$ such that

$$\mathbf{v}(\mathbf{r}+\mathbf{h}) - \mathbf{v}(\mathbf{r}) = \mathbf{h}.\nabla\mathbf{v} + \mathbf{h}.\mathbf{Z} \ , \tag{2.136}$$

where $|\mathbf{Z}| = \sqrt{\mathbf{Z}{:}\mathbf{Z}'} \to 0$ as $|\mathbf{h}| \to 0$,

then **v** is said to be differentiable at **r**, and $\nabla\mathbf{v}$ is called the **derivative** or **gradient** of **v** at **r**. If **v** is differentiable at all points of a domain $\mathscr{D}$, then $\nabla\mathbf{v}$ defines a second-order tensor field on $\mathscr{D}$.

The components of $\nabla\mathbf{v}$ in any Cartesian basis $(\mathbf{e}_1, \mathbf{e}_2, \mathbf{e}_3)$ may now be found. At any point $\mathbf{r} = x_i\mathbf{e}_i$, $\mathbf{v}(\mathbf{r}) = v_i(\mathbf{r})\mathbf{e}_i$. Let $\mathbf{h} = h\mathbf{e}_i$. Then
$\mathbf{v}(\mathbf{r}+h\mathbf{e}_i) - \mathbf{v}(\mathbf{r}) = h\mathbf{e}_i.\nabla\mathbf{v} + h\mathbf{e}_i.\mathbf{Z}$.

Therefore $$\mathbf{e}_i.\nabla\mathbf{v} = \lim_{h\to 0} \frac{\mathbf{v}(\mathbf{r}+h\mathbf{e}_i) - \mathbf{v}(\mathbf{r})}{h} = \frac{\partial\mathbf{v}}{\partial x_i} \ ,$$

hence $$\mathbf{e}_i.\nabla\mathbf{v}.\mathbf{e}_j = (\nabla\mathbf{v})_{ij} = \frac{\partial v_j}{\partial x_i} \ , \tag{2.137}$$

or $$\nabla\mathbf{v} = \frac{\partial v_j}{\partial x_i}\mathbf{e}_i \otimes \mathbf{e}_j \ . \tag{2.138}$$

Written in matrix form the gradient of **v** is:

$$\nabla\mathbf{v} = \begin{bmatrix} \partial v_1/\partial x_1 & \partial v_2/\partial x_1 & \partial v_3/\partial x_1 \\ \partial v_1/\partial x_2 & \partial v_2/\partial x_2 & \partial v_3/\partial x_2 \\ \partial v_1/\partial x_3 & \partial v_2/\partial x_3 & \partial v_3/\partial x_3 \end{bmatrix} . \tag{2.139}$$

From any of these forms it is easy to see that the divergence of **v** is a scalar invariant which is the trace of $\nabla\mathbf{v}$.

Thus $$\nabla.\mathbf{v} = \operatorname{div}\mathbf{v} = \operatorname{Tr}(\nabla\mathbf{v}) = \frac{\partial v_i}{\partial x_i} \ . \tag{2.140}$$

As we have defined $\nabla\mathbf{v}$ to be a second-order tensor we should naturally expect its components to transform according to the law given by (2.43). It is, however, instructive to obtain this result directly from the explicit expression of (2.137).

Thus $$\frac{\partial v_k^*}{\partial x_i^*} = \frac{\partial}{\partial x_i^*}(\theta_{km}v_m) ,$$

$$= \frac{\partial}{\partial x_n}(\theta_{km}v_m)\frac{\partial x_n}{\partial x_i^*} .$$

Now $$x_n = \theta_{in}x_i^* ,$$

therefore $$\frac{\partial x_n}{\partial x_i^*} = \theta_{in} .$$

Hence $$\frac{\partial v_k^*}{\partial x_i^*} = \theta_{in}\theta_{km}\frac{\partial v_m}{\partial x_n} . \tag{2.141}$$

Equation (2.141) is the transformation law for second-order tensors.

Now let us consider the case in which the vector field, **v**, is the gradient of the scalar field, ϕ, i.e. $\mathbf{v} = \nabla\phi$. We have $\nabla\mathbf{v} = \nabla(\nabla\phi)$ and div $\mathbf{v} = \nabla.(\nabla\phi)$. Now $\nabla.(\nabla\phi)$ is written as $\nabla^2\phi$ and is called the **Laplacian** of ϕ.

Clearly $$\nabla^2\phi = \frac{\partial^2\phi}{\partial x_i \partial x_i} . \tag{2.142}$$

It is sometimes useful to regard the symbol ∇ as a vector operator. Thus, formally:

$$\nabla = \mathbf{e}_i\frac{\partial}{\partial x_i} . \tag{2.143}$$

The expression (2.140) for $\nabla.\mathbf{v}$ then follows at once, but the tensor $\nabla\mathbf{v}$ implies a tensor product sign which is not written, i.e.

$$\nabla\mathbf{v} \equiv \nabla \otimes \mathbf{v} . \tag{2.144}$$

The expression on the right-hand side of (2.144) will not be used here although some authors have employed it.

Examples 2.15

(1) Given $\mathbf{v} = (x_2, x_1, x_1^2 + x_2^2)$, find (i) $\nabla\mathbf{v}$; (ii) $\nabla.\mathbf{v}$.

(i) We have, from (2.138) and (2.139):

$$\nabla\mathbf{v} = \frac{\partial v_j}{\partial x_i}\mathbf{e}_i \otimes \mathbf{e}_j = \begin{bmatrix} 0 & 1 & 2x_1 \\ 1 & 0 & 2x_2 \\ 0 & 0 & 0 \end{bmatrix} .$$

(ii) We have, from (2.140):

$$\nabla.\mathbf{v} = \mathrm{Tr}(\nabla\mathbf{v}) = 0 .$$

(2) Show that (i) $\nabla \mathbf{r} = \mathbf{1}$; (ii) $\nabla.\mathbf{r} = 3$.

(i) We have, from (2.138):

$$\nabla \mathbf{r} = \frac{\partial x_i}{\partial x_j} \mathbf{e}_i \otimes \mathbf{e}_j = \delta_{ij}\mathbf{e}_i \otimes \mathbf{e}_j = \mathbf{1} .$$

(ii) We have, from (2.140): $\nabla.\mathbf{r} = \mathrm{Tr}(\nabla \mathbf{r}) = 3$.

Problems 2.15

(1) Given $\mathbf{v} = (x_1^2 + x_2, 2x_2x_3, x_3^3)$, find (i) $\nabla \mathbf{v}$; (ii) $\nabla.\mathbf{v}$.

(2) Show that, for any scalar function $\phi(r)$ of r, and any vector $\mathbf{v}$:

(i) $\nabla[\phi(r)\mathbf{r}] = \nabla\phi \otimes \mathbf{r} + \phi(r)\mathbf{1}$;

(ii) $\nabla.[\phi(r)\mathbf{r}] = \mathbf{r}.\nabla\phi + 3\phi(r)$;

(iii) $\nabla[\phi(r)\mathbf{a}] = \nabla\phi \otimes \mathbf{a} + \phi(r)\nabla\mathbf{a}$;

(iv) $\nabla.[\phi(r)\mathbf{a}] = \mathbf{a}.\nabla\phi + \phi\nabla.\mathbf{a}$.

(3) Verify the results of question (2) when

$$\phi(r) = r, \qquad \mathbf{a} = x_1^2\mathbf{e}_1 .$$

(4) Verify the results of question 2 when

$$\phi(r) = 2r + 3r^2, \qquad \mathbf{a} = \mathbf{r} .$$

(5) Show by direct calculation that the components of the tensor $\nabla(\nabla\phi)$ obey the transformation law for second-order tensors.

CHAPTER 3

Third-Order Cartesian Tensors

3.1 TENSORS OF THE THIRD - ORDER

3.1.1 Definition

In Chapter 2 [Theorem 2.2.4(i)] we showed that a second-order tensor was defined uniquely by giving its value on three linearly independent vectors. This method of definition may be extended to define tensors of higher order; tensors of order n being used to define tensors of order $(n + 1)$. Thus if $\mathbf{v}_1, \mathbf{v}_2, \mathbf{v}_3$ are three linearly independent vectors and $\mathbf{T}_1, \mathbf{T}_2, \mathbf{T}_3$ are three second-order tensors, the relations

$$\mathbf{B}.\mathbf{v}_i = \mathbf{T}_i, i = 1, 2, 3 \ , \tag{3.1}$$

may be used to define a third-order tensor $\mathbf{B}$ once the dot product $\mathbf{B}.\mathbf{v}_i$ has been defined. To do this we proceed along the lines of section 2.1.2 and write:

$$(\mathbf{a} \otimes \mathbf{b} \otimes \mathbf{c}).\mathbf{v} = (\mathbf{a} \otimes \mathbf{b})(\mathbf{c}.\mathbf{v}) \ . \tag{3.2}$$

Equation (3.2) defines not only the dot product on the left-hand side but also the **tensor product of three vectors $\mathbf{a} \otimes \mathbf{b} \otimes \mathbf{c}$**. The right-hand side of (3.2) (being a dyad) is a second-order tensor. Hence, by (3.1), the tensor product $\mathbf{a} \otimes \mathbf{b} \otimes \mathbf{c}$ is a third-order tensor.

In section 2.2.1 we showed that any second-order tensor could be written as the sum of 9 dyads. We shall now show that any third-order tensor, such as is defined by (3.1), may be written as the sum of 27 tensor products of three vectors.

We note from (3.1) that if $\mathbf{B}$ is a third-order tensor, $\mathbf{B}.\mathbf{e}_m$ is a second-order tensor,

i.e.
$$\mathbf{B}.\mathbf{e}_m = B_{ikm}\, \mathbf{e}_i \otimes \mathbf{e}_k \ , \tag{3.3}$$

where
$$B_{ikm} = \mathbf{e}_i.(\mathbf{B}.\mathbf{e}_m).\mathbf{e}_k \ . \tag{3.4}$$

Equations (3.3) and (3.4) are satisfied if

$$\mathbf{B} = B_{ikm}\mathbf{e}_i \otimes \mathbf{e}_k \otimes \mathbf{e}_m \ . \tag{3.5}$$

Thus we have shown that it is possible to write **B** as the sum of tensor products of three vectors. The 27 numbers B_{ikm} are the components of the third-order tensor **B**. They are given explicitly by (3.4).

3.1.2 Products Involving Third-Order Tensors

The components of the post-multiplication inner product of a third-order tensor with a vector may be deduced from (3.3).

Thus
$$\mathbf{B.v} = \mathbf{B}.(v_m \mathbf{e}_m) \ , \tag{3.6}$$
$$= B_{ikm} v_m \mathbf{e}_i \otimes \mathbf{e}_k \ .$$

Therefore
$$(\mathbf{B.v})_{ik} = B_{ikm} v_m \ . \tag{3.7}$$

The components of the pre-multiplication product are defined to be:

$$(\mathbf{v.B})_{ik} = v_m B_{mik} \ . \tag{3.8}$$

The double inner product of a third-order tensor **B** and a second-order tensor **T** is a tensor of order $3 + 2 - 2 - 2 = 1$, i.e. a vector. The components of the vector are given by:

$$(\mathbf{B}:\mathbf{T})_i = B_{ikm} T_{mk} \ , \tag{3.9}$$

or
$$(\mathbf{T}:\mathbf{B})_i = T_{km} B_{mki} \ . \tag{3.10}$$

Finally, the **treble inner product** of two third-order tensors is a scalar given by:

$$\mathbf{B} \vdots \mathbf{C} = B_{ikm} C_{mki} \ . \tag{3.11}$$

3.1.3 The Transformation Law for Third-Order Tensors

The transformation law for the components of a third-order tensor may be deduced from the transformation law for vectors. Thus, if B^*_{jln} are the components of **B** in the transformed frame of reference, we have from (3.4):

$$B^*_{jln} = \mathbf{e}^*_j.(\mathbf{B}.\mathbf{e}^*_n).\mathbf{e}^*_l \ ,$$
$$= \theta_{ji} \mathbf{e}_i.(\mathbf{B}.\theta_{nm} \mathbf{e}_m).\theta_{lk} \mathbf{e}_k \ .$$

Hence
$$B^*_{jln} = \theta_{ji} \theta_{lk} \theta_{nm} B_{ikm} \ . \tag{3.12}$$

This expression should be compared with (2.43).

Example 3.1

(1) Vectors **a**, **b**, **c** are defined by **a** = (1, 0, 0); **b** = (1, 1, 0); **c** = (1, 1, 1). Write down the components of

(i) (**a** ⊗ **b** ⊗ **c**).**c**; (ii) **a**.(**a** ⊗ **b** ⊗ **c**); (iii) (**a** ⊗ **b** ⊗ **c**):(**a** ⊗ **b**);

(iv) (**a** ⊗ **b**):(**a** ⊗ **b** ⊗ **c**).

(v) Give the six non-zero components of **A** = **a** ⊗ **b** ⊗ **c**.

$$\text{(i) } (\mathbf{a} \otimes \mathbf{b} \otimes \mathbf{c}).\mathbf{c} = (\mathbf{a} \otimes \mathbf{b})(\mathbf{c}.\mathbf{c}) = 3 \begin{bmatrix} 1 & 1 & 0 \\ 0 & 0 & 0 \\ 0 & 0 & 0 \end{bmatrix} .$$

$$\text{(ii) } \mathbf{a}.(\mathbf{a} \otimes \mathbf{b} \otimes \mathbf{c}) = (\mathbf{a}.\mathbf{a})(\mathbf{b} \otimes \mathbf{c}) = \begin{bmatrix} 1 & 1 & 1 \\ 1 & 1 & 1 \\ 0 & 0 & 0 \end{bmatrix} .$$

(iii) $(\mathbf{a} \otimes \mathbf{b} \otimes \mathbf{c}):(\mathbf{a} \otimes \mathbf{b}) = (\mathbf{c}.\mathbf{a})(\mathbf{b}.\mathbf{b})\mathbf{a} = 2\mathbf{a} = (2, 0, 0)$.

(iv) $(\mathbf{a} \otimes \mathbf{b}):(\mathbf{a} \otimes \mathbf{b} \otimes \mathbf{c}) = (\mathbf{a}.\mathbf{b})^2 \mathbf{c} = \mathbf{c} = (1, 1, 1)$.

(v) The six non-zero components of $\mathbf{A} = \mathbf{a} \otimes \mathbf{b} \otimes \mathbf{c}$ are:

$$A_{111} = A_{121} = A_{112} = A_{122} = A_{113} = A_{123} = 1 .$$

Problem 3.1

(1) Vectors **a**, **b**, **c**, **d** are defined by

$\mathbf{a} = (3, 4, 1); \mathbf{b} = (1, 0, 1); \mathbf{c} = (2, 1, 0); \mathbf{d} = (3, 4, 5).$

(i) Write down $(\mathbf{a} \otimes \mathbf{b} \otimes \mathbf{c}).\mathbf{d}$ in matrix form.

(ii) Write down the 27 components of $\mathbf{a} \otimes \mathbf{b} \otimes \mathbf{c}$.

(iii) Let $\mathbf{A} = \mathbf{a} \otimes \mathbf{b} \otimes \mathbf{c}$; $\mathbf{B} = \mathbf{c} \otimes \mathbf{d}$. Show that $\mathbf{A}:\mathbf{B} = 40\mathbf{a}$.

(iv) Let $\mathbf{T} = \mathbf{a} \otimes \mathbf{d} \otimes \mathbf{d}$; $\mathbf{S} = \mathbf{b} \otimes \mathbf{c} \otimes \mathbf{c}$. Find all the components of $\mathbf{S}:\mathbf{T}$.

3.2 THE ALTERNATE TENSOR

3.2.1 Definition

The most useful third-order Cartesian tensor, **E**, with components ϵ_{ikm}, is defined as follows:

$$\mathbf{E}:\mathbf{a} \otimes \mathbf{b} = -\mathbf{a} \times \mathbf{b}, \qquad \forall \mathbf{a}, \mathbf{b} . \tag{3.13}$$

Thus $\epsilon_{ikm} a_m b_k = -(\mathbf{a} \times \mathbf{b})_i$.

The components ϵ_{ikm} are conveniently found by putting

$$\mathbf{a} = \mathbf{e}_p, \mathbf{b} = \mathbf{e}_q, \text{ so that } a_m = \delta_{mp}, b_k = \delta_{kq} .$$

Thus $\epsilon_{iqp} = -(\mathbf{e}_p \times \mathbf{e}_q)_i$,

therefore $\epsilon_{ikm} = [\mathbf{e}_i, \mathbf{e}_k, \mathbf{e}_m]$. (3.14)

The tensor **E** is known as the **alternate tensor** or **totally antisymmetric third-order tensor.**

From (3.14) we may deduce that the components of **E** have the following property:

$$\epsilon_{ikm} = \begin{cases} +1 \text{ if } i, k, m \text{ is an even permutation of the numbers } 1, 2, 3, \\ -1 \text{ if } i, k, m \text{ is an odd permutation of the numbers } 1, 2, 3, \\ 0 \text{ if any two of } i, k, m \text{ are equal.} \end{cases} \tag{3.15}$$

The function ϵ_{ikm} is sometimes known as the **Levi - Civita density function.**

Naturally we should expect the components of **E** to obey the transformation law (3.12). This fact may, however, be deduced independently using (3.14), which is clearly true for any other basis $(\mathbf{e}_j^*, \mathbf{e}_l^*, \mathbf{e}_n^*)$.

Thus
$$\begin{aligned} \epsilon_{jln}^* &= [\mathbf{e}_j^*, \mathbf{e}_l^*, \mathbf{e}_n^*] \ , \\ &= [\theta_{ji}\,\mathbf{e}_i, \theta_{lk}\mathbf{e}_k, \theta_{nm}\,\mathbf{e}_m] \ , \\ &= \theta_{ji}\theta_{lk}\theta_{nm}\,\epsilon_{ikm} \ . \end{aligned} \tag{3.16}$$

However, the fact that the components are defined by (3.15) in a basis-free way shows that they are invariant under rotations,

i.e.
$$\epsilon_{jln}^* = \epsilon_{jln} \ . \tag{3.17}$$

The property expressed by (3.17) is shared with the (second-order) unit tensor. Tensors whose components remain invariant under rotations are called **isotropic** (see section 4.2.1).

3.2.2 Applications of the Alternate Tensor

We have already seen, in section 3.2.1, that the vector product of two vectors may be written in terms of the alternate tensor. Thus, from (3.13) we have, using the result of Examples 3.2, No. 4:

$$\mathbf{a} \times \mathbf{b} = -\mathbf{E}{:}\mathbf{a} \otimes \mathbf{b} \ , \tag{3.18}$$

$$= \epsilon_{ikm} a_k b_m \mathbf{e}_i \ . \tag{3.19}$$

It follows that the triple scalar product $\mathbf{a} \times \mathbf{b.c}$ is given by:

$$\begin{aligned} \mathbf{a} \times \mathbf{b.c} &= \epsilon_{ikm} a_k b_m c_i \ , \\ &= \epsilon_{ikm} a_i b_k c_m \ . \end{aligned} \tag{3.20}$$

The definition (3.14) gives rise to an important result which is expressed by the following theorem:

Theorem 3.2.2 $\epsilon_{ikm}\epsilon_{iln} = \delta_{kl}\delta_{mn} - \delta_{kn}\delta_{ml}$.

Proof.

$$\begin{aligned}
\epsilon_{ikm}\epsilon_{iln} &= (\mathbf{e}_i.\mathbf{e}_k \times \mathbf{e}_m)(\mathbf{e}_i.\mathbf{e}_l \times \mathbf{e}_n) \,, \\
&= (\mathbf{e}_k \times \mathbf{e}_m).(\mathbf{e}_l \times \mathbf{e}_n) \,, \\
&= (\mathbf{e}_k \times \mathbf{e}_m) \times \mathbf{e}_l.\mathbf{e}_n \,, \\
&= [\mathbf{e}_m(\mathbf{e}_k.\mathbf{e}_l) - \mathbf{e}_k(\mathbf{e}_m.\mathbf{e}_l)].\mathbf{e}_n \,, \\
\epsilon_{ikm}\epsilon_{iln} &= \delta_{kl}\delta_{mn} - \delta_{kn}\delta_{ml} \,.
\end{aligned} \tag{3.21}$$

This result is useful for evaluating continued vector products (see for example Problems 3.2, No. 5) as well as a number of the differential vector functions considered in the next section.

3.2.3 Differential Functions involving the Alternate Tensor

In section 2.15.2 we showed how the symbol ∇ could be regarded as a vector operator.

Thus $$\nabla = \mathbf{e}_i \frac{\partial}{\partial x_i} \,. \tag{3.22}$$

In the same section we defined $\nabla.\mathbf{v}$ and $\nabla\mathbf{v}$ which are equivalent to $\mathbf{a}.\mathbf{b}$ and $\mathbf{a} \otimes \mathbf{b}$ respectively, where $\mathbf{a}$ and $\mathbf{b}$ are ordinary vectors. In equation (3.18) we saw that the cross product $\mathbf{a} \times \mathbf{b}$ could be defined in terms of the alternate tensor. It is logical therefore to define a differential function $\nabla \times \mathbf{v}$ as

$$\nabla \times \mathbf{v} = -\mathbf{E}:\nabla\mathbf{v} \,, \tag{3.23}$$

$$= \epsilon_{ikm} \frac{\partial v_m}{\partial x_k} \mathbf{e}_i \,. \tag{3.24}$$

$\nabla \times \mathbf{v}$ is a vector known as the **curl** of $\mathbf{v}$ and is alternatively written curl $\mathbf{v}$. It may also be expressed as a determinant:

$$\nabla \times \mathbf{v} = \begin{vmatrix} \mathbf{e}_1 & \mathbf{e}_2 & \mathbf{e}_3 \\ \partial/\partial x_1 & \partial/\partial x_2 & \partial/\partial x_3 \\ v_1 & v_2 & v_3 \end{vmatrix} \,. \tag{3.25}$$

By repeated application of (3.24) we find that

$$\begin{aligned}
\nabla \times (\nabla \times \mathbf{v}) &= \epsilon_{ikm}\epsilon_{iln} \frac{\partial^2 v_n}{\partial x_l \partial x_m} \mathbf{e}_k \,, \\
&= (\delta_{kl}\delta_{mn} - \delta_{kn}\delta_{ml}) \frac{\partial^2 v_n}{\partial x_l \partial x_m} \mathbf{e}_k \\
&= \left(\frac{\partial^2 v_m}{\partial x_k \partial x_m} - \frac{\partial^2 v_k}{\partial x_m \partial x_m} \right) \mathbf{e}_k \,, \\
&= \nabla(\nabla.\mathbf{v}) - \nabla^2\mathbf{v} \,,
\end{aligned}$$

or $$\text{curl curl } \mathbf{v} = \text{grad div } \mathbf{v} - \nabla^2\mathbf{v} \,. \tag{3.26}$$

Examples 3.2

(1) The numerical values of ϵ_{ikm} are:

$$\epsilon_{123} = \epsilon_{231} = \epsilon_{312} = 1, \epsilon_{321} = \epsilon_{132} = \epsilon_{213} = -1 .$$

The remaining 21 components of **E** are zero.

(2) The third scalar invariant of the second-order tensor **T** is $I_3 = \det \mathbf{T}$ (see section 2.6). Show that $I_3 = \epsilon_{jln} T_{1j} T_{2l} T_{3n}$.

We have $$I_3 = T_{11}(T_{22}T_{33} - T_{23}T_{32}) + T_{12}(T_{23}T_{31} - T_{21}T_{33}) + T_{13}(T_{21}T_{32} - T_{22}T_{31}) ,$$

$$= T_{1j}(\epsilon_{jln}T_{2l}T_{3n}) .$$

Hence $$I_3 = \epsilon_{jln}T_{1j}T_{2l}T_{3n} .$$

(3) Vectors **a**, **b** are defined by $\mathbf{a} = \mathbf{e}_1, \mathbf{b} = \mathbf{e}_1 + \mathbf{e}_2$.

Give the components of (i) **E.a ⊗ b**; (ii) **E:a ⊗ b** .

(i) Let **A** = **E.a ⊗ b**,

then $$A_{jln} = \epsilon_{jlp}a_p b_n = \epsilon_{jl1} b_n .$$

Therefore $$A_{jl1} = A_{jl2} = \epsilon_{jl1} ,$$

i.e. $$A_{321} = A_{322} = -1 ,$$

and $$A_{231} = A_{232} = +1 .$$

All other components of **A** are zero.

(ii) Let **d** = **E:a ⊗ b**,

then $$d_i = \epsilon_{ipq}a_q b_p = \epsilon_{ip1}b_p = \epsilon_{i21} .$$

Hence $$\mathbf{d} = (0, 0, -1) .$$

(4) For any i, k, m for which ϵ_{ikm} is defined:

$$\epsilon_{ikm} = \epsilon_{kmi} = \epsilon_{mik} = -\epsilon_{mki} = -\epsilon_{imk} = -\epsilon_{kim}$$

(5) Show that **(T.E).c** = **T.(E.c)** ; **c.(E.T)** = **(c.E).T** .

We have $$[(\mathbf{T.E}).\mathbf{c}]_{ik} = (\mathbf{T.E})_{ikp}c_p = T_{iq}\epsilon_{qkp}c_p ,$$

also $$[\mathbf{T}.(\mathbf{E.c})]_{ik} = T_{iq}(\mathbf{E.c})_{qk} = T_{iq}\epsilon_{qkp}c_p .$$

The second result is proved in a similar way.

Hence we may write **T.E.c** for **(T.E).c** or **T.(E.c)**, and **c.E.T** for **c.(E.T)** or **(c.E).T**.

(6) Prove that if $\nabla\mathbf{v} = \frac{1}{2}\mathbf{E}.\ \text{curl}\ \mathbf{v}$, then $\text{curl}\ \mathbf{v} = -\mathbf{E}:\nabla\mathbf{v}$. Is the reverse true?

We have $$(\nabla\mathbf{v})_{ik} = \tfrac{1}{2}\epsilon_{ikm}(\text{curl}\ \mathbf{v})_m ,$$

therefore $(\nabla\mathbf{v})_{ik}\epsilon_{iln} = \tfrac{1}{2}\epsilon_{ikm}\epsilon_{iln}(\text{curl }\mathbf{v})_m$,

$= \tfrac{1}{2}(\delta_{kl}\delta_{mn} - \delta_{kn}\delta_{ml})(\text{curl }\mathbf{v})_m$.

Therefore $(\nabla\mathbf{v})_{ik}\epsilon_{ikn} = (\text{curl }\mathbf{v})_n$,

thus $\text{curl }\mathbf{v} = -\mathbf{E}{:}\nabla\mathbf{v}$.

The reverse is not true for, suppose $\nabla\mathbf{v} = \mathbf{S} + \mathbf{A}$, where $\mathbf{S}$ is symmetric and $\mathbf{A}$ is antisymmetric. It is easy to show that $\mathbf{E}{:}\mathbf{S} = \mathbf{0}$ (see Problems 3.2, No. 4).

Therefore $\mathbf{E}{:}\nabla\mathbf{v} = \mathbf{E}{:}\mathbf{A}$,

and $\mathbf{E}{:}\mathbf{A} = -\text{curl }\mathbf{v}$,

or $\epsilon_{ikm}A_{mk} = -(\text{curl }\mathbf{v})_i$,

i.e. $A_{mk}\epsilon_{ikm}\epsilon_{iln} = -\epsilon_{iln}(\text{curl }\mathbf{v})_i$

therefore $A_{mk}(\delta_{kl}\delta_{mn} - \delta_{kn}\delta_{ml}) = -\epsilon_{iln}(\text{curl }\mathbf{v})_i$.

Hence $A_{nl} - A_{ln} = -\epsilon_{iln}(\text{curl }\mathbf{v})_i$.

But $\mathbf{A}$ is antisymmetric.

Hence $A_{ln} = \tfrac{1}{2}\epsilon_{iln}(\text{curl }\mathbf{v})_i$,

or $\mathbf{A} = \tfrac{1}{2}\mathbf{E}.\,\text{curl }\mathbf{v}$.

Thus $\text{curl }\mathbf{v} = -\mathbf{E}{:}\nabla\mathbf{v}$ implies only that the antisymmetric part of $\nabla\mathbf{v}$ is equal to $\tfrac{1}{2}\mathbf{E}.\,\text{curl }\mathbf{v}$.

(7) Show that, for any scalar ϕ, and any vector $\mathbf{v}$,

$$\nabla \times (\phi\mathbf{a}) = \nabla\phi \times \mathbf{a} + \phi\nabla \times \mathbf{a} .$$

We have
$$\nabla \times (\phi\mathbf{a}) = \epsilon_{ikm}\frac{\partial}{\partial x_k}(\phi a_m)\mathbf{e}_i,$$

$$= \left(\epsilon_{ikm}\frac{\partial\phi}{\partial x_k}a_m + \epsilon_{ikm}\phi\frac{\partial a_m}{\partial x_k}\right)\mathbf{e}_i ,$$

$$= \nabla\phi \times \mathbf{a} + \phi\nabla \times \mathbf{a} .$$

Problems 3.2

(1) Vectors $\mathbf{a}, \mathbf{b}, \mathbf{c}$ are defined by $\mathbf{a} = \mathbf{e}_1, \mathbf{b} = \mathbf{e}_1 + \mathbf{e}_2, \mathbf{c} = \mathbf{e}_1 + \mathbf{e}_2 + \mathbf{e}_3$. Find all the components of (i) $\mathbf{A} = \mathbf{a} \otimes \mathbf{c}.\mathbf{E}$; (ii) $\mathbf{d} = \mathbf{b} \otimes \mathbf{c}{:}\mathbf{E}$.

(2) Use Examples 3.2 No. 2 to show that

$$I_3 = (1/6)\epsilon_{ikm}\epsilon_{jln}T_{ij}T_{kl}T_{mn} .$$

(3) Show that, for any vector $\mathbf{u}$, $\mathbf{E}.\mathbf{u} = \mathbf{u}.\mathbf{E}$. Suppose the second-order tensor $\mathbf{T}$ is given by $\mathbf{T} = \mathbf{E}.\mathbf{u}$. Express $\mathbf{u}$ in terms of $\mathbf{T}$.

(4) Show that for any second-order tensor **T**, **E**:**T** = **T**:**E**. Show also that if **T** is symmetric then **E**:**T** = **0**.

(5) Use Theorem 3.2.2 to show that, for any four vectors **a**, **b**, **c**, **d**,

$$(\mathbf{a} \times \mathbf{b}) \times (\mathbf{c} \times \mathbf{d}) = \mathbf{b}[\mathbf{a}, \mathbf{c}, \mathbf{d}] - \mathbf{a}[\mathbf{b}, \mathbf{c}, \mathbf{d}] \; ,$$
$$= \mathbf{c}[\mathbf{a}, \mathbf{b}, \mathbf{d}] - \mathbf{d}[\mathbf{a}, \mathbf{b}, \mathbf{c}] \; .$$

(6) Prove that $\epsilon_{ikm}\epsilon_{ikn} = 2\delta_{mn}$.

(7) Use Examples 3.2 No. 2, and Equations (3.19), (3.20) to show that:

$$\text{(i) } \mathbf{a} \times \mathbf{b} = \begin{vmatrix} \mathbf{e}_1 & \mathbf{e}_2 & \mathbf{e}_3 \\ a_1 & a_2 & a_3 \\ b_1 & b_2 & b_3 \end{vmatrix} ; \quad \text{(ii) } \mathbf{a} \times \mathbf{b}.\mathbf{c} = \begin{vmatrix} a_1 & a_2 & a_3 \\ b_1 & b_2 & b_3 \\ c_1 & c_2 & c_3 \end{vmatrix} .$$

(8) Let A_{ik} be the cofactor of the element a_{ik} in the determinant

$$\Delta = \begin{vmatrix} a_{11} & a_{12} & a_{13} \\ a_{21} & a_{22} & a_{23} \\ a_{31} & a_{32} & a_{33} \end{vmatrix} .$$

Prove that $\Delta = (1/3)\, a_{ik}A_{ik}$ and $A_{ik} = \tfrac{1}{2}\epsilon_{ipr}\epsilon_{kqs}a_{pq}a_{rs}$.

(9) Show that, for any scalar ϕ and any vectors **u** and **v**:

(i) $\nabla.(\mathbf{u} \times \mathbf{v}) = \mathbf{v}.\nabla \times \mathbf{u} - \mathbf{u}.\nabla \times \mathbf{v}$;

(ii) $\nabla \times (\mathbf{u} \times \mathbf{v}) = \mathbf{u}\nabla.\mathbf{v} - \mathbf{v}\nabla.\mathbf{u} - \mathbf{u}.\nabla\mathbf{v} + \mathbf{v}.\nabla\mathbf{u}$;

(iii) $\nabla(\mathbf{u}.\mathbf{v}) = \mathbf{u}.\nabla\mathbf{v} + \mathbf{v}.\nabla\mathbf{u} + \mathbf{u} \times (\nabla \times \mathbf{v}) + \mathbf{v} \times (\nabla \times \mathbf{u})$;

(iv) $\nabla \times (\nabla\phi) = \mathbf{0}$;

(v) $\nabla.(\nabla \times \mathbf{u}) = 0$.

(10) Prove that $\tfrac{1}{2}\epsilon_{ipr}\epsilon_{jqs}\theta_{pq}\theta_{rs} = \theta_{ij}$.

[Hint: use (3.14) and (1.121)].

3.3 CROSS PRODUCTS OF TENSORS AND VECTORS

3.3.1 Definition

At the beginning of Chapter 2 we were able to give a meaning to the expression $(\mathbf{a} \otimes \mathbf{b}).\mathbf{c}$. We propose now to give a meaning to the expression $(\mathbf{a} \otimes \mathbf{b}) \times \mathbf{c}$, and hence to $\mathbf{T} \times \mathbf{c}$, where $\mathbf{T}$ is any second-order tensor. Let us write:

$$(\mathbf{a} \otimes \mathbf{b}) \times \mathbf{c} = \mathbf{a} \otimes (\mathbf{b} \times \mathbf{c}) \ . \tag{3.27}$$

Then $(\mathbf{a} \otimes \mathbf{b}) \times \mathbf{c}$ is a tensor whose components are given by:

$$\begin{aligned}[(\mathbf{a} \otimes \mathbf{b}) \times \mathbf{c}]_{ik} &= a_i(\mathbf{b} \times \mathbf{c})_k = a_i \epsilon_{kpq} b_p c_q \ , \\ &= a_i b_p c_q \epsilon_{kpq} \ .\end{aligned} \tag{3.28}$$

Hence $$(\mathbf{T} \times \mathbf{c})_{ik} = T_{ip} c_q \epsilon_{kpq} \ , \tag{3.29}$$

$$= -T_{ip} \epsilon_{pkq} c_q \ ,$$

i.e. $$\mathbf{T} \times \mathbf{c} = -\mathbf{T}.\mathbf{E}.\mathbf{c} \ . \tag{3.30}$$

Equations (3.29) and (3.30) define the cross product of $\mathbf{T}$ and $\mathbf{c}$ in which $\mathbf{T}$ is the pre-multiplier. A definition for the product $\mathbf{c} \times \mathbf{T}$, in which $\mathbf{T}$ is the post-multiplier, may be framed in a similar way. Thus we write:

$$\mathbf{c} \times (\mathbf{a} \otimes \mathbf{b}) = (\mathbf{c} \times \mathbf{a}) \otimes \mathbf{b} \ , \tag{3.31}$$

then $$[\mathbf{c} \times (\mathbf{a} \otimes \mathbf{b})]_{ik} = \epsilon_{ipq} c_p a_q b_k \ ,$$

and $$(\mathbf{c} \times \mathbf{T})_{ik} = \epsilon_{ipq} c_p T_{qk} \ . \tag{3.32}$$

Hence $$\mathbf{c} \times \mathbf{T} = -\mathbf{c}.\mathbf{E}.\mathbf{T} \ . \tag{3.33}$$

Let us consider the special case in which $\mathbf{T} = \mathbf{1}$ in (3.30) and (3.33). We obtain

$$\mathbf{1} \times \mathbf{c} = \mathbf{c} \times \mathbf{1} = -\mathbf{c}.\mathbf{E} = -\mathbf{E}.\mathbf{c} \tag{3.34}$$

(see Problems 3.2, No. 3).

The tensor $\mathbf{E}.\mathbf{c}$ or $-\mathbf{1} \times \mathbf{c}$ is what E.A. Milne calls the **tensor of a vector.**† It is written tens $\mathbf{c}$.

Thus $$\text{tens}\ \mathbf{c} = \mathbf{E}.\mathbf{c} \ . \tag{3.35}$$

The components of tens $\mathbf{c}$ are:

$$\text{tens}\ \mathbf{c} = \begin{bmatrix} 0 & c_3 & -c_2 \\ -c_3 & 0 & c_1 \\ c_2 & -c_1 & 0 \end{bmatrix} . \tag{3.36}$$

Suppose $\mathbf{T}$ = tens $\mathbf{c}$. The reverse process, namely finding $\mathbf{c}$ in terms of $\mathbf{T}$, leads to what is called the **vector of the tensor T** (see again Problems 3.2, No. 3), which is written vec $\mathbf{T}$. Thus

$$\mathbf{c} = \text{vec}\ \mathbf{T} = -\tfrac{1}{2}\mathbf{E}{:}\mathbf{T} \ ,$$

or $$\mathbf{c} = \tfrac{1}{2}(T_{23} - T_{32}, T_{31} - T_{13}, T_{12} - T_{21}) \ . \tag{3.37}$$

†E. A. Milne: *Vectorial Mechanics*, Methuen.

In the problems which follow [No. 1(iii)] we prove the result $\mathbf{a}.(\mathbf{b} \times \mathbf{T}) = (\mathbf{a} \times \mathbf{b}).\mathbf{T}$. Consider the special case in which $\mathbf{T} = \mathbf{1}$.

We have $$\mathbf{a} \times \mathbf{b} = \mathbf{a}.(\mathbf{b} \times \mathbf{1}) = -\mathbf{a}.\,\text{tens}\,\mathbf{b} \;. \tag{3.38}$$

Similarly $$\mathbf{a} \times \mathbf{b} = -\text{tens}\,\mathbf{a}.\mathbf{b} \;. \tag{3.39}$$

Thus we have been able to express the vector product of two vectors as the inner product of a tensor and a vector.

3.3.2 Rate of Change of a Tensor in a Rotating Frame of Reference

An immediate instance of a formula involving cross-products of tensors and vectors is afforded by the expression for the rate of change of a tensor in a rotating frame of reference. We shall follow the method adopted in section 1.11 for determining the rate of change of a vector in a rotating frame of reference.

Suppose the tensor $\mathbf{T}$ is a function of the time t, we may define the rate of change of $\mathbf{T}$ with respect to t in the usual way.

Thus $$\frac{d\mathbf{T}}{dt} = \lim_{\delta t \to 0} \frac{\mathbf{T}(t+\delta t) - \mathbf{T}(t)}{\delta t} \;. \tag{3.40}$$

The usual rules for the derivatives of products hold.

Thus $$\frac{d}{dt}(\mathbf{T}.\mathbf{v}) = \frac{d\mathbf{T}}{dt}.\mathbf{v} + \mathbf{T}.\frac{d\mathbf{v}}{dt} \;. \tag{3.41}$$

Also $$\frac{d}{dt}(\mathbf{T} \times \mathbf{v}) = \frac{d\mathbf{T}}{dt} \times \mathbf{v} + \mathbf{T} \times \frac{d\mathbf{v}}{dt} \;. \tag{3.42}$$

Now suppose $\mathbf{T} = T_{ij}\mathbf{e}_i^* \otimes \mathbf{e}_j^*$ is specified with respect to a frame of reference F^* rotating with angular velocity $\boldsymbol{\omega}$ with respect to a fixed frame F. Let $\partial\mathbf{T}/\partial t$ be the apparent rate of change of $\mathbf{T}$ with respect to the rotating frame F^*,

i.e. $$\frac{\partial \mathbf{T}}{\partial t} = \frac{dT_{ij}}{dt}\mathbf{e}_i^* \otimes \mathbf{e}_j^* \;. \tag{3.43}$$

Let $\dfrac{d\mathbf{T}}{dt}$ be the rate of change of $\mathbf{T}$ with respect to the fixed frame F.

Then
$$\begin{aligned}
\frac{d\mathbf{T}}{dt} &= \frac{d}{dt}(T_{ij}\mathbf{e}_i^* \otimes \mathbf{e}_j^*) \;, \\
&= \frac{dT_{ij}}{dt}\mathbf{e}_i^* \otimes \mathbf{e}_j^* + T_{ij}\frac{d}{dt}(\mathbf{e}_i^* \otimes \mathbf{e}_j^*) \;, \\
&= \frac{\partial \mathbf{T}}{\partial t} + T_{ij}[(\boldsymbol{\omega} \times \mathbf{e}_i^*) \otimes \mathbf{e}_j^* + \mathbf{e}_i^* \otimes (\boldsymbol{\omega} \times \mathbf{e}_j^*)] \;.
\end{aligned}$$

Hence $$\frac{d\mathbf{T}}{dt} = \frac{\partial \mathbf{T}}{\partial t} + \boldsymbol{\omega} \times \mathbf{T} - \mathbf{T} \times \boldsymbol{\omega} \ . \tag{3.44}$$

When **T** is a symmetric tensor (3.44) reduces to:

$$\frac{d\mathbf{T}}{dt} = \frac{\partial \mathbf{T}}{\partial t} + \boldsymbol{\omega} \times \mathbf{T} + (\boldsymbol{\omega} \times \mathbf{T})' \ , \tag{3.45}$$

(see Examples 3.3, No. 1).

Examples 3.3

(1) Prove that for any vector **c** and any symmetric tensor **T**:

$$(\mathbf{c} \times \mathbf{T})' = -\mathbf{T} \times \mathbf{c} \ .$$

We have $(\mathbf{c} \times \mathbf{T})_{ik} = \epsilon_{ipq} c_p T_{qk}$,

therefore
$$\begin{aligned} (\mathbf{c} \times \mathbf{T})'_{ik} &= (\mathbf{c} \times \mathbf{T})_{ki} \ , \\ &= \epsilon_{kpq} c_p T_{qi} \ , \\ &= + T_{iq} \epsilon_{qkp} c_p \ . \end{aligned}$$

Hence $(\mathbf{c} \times \mathbf{T})'_{ik} = -(\mathbf{T} \times \mathbf{c})_{ik}$.

(2) The vector vec **T** = **0** if and only if **T** is a symmetric tensor. This result follows at once from (3.37).

Problems 3.3

(1) Prove the following results:

(i) $\mathbf{a}.(\mathbf{T} \times \mathbf{b}) = (\mathbf{a}.\mathbf{T}) \times \mathbf{b}$;

(ii) $(\mathbf{a} \times \mathbf{T}).\mathbf{b} = \mathbf{a} \times (\mathbf{T}.\mathbf{b})$;

(iii) $\mathbf{a}.(\mathbf{b} \times \mathbf{T}) = (\mathbf{a} \times \mathbf{b}).\mathbf{T}$;

(iv) $(\mathbf{T} \times \mathbf{a}).\mathbf{b} = \mathbf{T}.(\mathbf{a} \times \mathbf{b})$;

(v) $\mathbf{T}.(\mathbf{a} \times \mathbf{b}) \neq (\mathbf{T}.\mathbf{a}) \times \mathbf{b}$;

(vi) $(\mathbf{a} \times \mathbf{b}).\mathbf{T} \neq \mathbf{a} \times (\mathbf{b}.\mathbf{T})$.

(2) Prove the following results:

(i) $\mathbf{a} \times (\mathbf{T} \times \mathbf{b}) = (\mathbf{a} \times \mathbf{T}) \times \mathbf{b}$;

(ii) $(\mathbf{T} \times \mathbf{a}) \times \mathbf{b} = -\mathbf{T}(\mathbf{a}.\mathbf{b}) + (\mathbf{T}.\mathbf{b}) \otimes \mathbf{a}$;

(iii) $\mathbf{a} \times (\mathbf{b} \times \mathbf{T}) = -(\mathbf{a}.\mathbf{b})\mathbf{T} + \mathbf{b} \otimes (\mathbf{a}.\mathbf{T})$;

(iv) $\mathbf{T} \times (\mathbf{a} \times \mathbf{b}) = -(\mathbf{T}.\mathbf{a}) \otimes \mathbf{b} + (\mathbf{T}.\mathbf{b}) \otimes \mathbf{a}$;

(v) $(\mathbf{a} \times \mathbf{b}) \times \mathbf{T} = -\mathbf{a} \otimes (\mathbf{b}.\mathbf{T}) + \mathbf{b} \otimes (\mathbf{a}.\mathbf{T})$.

(3) Prove that the only tensor **T** which satisfies the relation

$$\mathbf{T} \times \mathbf{c} = \mathbf{c} \times \mathbf{T}, \forall \mathbf{c} ,$$

is a multiple of the unit tensor **1**.

(4) Show that if $\mathbf{T} = \mathbf{a} \otimes \mathbf{b}$, then $\text{vec}\, \mathbf{T} = \tfrac{1}{2} \mathbf{a} \times \mathbf{b}$.

(5) Show that for any vector **c** and any tensor **T** then $\mathbf{T} = \text{tens}\, \mathbf{c} \Rightarrow \mathbf{c} = \text{vec}\, \mathbf{T}$ [i.e. $\text{vec}\,(\text{tens}\, \mathbf{c}) \equiv \mathbf{c}$] but that the converse is not true in general.

Find an expression for tens (vec **T**) and show that if $\mathbf{T} = \mathbf{A} + \mathbf{S}$ where **S** is symmetric and **A** is antisymmetric then:

(i) $\text{tens}\,(\text{vec}\, \mathbf{A}) = \text{tens}\,(\text{vec}\, \mathbf{T}) = \mathbf{A}$,

(ii) $\text{tens}\,(\text{vec}\, \mathbf{S}) = \mathbf{0}$.

3.4 THE GEOMETRY OF ROTATIONS

3.4.1 General Rotations

Rotations and the rotation matrix were first encountered in sections 1.10.2 and 1.10.3. The elements of the rotation matrix Θ were defined as the cosines of the angles between the coordinate axes before and after rotation [see equation (1.121)]. We have postponed until now any attempt at defining the elements of the rotation matrix in terms of (say) a rotation of an angle θ about a given line, because the expressions involve the components ϵ_{jkm} of the alternate tensor. We shall also seek a solution of the converse problem, that is of finding the angle and axis of rotation in terms of the elements of the rotation matrix.

We may frame the problem as follows. The position vector of a point P in a certain Cartesian system is **r**. The line OP is rotated through an angle θ in the positive sense about a line through the origin parallel to the unit vector **n**. It is required to write down **r**, the new position vector of P, and hence the rotation matrix corresponding to the rotation.

Let ON be the projection of OP on **n** (Fig. 3.1).

Then $$\vec{ON} = (\mathbf{r}.\mathbf{n})\mathbf{n} , \qquad (3.46)$$

and $$\vec{NP} = \mathbf{r} - (\mathbf{r}.\mathbf{n})\mathbf{n} . \qquad (3.47)$$

Now ON is also the projection of OP′ on **n**,

where $$\vec{NP}' = \vec{NP} \cos\theta + \mathbf{n} \times \vec{NP} \sin\theta \quad \text{(Fig. 3.2)} ,$$

therefore $$\mathbf{r}' = (\mathbf{r}.\mathbf{n})\mathbf{n} + [\mathbf{r} - (\mathbf{r}.\mathbf{n})\mathbf{n}] \cos\theta + \mathbf{n} \times \mathbf{r} \sin\theta . \qquad (3.48)$$

Now suppose the line OP remains fixed while the coordinate system is rotated through the same angle θ, in the positive sense, about ON. This has the effect of reversing the sign of θ in (3.48). Let **r*** be the new position vector of P.

Then $$\mathbf{r}^* = (\mathbf{r}.\mathbf{n})\mathbf{n} + [\mathbf{r} - (\mathbf{r}.\mathbf{n})\mathbf{n}] \cos\theta - \mathbf{n} \times \mathbf{r} \sin\theta . \qquad (3.49)$$

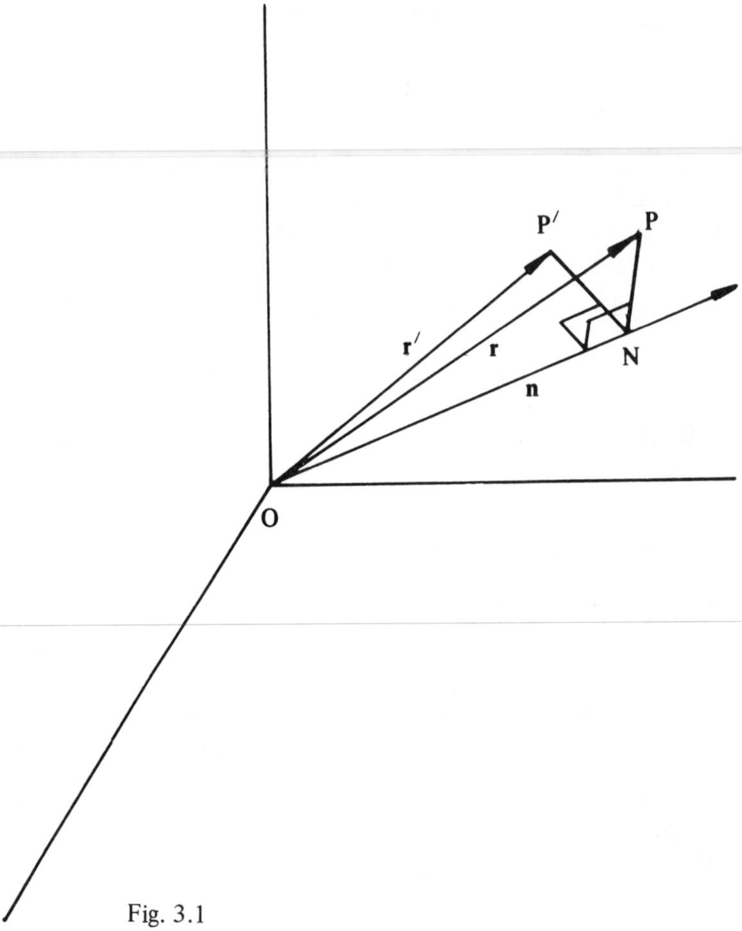

Fig. 3.1

In component form (3.49) becomes:

$$x_i^* = x_j n_j n_i + (x_i - x_j n_j n_i)\cos\theta = \epsilon_{ijk} n_j x_k \sin\theta$$
$$= [n_i n_j + (\delta_{ij} - n_i n_j)\cos\theta - \epsilon_{ikj} n_k \sin\theta]\,x_j \ .$$

But $x_i^* = \theta_{ij} x_j$,

therefore $$\theta_{ij} = n_i n_j (1 - \cos\theta) + \cos\theta\,\delta_{ij} + \epsilon_{ijk} n_k \sin\theta \ . \qquad (3.50)$$

Equation (3.50) gives the elements of the rotation matrix corresponding to a rotation of angle θ in the positive sense about a line in the direction of the unit vector **n**.

We come now to the converse problem, that is of finding the angle and axis

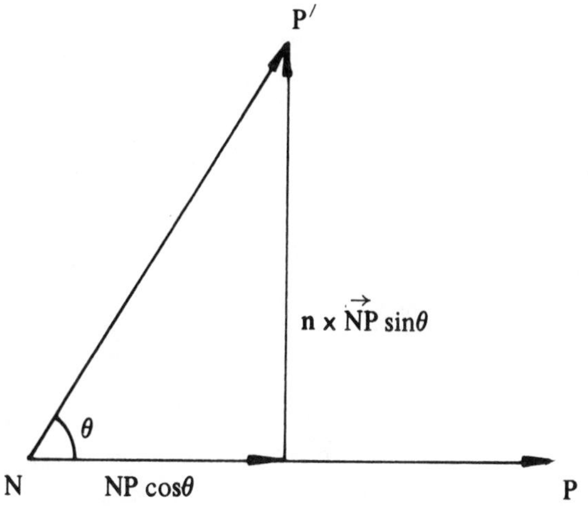

Fig. 3.2

of rotation. From (3.50) we have:

$$\theta_{ij} - \theta_{ji} = (\epsilon_{ijk} - \epsilon_{jik})n_k \sin\theta \ ,$$
$$= 2\epsilon_{ijk}n_k \sin\theta \ . \qquad (3.51)$$

Hence $$\mathbf{n} = \frac{1}{2\sin\theta}(\theta_{23} - \theta_{32}, \theta_{31} - \theta_{13}, \theta_{12} - \theta_{21}) \ . \qquad (3.52)$$

Equation (3.52) is sufficient to determine both **n** and $\sin\theta$ uniquely except for their signs. It should be remembered, however, that the rotation specified by $(\mathbf{n}, \theta)$ is precisely the same as that specified by $(-\mathbf{n}, -\theta)$.

Equation (3.51) can also be made to yield an explicit expression for $\sin\theta$. We have:

$$\epsilon_{ijk}\epsilon_{ijp}n_k n_p \sin\theta = \tfrac{1}{2}(\theta_{ij} - \theta_{ji})\epsilon_{ijp}n_p \ . \qquad (3.53)$$

Hence, using the result of Problems 3.2, No. 6:

$$2\delta_{kp}n_k n_p \sin\theta = \theta_{ij}\epsilon_{ijp}n_p \ ,$$

or $$\sin\theta = \tfrac{1}{2}\epsilon_{ijp}\theta_{ij}n_p \ . \qquad (3.54)$$

An explicit expression for $\cos\theta$ is found by letting $i = j$ in (3.50):

$$\theta_{ii} = 1 + 2\cos\theta \ .$$

Hence $$\cos\theta = \tfrac{1}{2}(\theta_{ii} - 1) \ . \qquad (3.55)$$

We have now determined the rotation uniquely.

It is instructive to look for the eigenvalues and eigenvectors of the rotation matrix. If λ is an eigenvalue and **u** is an eigenvector, then:

$$\theta_{ij}u_j = \lambda u_i \ . \tag{3.56}$$

Hence, from (3.50):

$$n_i n_j u_j(1 - \cos\theta) + \cos\theta\,\delta_{ij}u_j + \epsilon_{ijk}u_j n_k \sin\theta = \lambda u_i \ , \tag{3.57}$$

or, remembering that λ and **u** may be complex,

$$\mathbf{n}(\bar{\mathbf{u}}.\mathbf{n})\,(1 - \cos\theta) + \mathbf{u}\cos\theta + \bar{\mathbf{u}} \times \bar{\mathbf{n}}\sin\theta = \lambda\mathbf{u} \ . \tag{3.58}$$

If we now form the scalar product of (3.58) with $\bar{\mathbf{n}}$ and $\bar{\mathbf{u}}$ in turn, we obtain:

$$\bar{\mathbf{u}}.\mathbf{n} = \lambda\bar{\mathbf{u}}.\mathbf{n} \ , \tag{3.59}$$

$$(\bar{\mathbf{u}}.\mathbf{n})^2\,(1 - \cos\theta) = (\bar{\mathbf{u}}.\mathbf{u})\,(\lambda - \cos\theta) \ . \tag{3.60}$$

From (3.59) we conclude that either $\lambda = 1$ or $\bar{\mathbf{u}}.\mathbf{n} = 0$. When $\lambda = 1$, (3.60) yields $\mathbf{u} \parallel \mathbf{n}$. We have shown that $\lambda = 1$ is the eigenvalue with corresponding eigenvector **n**. The other two eigenvalues and their corresponding eigenvectors are complex and will be found in section 4.2.5 and Problems 4.2, No. 4.

3.4.2 Rotations about the Coordinate Axes

Let us substitute the expression (3.50) for θ_{ji} into the transformation law (2.43) to obtain:

$$\left.\begin{aligned} T^*_{jl} &= [n_j n_l n_p n_q(1 - \cos\theta)^2 + \epsilon_{jpr}\epsilon_{lqs}n_r n_s \sin^2\theta \\ &\quad + (\epsilon_{jpr}n_l n_r n_q + \epsilon_{lqr}n_j n_r n_p)\sin\theta\,(1 - \cos\theta)]\,T_{pq} \\ &\quad + \cos\theta\,[\epsilon_{lqr}n_r \sin\theta + (1 - \cos\theta)n_l n_q]\,T_{jq} \\ &\quad + \cos\theta\,[\epsilon_{jpr}n_r \sin\theta + (1 - \cos\theta)n_j n_p]\,T_{pl} + \cos^2\theta\,T_{jl} \ . \end{aligned}\right\} \tag{3.61}$$

Expression (3.61) gives the components of the transformed tensor in terms of the components of the original tensor and a rotation specified by an angle θ about a line in the direction of the unit vector **n**. Suppose now that **n** is in the direction of a coordinate axis, e.g. $\mathbf{n} = \mathbf{e}_3$. Equation (3.61) becomes:

$$\begin{aligned} T^*_{jl} &= \delta_{j3}\delta_{l3}(1 - \cos\theta)^2\,T_{33} + \epsilon_{jp3}\epsilon_{lq3}\sin^2\theta\,T_{pq} \\ &\quad + \sin\theta(1 - \cos\theta)\,(\epsilon_{jp3}\delta_{l3}T_{p3} + \epsilon_{lq3}\delta_{j3}T_{3q}) \\ &\quad + \cos\theta\,\sin\theta(\epsilon_{lq3}T_{jq} + \epsilon_{jp3}T_{pl}) \\ &\quad + \cos\theta(1 - \cos\theta)\,(\delta_{l3}T_{j3} + \delta_{j3}T_{3l}) + \cos^2\theta\,T_{jl} \ . \end{aligned} \tag{3.62}$$

From (3.62) we may obtain the components of **T** with respect to axes rotated about $\mathbf{e}_3$ through an angle θ in the positive sense:

$$\left.\begin{aligned}
T^*_{11} &= \cos^2\theta\, T_{11} + \sin^2\theta\, T_{22} + \cos\theta \sin\theta\,(T_{12} + T_{21}) ,\\
T^*_{12} &= \cos^2\theta\, T_{12} - \sin^2\theta\, T_{21} + \cos\theta \sin\theta\,(T_{22} - T_{11}) ,\\
T^*_{13} &= T_{13} + \sin\theta\, T_{23} ,\\
T^*_{21} &= \cos^2\theta\, T_{21} - \sin^2\theta\, T_{12} + \cos\theta \sin\theta\,(T_{22} - T_{11}) ,\\
T^*_{22} &= \cos^2\theta\, T_{22} + \sin^2\theta\, T_{11} - \cos\theta \sin\theta\,(T_{12} + T_{21}) ,\\
T^*_{23} &= T_{23} - \sin\theta\, T_{13} ,\\
T^*_{31} &= T_{31} + \sin\theta\, T_{32} ,\\
T^*_{32} &= T_{32} - \sin\theta\, T_{31} ,\\
T^*_{33} &= T_{33} .
\end{aligned}\right\} \qquad (3.63)$$

If the tensor **T** is symmetric and in diagonal form with respect to the original set of axes, (3.63) becomes:

$$\left.\begin{aligned}
T^*_{11} &= \cos^2\theta\, T_{11} + \sin^2\theta\, T_{22} ,\\
T^*_{12} &= \cos\theta \sin\theta\,(T_{22} - T_{11}) ,\\
T^*_{13} &= 0 ,\\
T^*_{21} &= \cos\theta \sin\theta\,(T_{22} - T_{11}) ,\\
T^*_{22} &= \cos^2\theta\, T_{22} + \sin^2\theta\, T_{11} ,\\
T^*_{23} &= 0 ,\\
T^*_{31} &= 0 ,\\
T^*_{32} &= 0 ,\\
T^*_{33} &= T_{33} .
\end{aligned}\right\} \qquad (3.64)$$

It will be noticed that the first two equations of (3.64) may be written:

$$\left.\begin{aligned}
T^*_{11} &= \tfrac{1}{2}(T_{11} + T_{22}) + \tfrac{1}{2}(T_{11} - T_{22}) \cos 2\theta \\
T^*_{12} &= \tfrac{1}{2}(T_{22} - T_{11}) \sin 2\theta .
\end{aligned}\right\} \qquad (3.65)$$

The equations (3.65) are the parametric equations of a circle, a fact which will be referred to in Examples 6.1, No. 6.

Examples 3.4

(1) A set of Cartesian axes is rotated by an angle α in the positive sense about the x_3-axis. Use equation (3.50) to write down the rotation matrix. Compare your result with Examples 1.10, No. 1.

We have $\theta_{ij} = \delta_{3i}\delta_{3j}(1 - \cos\alpha) + \delta_{ij}\cos\alpha + \epsilon_{ij3}\sin\alpha$.

Hence
$$\Theta = \begin{bmatrix} \cos\alpha & \sin\alpha & 0 \\ -\sin\alpha & \cos\alpha & 0 \\ 0 & 0 & 1 \end{bmatrix} .$$

(2) A set of Cartesian axes is rotated by an angle $\pi/6$ in the positive sense about a line through the origin in the direction of the vector (1, 2, 2). Write down the rotation matrix.

The unit normal along the axis of rotation is given by $\mathbf{n} = \frac{1}{3}(1, 2, 2)$. Hence, from (3.50), we have:

$$\theta_{ij} = n_i n_j(1 - \cos\theta) + \cos\theta\,\delta_{ij} + \epsilon_{ijk}n_k\sin\theta ,$$

or
$$\Theta = 1/9 \begin{bmatrix} 1 + 4\sqrt{3} & 5 - \sqrt{3} & -(1 + \sqrt{3}) \\ -(1 + \sqrt{3}) & 4 + 5\sqrt{3}/2 & 5{\cdot}5 - 2\sqrt{3} \\ 5 - \sqrt{3} & 2{\cdot}5 - 2\sqrt{3} & 4 + 5\sqrt{3}/2 \end{bmatrix} .$$

As a check on this result we note from (3.52) that

$$\mathbf{u} = \frac{1}{6\sin\theta}(1, 2, 2) .$$

Thus $\mathbf{u}$, the axis of rotation, is parallel to $\mathbf{n}$.
From (3.55) we obtain:

$$\cos\theta = \tfrac{1}{2}(\theta_{ii} - 1) = \sqrt{3}/2 .$$

Finally, from (3.54) we obtain:

$$\sin\theta = \tfrac{1}{2}\epsilon_{ijk}\theta_{ij}n_k ,$$
$$= \tfrac{1}{2}[(\theta_{23} - \theta_{32})n_1 + (\theta_{31} - \theta_{13})n_2 + (\theta_{12} - \theta_{21})n_3] ,$$

i.e. $\sin\theta = \tfrac{1}{2}$.

(3) *Small Rotations*

If the angle of rotation θ is small (3.50) reduces approximately to:

$$\theta_{ij} \cong \delta_{ij} + \epsilon_{ijk}n_k\sin\theta . \tag{3.66}$$

This result, although immediately obvious from (3.50), is of great importance when establishing the uniqueness of the unit tensor as a second-order isotropic tensor (see section 4.2.2).

Problems 3.4

(1) A set of Cartesian axes is rotated by an angle $\pi/4$ in the positive sense about a line through the origin in the direction of the vector (2, 3, 6). Write down the rotation matrix and check your result using (3.52), (3.54) and (3.55).

(2) Repeat No. 1 for an angle of $\pi/3$ about a line in the direction of the vector (1, 4, 8).

(3) A rotation is specified by the matrix

$$\begin{bmatrix} 1/\sqrt{3} & 1/\sqrt{3} & 1/\sqrt{3} \\ 0 & 1/\sqrt{2} & -1/\sqrt{2} \\ -2/\sqrt{6} & 1/\sqrt{6} & 1/\sqrt{6} \end{bmatrix} .$$

Show that the rotation is by an angle θ about an axis parallel to the vector

$$\mathbf{u} = (\sqrt{3}+1, 2+\sqrt{2}, -\sqrt{2}) ,$$

where $12 \cos\theta = (2\sqrt{3} + 3\sqrt{2} + \sqrt{6} - 6)$,

and $$6 \sin\theta = -\frac{(6\sqrt{6} + 3\sqrt{2} + 4\sqrt{3})}{\sqrt{(12 + 2\sqrt{3} + 4\sqrt{2})}} .$$

(4) Prove that for any rotation matrix Θ with elements θ_{ij}:

$$\det(\theta_{ij} - \delta_{ij}) = 0 ,$$

(see Problems 1. 10, No. 2).

(5) Write down the rotation matrix for a rotation of $\pi/4$ in the positive sense about (i) the x_1-axis,; (ii) the x_2-axis; (iii) the x_3-axis. Repeat (i) for a rotation in the negative sense.

(6) Repeat No. 5 for rotations of $\pi/2$ and π.

(7) Generalize the results of Nos. 5, 6, to show that for a rotation about the axis $\mathbf{e}_m$ of (i) $\pi/4$,; (ii) $\pi/2$; and (iii) π, the coefficients of the rotation matrix are given by:

(i) $$\theta_{ij} = \delta_{im}\delta_{jm}(1-\sqrt{2}/2) + \delta_{ij}\sqrt{2}/2 \pm \epsilon_{ijm}\sqrt{2}/2 ;$$

(ii) $$\theta_{ij} = \delta_{im}\delta_{jm} \pm \epsilon_{ijm} ;$$

(iii) $$\theta_{ij} = 2\delta_{im}\delta_{jm} - \delta_{ij} ,$$

where the summation is in no case carried out over the index m, and the negative sign in (i) and (ii) corresponds to a rotation in the negative sense.

(8) $\mathbf{T}$ is a symmetric tensor. The axes are rotated by an angle θ in the positive sense about the x_3-axis. Show that there are two values of θ for which T^*_{12} is

zero, given by

$$\tan 2\theta = \frac{2T_{12}}{T_{11} - T_{22}} .$$

Find also the value of θ for which $\mathbf{T}^*_{11}$ is (i) a maximum and (ii) a minimum.

(9) Show that the displacement of the point whose position vector is **r**, caused by a rotation of a *small* angle θ about a line in the direction of the unit vector **n**, is given approximately by

$$\delta\mathbf{r} = \theta\mathbf{n} \times \mathbf{r} .$$

3.5 INTEGRAL THEOREMS IN TENSOR FORM

3.5.1 The Tensor Form of Green's Theorem

We shall now seek to extend the Divergence Theorem to include tensor quantities of arbitrary order. The **Divergence Theorem** may be stated as follows:

Let S be the surface of any volume V and **n** the outward normal to any surface element dS. Then, if **v** is any vector function of position defined within V:

$$\int_S \mathbf{v}.\mathbf{n}\mathrm{d}S = \int_V \mathrm{div}\ \mathbf{v}\,\mathrm{d}V . \tag{3.67}$$

Now let $\mathbf{v} = \phi_{ikm} \ldots \mathbf{e}_p$ where $\phi_{ikm} \ldots$ is a component of a tensor function of arbitrary order. Equation (3.67) becomes:

$$\int_S \phi_{ikm} \ldots n_p \mathrm{d}S = \int_V \frac{\partial \phi_{ikm \cdots}}{\partial x_p} \mathrm{d}V . \tag{3.68}$$

The result (3.68) is known as **Green's Theorem**.

3.5.2 Special Cases of Green's Theorem

(i) $\phi_{ikm} \ldots = v_i$, the i-component of the vector **v**. Then (3.68) becomes:

$$\int_S v_i n_p \mathrm{d}S = \int_V \frac{\partial v_i}{\partial x_p} \mathrm{d}V , \tag{3.69}$$

i.e.

$$\int_S \mathrm{d}\mathbf{S} \otimes \mathbf{v} = \int_V \nabla \mathbf{v} \mathrm{d}V . \tag{3.70}$$

Of course the contraction of (3.70) is simply the Divergence Theorem (3.67).

(ii) $\phi_{ikm\,\ldots} = \phi$, a scalar. Then (3.68) becomes:

$$\int_S \phi n_p \mathrm{d}S = \int_V \frac{\partial \phi}{\partial x_p} \mathrm{d}V \ , \tag{3.71}$$

or

$$\int_S \phi \mathrm{d}\mathbf{S} = \int_V \nabla \phi \mathrm{d}V \ . \tag{3.72}$$

(iii) $\phi_{ikm\,\ldots} = T_{ik}$, the (i,k)-component of the tensor $\mathbf{T}$. Then (3.68) becomes:

$$\int_S T_{ik} n_p \mathrm{d}S = \int_V \frac{\partial T_{ik}}{\partial x_p} \mathrm{d}V \ . \tag{3.73}$$

If we contract (3.73) by letting $p = i$ we obtain:

$$\int_S \mathrm{d}\mathbf{S}.\mathbf{T} = \int_V \mathrm{div}\, \mathbf{T} \mathrm{d}V \ . \tag{3.74}$$

3.5.3 The Tensor Form of Stokes's Theorem

Stokes's Theorem may be stated as follows: Let S be an open surface bounded by the curve C. Then for any vector function of position $\mathbf{v}$, defined over S:

$$\oint_C \mathbf{v}.\mathrm{d}\mathbf{r} = \int_S \mathrm{curl}\, \mathbf{v}.\mathrm{d}\mathbf{S} \ . \tag{3.75}$$

Now let $v_i = \phi_{ikm\,\ldots}$ where $\phi_{ikm\,\ldots}$ is a component of a tensor function of arbitrary order. Then (3.75) becomes:

$$\oint_C \phi_{ikm\,\ldots} \mathrm{d}x_i = \int_S \epsilon_{rpq} \frac{\partial \phi_{qkm\,\ldots}}{\partial x_p} \mathrm{d}S_r \ . \tag{3.76}$$

The result (3.76) is the tensor form of Stokes's Theorem.

3.5.4 Special Cases of Stokes's Theorem

(i) $\phi_{ikm\,\ldots} = v_i$, the i-component of the vector $\mathbf{v}$. Then (3.76) becomes (3.75).

(ii) $\phi_{ikm\,\ldots} = \phi\delta_{ik}$.

Then

$$\oint_C \phi \delta_{ik} \mathrm{d}x_i = \int_S \epsilon_{rpq} \frac{\partial \phi}{\partial x_p} \delta_{qk} \mathrm{d}S_r \ , \tag{3.77}$$

i.e.

$$\oint_C \phi \mathrm{d}\mathbf{r} = \int_S \mathrm{d}\mathbf{S} \times \nabla \phi \ . \tag{3.78}$$

(iii) $\phi_{ikm} \ldots = \epsilon_{ikm} v_k$.

Then
$$\oint_C \epsilon_{ikm} v_k \mathrm{d}x_i = \int_S \epsilon_{rpq} \epsilon_{qkm} \frac{\partial v_k}{\partial x_p} \mathrm{d}S_r \ ,$$

$$= \int_S (\delta_{rk}\delta_{pm} - \delta_{rm}\delta_{pk}) \frac{\partial v_k}{\partial x_p} \mathrm{d}S_r \ ,$$

$$= \int_S \left(\frac{\partial v_r}{\partial x_m} - \frac{\partial v_p}{\partial x_p} \delta_{rm} \right) \mathrm{d}S_r \ ,$$

i.e.
$$\oint_C \mathrm{d}\mathbf{r} \times \mathbf{v} = \int_S \nabla\mathbf{v}.\mathrm{d}\mathbf{S} - \int_S \mathrm{div}\,\mathbf{v}\mathrm{d}\mathbf{S} \ . \tag{3.79}$$

Problems 3.5

(1) By choosing an appropriate form for $\phi_{ikm} \ldots$ in (3.68) prove that:

(i) $\int_S \mathrm{d}\mathbf{S} \times \mathbf{v} = \int_V \mathrm{curl}\,\mathbf{v}\mathrm{d}V$;

(ii) $\int_S (\mathbf{v} \times \mathbf{T}).\mathrm{d}\mathbf{S} = \int_V [\mathbf{v} \times \mathrm{div}\,\mathbf{T}' - 2\,\mathrm{vec}(\mathbf{T}.\nabla\mathbf{v})]\,\mathrm{d}V$;

(iii) $\int_S (\mathbf{r} \times \mathbf{T}).\mathrm{d}\mathbf{S} = \int_V (\mathbf{r} \times \mathrm{div}\,\mathbf{T}' - 2\,\mathrm{vec}\,\mathbf{T})\mathrm{d}V$.

(2) S is a portion of a plane surface bounded by the closed curve C. Prove that

$$\int_S \mathrm{div}\,\mathbf{v}\mathrm{d}S = \oint_C \mathbf{v}.\mathbf{n}\mathrm{d}s \ ,$$

where the vectors **v** and **n**, the unit outward normal to C, both lie in the plane of S.

3.6 CURVILINEAR COORDINATES

3.6.1 Introduction

Let u_1, u_2, u_3, where $u_i = u_i(x_1, x_2, x_3)$, be a system of **curvilinear coordinates.** Then, provided the Jacobian $J = |(\partial u_i/\partial x_j)| \neq 0$, we may write $x_i = x_i(u_1, u_2, u_3)$.

If $\mathbf{r}(u_1, u_2, u_3)$ is the position vector of a point, then

$$\delta\mathbf{r} = \sum_{i=1}^{3} h_i \delta u_i \mathbf{e}_i \ , \tag{3.80}$$

where $\mathbf{e}_i$ is a unit vector in the direction of the tangent to the u_i **coordinate line** (along which the other two u's are constant). The quantity

$$h_i = | \, \partial\mathbf{r}/\partial u_i \, | \tag{3.81}$$

is known as a **scale factor**.

Whenever scale factors are used the Summation Convention, as defined in section 1.9.1, breaks down. In equation (3.80) a summation is required over the suffix i. However, the presence of h_i means that this suffix occurs three times. For this reason we are forced to abandon the Convention when dealing with curvilinear coordinates.

We shall, in the next section, express the Cartesian tensor $\nabla\mathbf{v}$ in terms of curvilinear coordinates. The procedure is complicated by the fact that the unit vectors $\mathbf{e}_i$ are no longer fixed in direction. We therefore require an expression for $\partial\mathbf{e}_i/\partial u_j$, and it is this quantity which we now seek to evaluate. We have, from equation (3.80):

$$\mathbf{e}_i = \frac{1}{h_i}\frac{\partial\mathbf{r}}{\partial u_i} \ , \tag{3.82}$$

where no summation is implied over the index i.

Therefore $$h_i \frac{\partial\mathbf{e}_i}{\partial u_j} = \frac{\partial^2\mathbf{r}}{\partial u_i \partial u_j} - \frac{\partial h_i}{\partial u_j}\mathbf{e}_i \ ,$$

and $$h_j \frac{\partial\mathbf{e}_j}{\partial u_i} = \frac{\partial^2\mathbf{r}}{\partial u_j \partial u_i} - \frac{\partial h_j}{\partial u_i}\mathbf{e}_j \ ,$$

hence $$h_i \frac{\partial\mathbf{e}_i}{\partial u_j} - h_j \frac{\partial\mathbf{e}_j}{\partial u_i} = \frac{\partial h_j}{\partial u_i}\mathbf{e}_j - \frac{\partial h_i}{\partial u_j}\mathbf{e}_i \ . \tag{3.83}$$

Now let us suppose that $\mathbf{e}_i.\mathbf{e}_j = \delta_{ij}$. The coordinates (u_1, u_2, u_3) are then said to be **orthogonal curvilinear coordinates**. We know too that, since $\mathbf{e}_i$ is a unit vector, $\partial\mathbf{e}_i/\partial u_j$ is a vector perpendicular to $\mathbf{e}_i$, also $\partial^2\mathbf{r}/\partial u_i \partial u_j$ is a vector in the plane of $\mathbf{e}_i$ and $\mathbf{e}_j$ (see Problems 3.6 No. 6). Hence, for (3.83) to be true when $i \neq j$ we must have:

$$\frac{\partial\mathbf{e}_i}{\partial u_j} = \frac{1}{h_i}\frac{\partial h_j}{\partial u_i}\mathbf{e}_j, \, i \neq j \ . \tag{3.84}$$

To evaluate $\partial \mathbf{e}_i/\partial u_j$ when $i = j$, let us consider:

$$\frac{\partial \mathbf{e}_1}{\partial u_1} = \frac{\partial}{\partial u_1}(\mathbf{e}_2 \times \mathbf{e}_3) \; ,$$

$$= \frac{\partial \mathbf{e}_2}{\partial u_1} \times \mathbf{e}_3 + \mathbf{e}_2 \times \frac{\partial \mathbf{e}_3}{\partial u_1} \; ,$$

$$= \frac{1}{h_2}\frac{\partial h_1}{\partial u_2}\mathbf{e}_1 \times \mathbf{e}_3 + \mathbf{e}_2 \times \left(\frac{1}{h_3}\frac{\partial h_1}{\partial u_3}\mathbf{e}_1\right),$$

$$= -\frac{1}{h_2}\frac{\partial h_1}{\partial u_2}\mathbf{e}_2 - \frac{1}{h_3}\frac{\partial h_1}{\partial u_3}\mathbf{e}_3 \; . \tag{3.85}$$

Combining (3.84) and (3.85), we see that an expression which is true for all i and j is:

$$\frac{\partial \mathbf{e}_i}{\partial u_j} = \frac{1}{h_i}\frac{\partial h_j}{\partial u_i}\mathbf{e}_j - \sum_{k=1}^{3} \delta_{ij}\frac{1}{h_k}\frac{\partial h_j}{\partial u_k}\mathbf{e}_k \; . \tag{3.86}$$

3.6.2 Cartesian Tensors Referred to Orthogonal Curvilinear Coordinates

Many problems in mathematical physics are, because of their boundary conditions or some special symmetry, more conveniently expressed in coordinates other than Cartesian coordinates. The equations describing the problem, if true when expressed in terms of Cartesian coordinates, are equally true when all vector and tensor quantities are expressed in terms of curvilinear coordinates, according to the rules which we are about to develop. According to these rules dimensionality will be preserved. The components of the stress tensor, for example, which are stresses in the Cartesian representation, will also be stresses in the curvilinear representation. These components are said to be the **physical components** of the tensor. If all quantities are converted to a curvilinear system and the resulting equations solved with appropriate boundary conditions there is little more to be said on the subject.

Nevertheless, it should be pointed out that the components of the resulting quantities expressed in terms of curvilinear coordinates do not transform from one curvilinear system to another according to the laws which we have developed for Cartesian tensors. It is, of course, perfectly possible to write down quantities whose components *do* transform according to these laws, but these components are no longer stresses, strains etc. as the case may be.

Furthermore it is necessary to distinguish between the contravariant and covariant components of vectors and tensors, a distinction which we have hitherto been able to avoid by confining our attention to Cartesian tensors. These matters will be dealt with briefly in the Appendix. The remainder of the present section will be devoted to a discussion of methods used to calculate the physical components of tensors referred to orthogonal curvilinear coordinates.

The gradient operator ∇ is defined in curvilinear coordinates by:

$$\nabla = \sum_{i=1}^{3} \mathbf{e}_i \frac{1}{h_i} \frac{\partial}{\partial u_i} \; . \tag{3.87}$$

(see Problems 3.6, No. 1). Thus, the gradient of the scalar function Φ is given by:

$$\nabla\Phi = \sum_{i=1}^{3} \frac{1}{h_i} \frac{\partial \Phi}{\partial u_i} \mathbf{e}_i \; . \tag{3.88}$$

The gradient of the vector $\mathbf{v}$ is given by:

$$\nabla\mathbf{v} = \sum_{i=1}^{3} \sum_{j=1}^{3} \frac{1}{h_i} \mathbf{e}_i \otimes \frac{\partial}{\partial u_i} (v_j \mathbf{e}_j) \; , \tag{3.89}$$

which, by the use of (3.86), becomes

$$\nabla\mathbf{v} = \sum_{i=1}^{3} \sum_{j=1}^{3} \frac{1}{h_i} \mathbf{e}_i \otimes \left(\mathbf{e}_j \frac{\partial v_j}{\partial u_i} + \frac{v_j}{h_j} \frac{\partial h_i}{\partial u_j} \mathbf{e}_i - \sum_{k=1}^{3} v_j \delta_{ij} \frac{1}{h_k} \frac{\partial h_i}{\partial u_k} \mathbf{e}_k \right) ,$$

$$= \sum_{i=1}^{3} \sum_{j=1}^{3} \left[\frac{v_j}{h_i h_j} \frac{\partial h_i}{\partial u_j} \mathbf{e}_i \otimes \mathbf{e}_i + \frac{1}{h_i} \left(\frac{\partial v_j}{\partial u_i} - \frac{v_i}{h_j} \frac{\partial h_i}{\partial u_j} \right) \mathbf{e}_i \otimes \mathbf{e}_j \right] . \tag{3.90}$$

It is useful for problems in elasticity and fluid mechanics to write down expressions for the symmetric and antisymmetric parts of $\nabla\mathbf{v}$. This we shall now do The symmetric part of (3.90) is given by:

$$\mathbf{e} = \tfrac{1}{2}[(\nabla\mathbf{v})' + \nabla\mathbf{v}] \; ,$$

$$= \sum_{i=1}^{3} \sum_{j=1}^{3} \left[\frac{v_j}{h_i h_j} \frac{\partial h_i}{\partial u_j} \mathbf{e}_i \otimes \mathbf{e}_i + \frac{1}{2} \frac{h_i}{h_j} \frac{\partial}{\partial u_j} \left(\frac{v_i}{h_i} \right) (\mathbf{e}_i \otimes \mathbf{e}_j + \mathbf{e}_j \otimes \mathbf{e}_i) \right] . \tag{3.91}$$

The antisymmetric part of (3.90) is given by:

$$-\boldsymbol{\xi} = \tfrac{1}{2}[\nabla\mathbf{v} - (\nabla\mathbf{v})'] \; ,$$

$$= \sum_{i=1}^{3} \sum_{j=1}^{3} \frac{1}{2h_i h_j} \frac{\partial}{\partial u_i} (v_j h_j) (\mathbf{e}_i \otimes \mathbf{e}_j - \mathbf{e}_j \otimes \mathbf{e}_i) \; . \tag{3.92}$$

The expressions (3.90) - (3.92) may now be used to derive expressions for the other differential functions of $\mathbf{v}$. For example the divergence $\nabla . \mathbf{v}$ is found by contracting (3.91) over the indices i and j.

Thus $\nabla . \mathbf{v} = (\nabla \mathbf{v})_{ii} = e_{ii}$.

Unfortunately the result obtained by putting $j = i$ everywhere is not correct, as we shall see. Consider for example e_{11}

$$e_{11} = \sum_{j=1}^{3}\left(\frac{v_j}{h_1 h_j}\frac{\partial h_1}{\partial u_j}\right) + \frac{\partial}{\partial u_1}\left(\frac{v_1}{h_1}\right) .$$

Hence $$\nabla . \mathbf{v} = e_{ii} = \sum_{i=1}^{3}\left[\sum_{j=1}^{3}\left(\frac{v_j}{h_i h_j}\frac{\partial h_i}{\partial u_j}\right) + \frac{\partial}{\partial u_i}\left(\frac{v_i}{h_i}\right)\right] . \tag{3.93}$$

The expression (3.93) written out in full is probably more useful in practice (see Examples 3.6, No. 1).

In order to calculate $\nabla \times \mathbf{v}$ we note from (3.23) and Examples 3.2, No. 6 that:

$$\nabla \times \mathbf{v} = -\mathbf{E}{:}\nabla\mathbf{v} = \mathbf{E}{:}\boldsymbol{\xi} = -\epsilon_{ijk}\xi_{ij}\mathbf{e}_k , \tag{3.94}$$

where $\boldsymbol{\xi}$ is minus the antisymmetric part of $\nabla\mathbf{v}$ defined by (3.92).

Hence $$\nabla \times \mathbf{v} = \sum_{i=1}^{3}\sum_{j=1}^{3}\sum_{k=1}^{3}\frac{1}{h_i h_j}\frac{\partial}{\partial u_i}(v_j h_j)\epsilon_{ijk}\mathbf{e}_k . \tag{3.95}$$

The function $\mathbf{n}.\nabla\mathbf{v}$ is given by:

$$\mathbf{n}.\nabla\mathbf{v} = n_i(\nabla\mathbf{v})_{ij}\mathbf{e}_j ,$$

which, by virtue of (3.90), becomes:

$$\mathbf{n}.\nabla\mathbf{v} = \sum_{i=1}^{3}\sum_{j=1}^{3}\left(\frac{n_i}{h_i}\frac{\partial v_j}{\partial u_i}\mathbf{e}_j - \frac{n_i v_i}{h_i h_j}\frac{\partial h_i}{\partial u_j}\mathbf{e}_j + \sum_{k=1}^{3}\frac{n_i v_k}{h_i h_k}\frac{\partial h_i}{\partial u_k}\mathbf{e}_i\right) ,$$

$$= \sum_{i=1}^{3}\left[(\mathbf{n}.\nabla v_i)\mathbf{e}_i + \sum_{j=1}^{3}\frac{v_j}{h_i h_j}\left(n_i\frac{\partial h_i}{\partial u_j} - n_j\frac{\partial h_j}{\partial u_i}\right)\mathbf{e}_i\right] . \tag{3.96}$$

The Laplacian $\nabla^2\Phi$ is readily obtained from (3.93) by replacing v_i with $\dfrac{1}{h_i}\dfrac{\partial\Phi}{\partial u_i}$:

$$\nabla^2\Phi = \sum_{i=1}^{3}\left[\sum_{j=1}^{3}\left(\frac{1}{h_i^2 h_j}\frac{\partial\Phi}{\partial u_i}\frac{\partial h_j}{\partial u_i}\right) + \frac{\partial}{\partial u_i}\left(\frac{1}{h_i^2}\frac{\partial\Phi}{\partial u_i}\right)\right] . \tag{3.97}$$

Finally the function $\nabla^2 \mathbf{v}$ is probably best derived from (3.26) as required.

3.6.3 Spherical Polar Coordinates

We consider, as an example of a curvilinear coordinate system, the spherical polar coordinates $u_1 = r$, $u_2 = \theta$, $u_3 = \phi$. The Cartesian components of the vector $\mathbf{r}$ are:

$$\mathbf{r} = (r \sin\theta \cos\phi, r \sin\theta \sin\phi, r \cos\theta) \ . \tag{3.98}$$

$$\text{Therefore}\quad \left.\begin{aligned} \partial\mathbf{r}/\partial r &= (\sin\theta\cos\phi, \sin\theta\sin\phi, \cos\theta) \ , \\ \partial\mathbf{r}/\partial\theta &= r(\cos\theta\cos\phi, \cos\theta\sin\phi, -\sin\theta) \ , \\ \partial\mathbf{r}/\partial\phi &= r(-\sin\theta\sin\phi, \sin\theta\cos\phi, 0) \ . \end{aligned}\right\} \tag{3.99}$$

The scale factors, calculated from (3.81), are:

$$h_1 = 1, h_2 = r, h_3 = r \sin\theta \ . \tag{3.100}$$

Substituting these values into (3.88), we find:

$$\nabla\Phi = \left(\frac{\partial\Phi}{\partial r}, \frac{1}{r}\frac{\partial\Phi}{\partial\theta}, \frac{1}{r\sin\theta}\frac{\partial\Phi}{\partial\phi}\right) \ . \tag{3.101}$$

Equation (3.101) gives the physical components of $\nabla\Phi$ in the directions r increasing, θ increasing and ϕ increasing.

From (3.90) the physical components of $\nabla\mathbf{v}$ are:

$$\left.\begin{aligned}
(\nabla\mathbf{v})_{11} &= \frac{\partial v_1}{\partial r}, \qquad (\nabla\mathbf{v})_{22} = \frac{v_1}{r} + \frac{1}{r}\frac{\partial v_2}{\partial\theta}, \\
(\nabla\mathbf{v})_{33} &= \frac{v_1}{r} + \frac{v_2\cot\theta}{r} + \frac{1}{r\sin\theta}\frac{\partial v_3}{\partial\phi}, \\
(\nabla\mathbf{v})_{23} &= \frac{1}{r}\frac{\partial v_3}{\partial\theta}, \qquad (\nabla\mathbf{v})_{32} = \frac{1}{r\sin\theta}\frac{\partial v_2}{\partial\phi} - \frac{v_3\cot\theta}{r}, \\
(\nabla\mathbf{v})_{31} &= \frac{1}{r\sin\theta}\frac{\partial v_1}{\partial\phi} - \frac{v_3}{r}, \qquad (\nabla\mathbf{v})_{13} = \frac{\partial v_3}{\partial r}, \\
(\nabla\mathbf{v})_{12} &= \frac{\partial v_2}{\partial r}, \qquad (\nabla\mathbf{v})_{21} = \frac{1}{r}\frac{\partial v_1}{\partial\theta} - \frac{v_2}{r} \ .
\end{aligned}\right\} \tag{3.102}$$

The physical components of $\mathbf{e} = \frac{1}{2}[\nabla\mathbf{v} + (\nabla\mathbf{v})']$ are:

$$\left.\begin{aligned}
e_{11} &= \frac{\partial v_1}{\partial r}, \quad e_{22} = \frac{v_1}{r} + \frac{1}{r}\frac{\partial v_2}{\partial \theta},\\
e_{33} &= \frac{v_1}{r} + \frac{v_2 \cot\theta}{r} + \frac{1}{r\sin\theta}\frac{\partial v_3}{\partial \phi},\\
e_{23} &= e_{32} = \frac{1}{2}\left[\frac{1}{r\sin\theta}\frac{\partial v_2}{\partial \phi} + \frac{\sin\theta}{r}\frac{\partial}{\partial\theta}\left(\frac{v_3}{\sin\theta}\right)\right],\\
e_{31} &= e_{13} = \frac{1}{2}\left[\frac{1}{r\sin\theta}\frac{\partial v_1}{\partial \phi} + r\frac{\partial}{\partial r}\left(\frac{v_3}{r}\right)\right],\\
e_{12} &= e_{21} = \frac{1}{2}\left[\frac{1}{r}\frac{\partial v_1}{\partial \theta} + r\frac{\partial}{\partial r}\left(\frac{v_2}{r}\right)\right].
\end{aligned}\right\} \tag{3.103}$$

The non-zero physical components of $\boldsymbol{\xi} = \frac{1}{2}[(\nabla\mathbf{v})' - \nabla\mathbf{v}]$ are:

$$\left.\begin{aligned}
\xi_{23} &= -\xi_{32} = \frac{1}{2r\sin\theta}\left[\frac{\partial v_2}{\partial\theta} - \frac{\partial}{\partial\theta}(v_3\sin\theta)\right],\\
\xi_{31} &= -\xi_{13} = \frac{1}{2}\left[\frac{-1}{r\sin\theta}\frac{\partial v_1}{\partial\phi} + \frac{1}{r}\frac{\partial}{\partial r}(r v_3)\right],\\
\xi_{12} &= -\xi_{21} = \frac{1}{2}\left[\frac{1}{r}\frac{\partial v_1}{\partial\theta} - \frac{1}{r}\frac{\partial}{\partial r}(r v_2)\right].
\end{aligned}\right\} \tag{3.104}$$

The divergence, $\nabla.\mathbf{v} = (\nabla\mathbf{v})_{ii}$ is given by:

$$\nabla.\mathbf{v} = \frac{1}{r^2}\frac{\partial}{\partial r}(r^2 v_1) + \frac{1}{r\sin\theta}\frac{\partial}{\partial\theta}(v_2\sin\theta) + \frac{1}{r\sin\theta}\frac{\partial v_3}{\partial\phi}. \tag{3.105}$$

The curl, $\nabla\times\mathbf{v}$ is obtained from (3.94):

$$\nabla\times\mathbf{v} = -\epsilon_{ijk}\xi_{ij}\mathbf{e}_k .$$

Therefore

$$\left.\begin{aligned}
(\nabla\times\mathbf{v})_1 &= 2\xi_{32} = \frac{1}{r\sin\theta}\left[\frac{\partial}{\partial\theta}(v_3\sin\theta) - \frac{\partial v_2}{\partial\phi}\right],\\
(\nabla\times\mathbf{v})_2 &= 2\xi_{13} = \frac{1}{r\sin\theta}\frac{\partial v_1}{\partial\theta} - \frac{1}{r}\frac{\partial}{\partial r}(r v_3),\\
(\nabla\times\mathbf{v})_3 &= 2\xi_{21} = \frac{1}{r}\frac{\partial}{\partial r}(r v_2) - \frac{1}{r}\frac{\partial v_1}{\partial\theta}.
\end{aligned}\right\} \tag{3.106}$$

The Laplacian $\nabla^2\Phi$ is given by (3.97) or, more conveniently, by (3.101) and (3.105):

$$\nabla^2\Phi = \frac{1}{r^2}\frac{\partial}{\partial r}\left(r^2\frac{\partial\Phi}{\partial r}\right) + \frac{1}{r^2\sin\theta}\frac{\partial}{\partial\theta}\left(\sin\theta\frac{\partial\Phi}{\partial\theta}\right) + \frac{1}{r^2\sin^2\theta}\frac{\partial^2\Phi}{\partial\phi^2} \quad . \tag{3.107}$$

Finally, to write down $\nabla^2\mathbf{v}$ from (3.26) we first write down the physical components of $\nabla(\nabla.\mathbf{v})$:

$$\left.\begin{aligned}
\nabla(\nabla.\mathbf{v})_1 &= \frac{\partial}{\partial r}\left[\frac{1}{r^2}\frac{\partial}{\partial r}(r^2v_1) + \frac{1}{r\sin\theta}\frac{\partial}{\partial\theta}(v_2\sin\theta) + \frac{1}{r\sin\theta}\frac{\partial v_3}{\partial\phi}\right]\\
\nabla(\nabla.\mathbf{v})_2 &= \frac{1}{r}\frac{\partial}{\partial\theta}\left[\frac{1}{r^2}\frac{\partial}{\partial r}(r^2v_1) + \frac{1}{r\sin\theta}\frac{\partial}{\partial\theta}(v_2\sin\theta) + \frac{1}{r\sin\theta}\frac{\partial v_3}{\partial\phi}\right]\\
\nabla(\nabla.\mathbf{v})_3 &= \frac{1}{r\sin\theta}\frac{\partial}{\partial\phi}\left[\frac{1}{r^2}\frac{\partial}{\partial r}(r^2v_1) + \frac{1}{r\sin\theta}\frac{\partial}{\partial\theta}(v_2\sin\theta) + \frac{1}{r\sin\theta}\frac{\partial v_3}{\partial\phi}\right]
\end{aligned}\right\} \tag{3.108}$$

The physical components of $\nabla\times(\nabla\times\mathbf{v})$ are:

$$\left.\begin{aligned}
\nabla\times(\nabla\times\mathbf{v})_1 &= \frac{1}{r\sin\theta}\left[\frac{\partial}{\partial\theta}\left(\frac{\sin\theta}{r}\frac{\partial}{\partial r}(rv_2) - \frac{\sin\theta}{r}\frac{\partial v_1}{\partial\theta}\right)\right.\\
&\qquad \left. - \frac{\partial}{\partial\phi}\left(\frac{1}{r\sin\theta}\frac{\partial v_1}{\partial\phi} - \frac{1}{r}\frac{\partial}{\partial r}(rv_3)\right)\right],\\
\nabla\times(\nabla\times\mathbf{v})_2 &= \frac{1}{r\sin\theta}\frac{\partial}{\partial\phi}\left[\frac{1}{r\sin\theta}\left(\frac{\partial}{\partial\theta}(v_3\sin\theta) - \frac{\partial v_2}{\partial\phi}\right)\right]\\
&\qquad - \frac{1}{r}\frac{\partial}{\partial r}\left(\frac{\partial}{\partial r}(rv_2) - \frac{\partial v_1}{\partial\theta}\right),\\
\nabla\times(\nabla\times\mathbf{v})_3 &= \frac{1}{r}\frac{\partial}{\partial r}\left[\frac{1}{\sin\theta}\frac{\partial v_1}{\partial\phi} - \frac{\partial}{\partial r}(rv_3)\right]\\
&\qquad - \frac{1}{r}\frac{\partial}{\partial\theta}\left[\frac{1}{r\sin\theta}\left(\frac{\partial}{\partial\theta}(v_3\sin\theta) - \frac{\partial v_2}{\partial\phi}\right)\right].
\end{aligned}\right\} \tag{3.109}$$

Hence the physical components of $\nabla^2\mathbf{v}$ are:

$$\left.\begin{aligned}
(\nabla^2\mathbf{v})_1 &= \nabla^2 v_1 - \frac{2v_1}{r^2} - \frac{2}{r^2\sin\theta}\frac{\partial}{\partial\theta}(v_2\sin\theta) - \frac{2}{r^2\sin\theta}\frac{\partial v_3}{\partial\phi},\\
(\nabla^2\mathbf{v})_2 &= \nabla^2 v_2 + \frac{2}{r^2}\frac{\partial v_1}{\partial\theta} - \frac{v_2}{r^2\sin^2\theta} - \frac{2\cos\theta}{r^2\sin^2\theta}\frac{\partial v_3}{\partial\phi},\\
(\nabla^2\mathbf{v})_3 &= \nabla^2 v_3 + \frac{2}{r^2\sin^2\theta}\frac{\partial v_1}{\partial\phi} + \frac{2\cos\theta}{r^2\sin^2\theta}\frac{\partial v_2}{\partial\phi} - \frac{v_3}{r^2\sin^2\theta}.
\end{aligned}\right\} \tag{3.110}$$

3.6.4 Cylindrical Polar Coordinates

We shall now repeat the process of the previous section for the cylindrical co-ordinate system (r, θ, z), quoting only the results. The Cartesian components of the vector **r** are:

$$\mathbf{r} = (r\cos\theta, r\sin\theta, z) \ . \tag{3.111}$$

The scale factors are:

$$h_1 = 1, h_2 = r, h_3 = 1 \ . \tag{3.112}$$

The gradient, $\nabla\Phi$, is given by:

$$\nabla\Phi = \left(\frac{\partial\Phi}{\partial r}, \frac{1}{r}\frac{\partial\Phi}{\partial\theta}, \frac{\partial\Phi}{\partial z}\right) . \tag{3.113}$$

The physical components of $\nabla\mathbf{v}$ are:

$$\left.\begin{aligned}
(\nabla\mathbf{v})_{11} &= \frac{\partial v_1}{\partial r}, \quad (\nabla\mathbf{v})_{22} = \frac{v_1}{r} + \frac{1}{r}\frac{\partial v_2}{\partial\theta}, \quad (\nabla\mathbf{v})_{33} = \frac{\partial v_3}{\partial z}, \\
(\nabla\mathbf{v})_{23} &= \frac{1}{r}\frac{\partial v_3}{\partial\theta}, \quad (\nabla\mathbf{v})_{32} = \frac{\partial v_2}{\partial z}, \quad (\nabla\mathbf{v})_{31} = \frac{\partial v_1}{\partial z}, \\
(\nabla\mathbf{v})_{13} &= \frac{\partial v_3}{\partial r}, \quad (\nabla\mathbf{v})_{12} = \frac{\partial v_2}{\partial r}, \quad (\nabla\mathbf{v})_{21} = \frac{1}{r}\frac{\partial v_1}{\partial\theta} - \frac{v_2}{r} .
\end{aligned}\right\} \tag{3.114}$$

The physical components of $\mathbf{e} = \tfrac{1}{2}[(\nabla\mathbf{v}) + (\nabla\mathbf{v})']$ are:

$$\left.\begin{aligned}
e_{11} &= \frac{\partial v_1}{\partial r}, \quad e_{22} = \frac{v_1}{r} + \frac{1}{r}\frac{\partial v_2}{\partial\theta}, \quad e_{33} = \frac{\partial v_3}{\partial z} \\
e_{23} &= e_{32} = \tfrac{1}{2}\left(\frac{\partial v_2}{\partial z} + \frac{1}{r}\frac{\partial v_3}{\partial\theta}\right), \quad e_{31} = e_{13} = \tfrac{1}{2}\left(\frac{\partial v_1}{\partial z} + \frac{\partial v_3}{\partial r}\right), \\
e_{12} &= e_{21} = \tfrac{1}{2}\left[\frac{1}{r}\frac{\partial v_1}{\partial\theta} + r\frac{\partial}{\partial r}\left(\frac{v_2}{r}\right)\right] .
\end{aligned}\right\} \tag{3.115}$$

The non-zero physical components of $\boldsymbol{\xi} = \tfrac{1}{2}[(\nabla\mathbf{v})' - \nabla\mathbf{v}]$ are:

$$\left.\begin{aligned}
\xi_{23} &= -\xi_{32} = \tfrac{1}{2}\left(\frac{\partial v_2}{\partial z} - \frac{1}{r}\frac{\partial v_3}{\partial\theta}\right), \quad \xi_{31} = -\xi_{13} = \tfrac{1}{2}\left(\frac{\partial v_3}{\partial r} - \frac{\partial v_1}{\partial z}\right) \\
\xi_{12} &= -\xi_{21} = \frac{1}{2r}\left[\frac{\partial v_1}{\partial\theta} - \frac{\partial}{\partial r}(rv_2)\right] .
\end{aligned}\right\} \tag{3.116}$$

The divergence is given by:

$$\nabla.\mathbf{v} = \frac{1}{r}\frac{\partial}{\partial r}(rv_1) + \frac{1}{r}\frac{\partial v_2}{\partial\theta} + \frac{\partial v_3}{\partial z} \ . \tag{3.117}$$

The physical components of curl are given by:

$$\left.\begin{aligned}(\nabla\times\mathbf{v})_1 &= \frac{1}{r}\frac{\partial v_3}{\partial\theta}-\frac{\partial v_2}{\partial z}\ ,\\ (\nabla\times\mathbf{v})_2 &= \frac{\partial v_1}{\partial z}-\frac{\partial v_3}{\partial r}\ ,\\ (\nabla\times\mathbf{v})_3 &= \frac{1}{r}\left[\frac{\partial}{\partial r}(rv_2)-\frac{\partial v_1}{\partial\theta}\right]\ .\end{aligned}\right\}\tag{3.118}$$

The Laplacian is given by:

$$\nabla^2\Phi = \frac{1}{r}\frac{\partial}{\partial r}\left(r\frac{\partial\Phi}{\partial r}\right)+\frac{1}{r^2}\frac{\partial}{\partial\theta}\left(\frac{\partial\Phi}{\partial\theta}\right)+\frac{\partial^2\Phi}{\partial z^2}\ .\tag{3.119}$$

Finally the physical components of $\nabla^2\mathbf{v}$ are:

$$\left.\begin{aligned}(\nabla^2\mathbf{v})_1 &= \nabla^2 v_1-\frac{v_1}{r^2}-\frac{2}{r^2}\frac{\partial v_2}{\partial\theta}\ ,\\ (\nabla^2\mathbf{v})_2 &= \nabla^2 v_2+\frac{2}{r^2}\frac{\partial v_1}{\partial\theta}-\frac{v_2}{r^2}\ ,\\ (\nabla^2\mathbf{v})_3 &= \nabla^2 v_3\ .\end{aligned}\right\}\tag{3.120}$$

Examples 3.6

(1) The expression (3.93) for $\nabla.\mathbf{v}$ written out in full is:

$$\nabla.\mathbf{v} = \frac{1}{h_1h_2h_3}\left[\frac{\partial}{\partial u_1}(v_1h_2h_3)+\frac{\partial}{\partial u_2}(v_2h_1h_3)+\frac{\partial}{\partial u_3}(v_3h_1h_2)\right]\ ;$$

for $\displaystyle\sum_{i=1}^{3}\left[\sum_{j=1}^{3}\left(\frac{v_j}{h_ih_j}\frac{\partial h_i}{\partial u_j}\right)+\frac{\partial}{\partial u_i}\left(\frac{v_i}{h_i}\right)\right]$.

$$= \frac{v_1}{h_1^2}\frac{\partial h_1}{\partial u_1}+\frac{v_1}{h_1h_2}\frac{\partial h_2}{\partial u_1}+\frac{v_1}{h_1h_3}\frac{\partial h_3}{\partial u_1}+\frac{1}{h_1}\frac{\partial v_1}{\partial u_1}-\frac{v_1}{h_1^2}\frac{\partial h_1}{\partial u_1}$$

$$+\frac{v_2}{h_1h_2}\frac{\partial h_1}{\partial u_2}+\text{etc.}+\frac{v_3}{h_1h_3}\frac{\partial h_1}{\partial u_3}+\text{etc.},$$

$$= \frac{1}{h_1h_2h_3}\left[\frac{\partial}{\partial u_1}(v_1h_2h_3)+\text{etc.}\right]\ .$$

(2) From equation (3.80) we have:

$$\delta\mathbf{r}^2 = \sum_{i=1}^{3} h_i^2\,\delta u_i^2\ .$$

Thus, for spherical polar coordinates we have, from (3.98):

$$\delta \mathbf{r}^2 = \delta r^2 + r^2 \delta\theta^2 + r^2 \sin^2\theta\, \delta\phi^2 \,,$$

leading to $h_1 = 1, h_2 = r, h_3 = r \sin\theta$. This provides an alternative to (3.81) for calculating the scale factors.

Problems 3.6

(1) Derive the expression (3.87) for the gradient operator from the Cartesian definition (2.143) and equation (3.82).

(2) Derive from (3.95) the determinantal expression for $\nabla \times \mathbf{v}$ in curvilinear coordinates:

$$\nabla \times \mathbf{v} = \frac{1}{h_1 h_2 h_3} \begin{vmatrix} h_1 \mathbf{e}_1 & h_2 \mathbf{e}_2 & h_2 \mathbf{e}_3 \\ \partial/\partial u_1 & \partial/\partial u_2 & \partial/\partial u_3 \\ h_1 v_1 & h_2 v_2 & h_3 v_3 \end{vmatrix} .$$

(3) Use (3.97) to show that the Laplacian $\nabla^2 \Phi$ written out in full is:

$$\nabla^2 \Phi = \frac{1}{h_1 h_2 h_3} \left[\frac{\partial}{\partial u_1} \left(\frac{h_2 h_3}{h_1} \frac{\partial \Phi}{\partial u_1} \right) + \frac{\partial}{\partial u_2} \left(\frac{h_3 h_1}{h_2} \frac{\partial \Phi}{\partial u_2} \right) + \frac{\partial}{\partial u_3} \left(\frac{h_1 h_2}{h_3} \frac{\partial \Phi}{\partial u_3} \right) \right] .$$

(4) Use (3.96) to show that the expression for $\mathbf{n}.\nabla\mathbf{v}$ written out in full is:

$$(\mathbf{n}.\nabla\mathbf{v})_1 = \mathbf{n}.\nabla v_1 + \frac{1}{h_1 h_2 h_3} \left[v_2 h_3 \left(n_1 \frac{\partial h_1}{\partial u_2} - n_2 \frac{\partial h_2}{\partial u_1} \right) \right.$$
$$\left. + v_3 h_2 \left(n_1 \frac{\partial h_1}{\partial u_3} - n_3 \frac{\partial h_3}{\partial u_1} \right) \right] ,$$

together with similar expression for $(\mathbf{n}.\nabla\mathbf{v})_2$ and $(\mathbf{n}.\nabla\mathbf{v})_3$.

(5) Show that, for the cylindrical coordinates (r, θ, z),

$$\delta \mathbf{r}^2 = \delta r^2 + r^2 \delta\theta^2 + \delta z^2 \ .$$

Hence write down the scale factors.

(6) Prove that if (u_1, u_2, u_3) are orthogonal curvilinear coordinates:

$$\frac{\partial^2 \mathbf{r}}{\partial u_i \partial u_k} \cdot \frac{\partial \mathbf{r}}{\partial u_j} + \frac{\partial^2 \mathbf{r}}{\partial u_i \partial u_j} \cdot \frac{\partial \mathbf{r}}{\partial u_k} = 0, (i \neq j \neq k) \ ,$$

together with two similar expressions. Hence show that $\dfrac{\partial^2 \mathbf{r}}{\partial u_i \partial u_j}$ is a vector in the plane of $\mathbf{e}_i$ and $\mathbf{e}_j$.

3.7 THIRD-ORDER TENSOR FIELDS

We first define the derivative of a second-order Cartesian tensor. Let $\mathbf{T}$ be a tensor field. If there exists a third-order tensor $\nabla\mathbf{T}$ such that:

$$\mathbf{T}(\mathbf{r}+\mathbf{h}) - \mathbf{T}(\mathbf{r}) = \mathbf{h}.\nabla\mathbf{T} + \mathbf{h}.\mathbf{Z} \ , \qquad (3.121)$$

where $|\mathbf{Z}| = \sqrt{(\mathbf{Z}:\mathbf{Z})} \to 0$ as $|\mathbf{h}| \to 0$, then $\mathbf{T}$ is said to be differentiable at $\mathbf{r}$ and $\nabla\mathbf{T}$ is called the **derivative** or **gradient** of $\mathbf{T}$ at $\mathbf{r}$. If $\mathbf{T}$ is differentiable at all points of a domain $\mathscr{D}$, then $\nabla\mathbf{T}$ defines a third-order tensor field on $\mathscr{D}$.

The components of $\nabla\mathbf{T}$ in any Cartesian basis may now be found. At any point $\mathbf{r} = x_i\mathbf{e}_i$, $\mathbf{T}(\mathbf{r}) = T_{ik}(\mathbf{r})\mathbf{e}_i \otimes \mathbf{e}_k$. Let $\mathbf{h} = h\mathbf{e}_i$. Then
$\mathbf{T}(\mathbf{r}+h\mathbf{e}_i) - \mathbf{T}(\mathbf{r}) = h\mathbf{e}_i.\nabla\mathbf{T} + h\mathbf{e}_i.\mathbf{Z}$.

$$\text{Therefore} \quad \mathbf{e}_i.\nabla\mathbf{T} = \lim_{h\to 0} \frac{\mathbf{T}(\mathbf{r}+h\mathbf{e}_i) - \mathbf{T}(\mathbf{r})}{h} = \frac{\partial \mathbf{T}}{\partial x_i} \ ,$$

$$\text{hence} \qquad (\nabla\mathbf{T})_{ikm} = \frac{\partial T_{km}}{\partial x_i} \ , \qquad (3.122)$$

$$\text{or} \qquad \nabla\mathbf{T} = \frac{\partial T_{km}}{\partial x_i}\, \mathbf{e}_i \otimes \mathbf{e}_k \otimes \mathbf{e}_m \ . \qquad (3.123)$$

Equation (3.123) may be contracted over the indices i and k to give:

$$\nabla.\mathbf{T} = \operatorname{div}\mathbf{T} = \frac{\partial T_{ik}}{\partial x_i}\,\mathbf{e}_k \ . \qquad (3.124)$$

In curvilinear coordinates we have:

$$\nabla\mathbf{T} = \sum_{i=1}^{3}\sum_{k=1}^{3}\sum_{m=1}^{3} \frac{1}{h_i}\,\mathbf{e}_i \otimes \frac{\partial}{\partial u_i}(T_{km}\mathbf{e}_k \otimes \mathbf{e}_m) \ ,$$

$$= \sum_{i=1}^{3}\sum_{k=1}^{3}\sum_{m=1}^{3} \frac{1}{h_i}\,\mathbf{e}_i \otimes \Bigg[\mathbf{e}_k \otimes \mathbf{e}_m \frac{\partial T_{km}}{\partial u_i}$$

$$+ T_{km}\left(\frac{1}{h_k}\frac{\partial h_i}{\partial u_k}\,\mathbf{e}_i - \delta_{ik}\sum_{p=1}^{3} \frac{1}{h_p}\frac{\partial h_i}{\partial u_p}\,\mathbf{e}_p \right) \otimes \mathbf{e}_m$$

$$+ T_{km}\mathbf{e}_k \otimes \left(\frac{1}{h_m}\frac{\partial h_i}{\partial u_m}\,\mathbf{e}_i - \delta_{im}\sum_{p=1}^{3} \frac{1}{h_p}\frac{\partial h_i}{\partial u_p}\,\mathbf{e}_p \right) \Bigg] . \qquad (3.125)$$

$$(\nabla\mathbf{T})_{ikm} = \frac{1}{h_i}\frac{\partial T_{km}}{\partial u_i} + \sum_{p=1}^{3} T_{pm}\frac{1}{h_i h_p}\frac{\partial h_i}{\partial u_p}\delta_{ik}$$

$$+ \sum_{p=1}^{3} T_{kp}\frac{1}{h_i h_p}\frac{\partial h_i}{\partial u_p}\delta_{im} - T_{im}\frac{1}{h_i h_k}\frac{\partial h_i}{\partial u_k} - T_{ki}\frac{1}{h_i h_m}\frac{\partial h_i}{\partial u_m} \,. \qquad (3.126)$$

Contracting over the indices i and k we have:

$$(\nabla.\mathbf{T})_k = \sum_{i=1}^{3}\sum_{m=1}^{3}\left(\frac{1}{h_i}\frac{\partial T_{ik}}{\partial u_i} + T_{mk}\frac{1}{h_i h_m}\frac{\partial h_i}{\partial u_m} + \frac{T_{km}}{h_k h_m}\frac{\partial h_k}{\partial u_m}\right.$$

$$\left. - T_{ik}\frac{1}{h_i^2}\frac{\partial h_i}{\partial u_i} - T_{ii}\frac{1}{h_i h_k}\frac{\partial h_i}{\partial u_k}\right) . \qquad (3.127)$$

Examples 3.7

(1) In the case of cylindrical polar coordinates (3.127) yields:

$$(\nabla.\mathbf{T})_k = \frac{\partial T_{1k}}{\partial r} + \frac{1}{r}\frac{\partial T_{2k}}{\partial \theta} + \frac{\partial T_{3k}}{\partial z} + \frac{T_{1k}}{r} + \frac{T_{21}}{r}\delta_{2k} - \frac{T_{22}}{r}\delta_{1k} \,,$$

$$(\nabla.\mathbf{T})_1 = \frac{\partial T_{11}}{\partial r} + \frac{1}{r}\frac{\partial T_{21}}{\partial \theta} + \frac{\partial T_{31}}{\partial z} + \frac{T_{11}}{r} - \frac{T_{22}}{r} \,,$$

$$(\nabla.\mathbf{T})_2 = \frac{\partial T_{12}}{\partial r} + \frac{1}{r}\frac{\partial T_{22}}{\partial z} + \frac{\partial T_{32}}{\partial z} + \frac{T_{12}}{r} + \frac{T_{21}}{r} \,,$$

$$(\nabla.\mathbf{T})_3 = \frac{\partial T_{13}}{\partial r} + \frac{1}{r}\frac{\partial T_{23}}{\partial \theta} + \frac{\partial T_{33}}{\partial z} + \frac{T_{13}}{r} \,.$$

(2) In the case of spherical polar coordinates (3.127) yields:

$$(\nabla.\mathbf{T})_k = \frac{\partial T_{1k}}{\partial r} + \frac{1}{r}\frac{\partial T_{2k}}{\partial \theta} + \frac{1}{r\sin\theta}\frac{\partial T_{3k}}{\partial \theta} + \frac{2T_{1k}}{r} + \frac{T_{2k}\cot\theta}{r} + \frac{T_{21}}{r}\delta_{2k}$$

$$+\left(\frac{T_{31}}{r} + \frac{T_{32}\cot\theta}{r}\right)\delta_{3k} - \left(\frac{T_{22}}{r} + \frac{T_{33}}{r}\right)\delta_{1k} - \frac{T_{33}\cot\theta}{r}\delta_{2k} \,,$$

$$(\nabla.\mathbf{T})_1 = \frac{1}{r^2}\frac{\partial}{\partial r}(r^2 T_{11}) + \frac{1}{r\sin\theta}\left[\frac{\partial}{\partial\theta}(T_{21}\sin\theta) + \frac{\partial T_{31}}{\partial\phi}\right] - \frac{1}{r}(T_{22} + T_{33}),$$

$$(\nabla.\mathbf{T})_2 = \frac{1}{r^2}\frac{\partial}{\partial r}(r^2 T_{12}) + \frac{1}{r\sin\theta}\left[\frac{\partial}{\partial\theta}(T_{22}\sin\theta) + \frac{\partial T_{32}}{\partial\phi}\right] - \frac{1}{r}(T_{21} - T_{33}\cot\theta),$$

$$(\nabla.\mathbf{T})_3 = \frac{1}{r}\frac{\partial}{\partial r}(r^2 T_{13}) + \frac{1}{r\sin\theta}\left[\frac{\partial}{\partial\theta}(T_{23}\sin\theta) + \frac{\partial T_{33}}{\partial\phi}\right] + \frac{1}{r}(T_{31} + T_{32}\cot\theta) \,.$$

CHAPTER 4

Fourth-Order Cartesian Tensors

4.1 TENSORS OF THE FOURTH-ORDER - DEFINITION

The highest order of Cartesian tensor which we shall treat explicitly is the fourth-order tensor. Proceeding along the lines of section 3.1.1, which dealt with third-order tensors, we write:

$$\mathbf{C}.\mathbf{v}_i = \mathbf{B}_i, \qquad i = 1, 2, 3 \ , \tag{4.1}$$

where $\mathbf{v}_1, \mathbf{v}_2, \mathbf{v}_3$ are three linearly independent vectors and $\mathbf{B}_1, \mathbf{B}_2, \mathbf{B}_3$ are three third-order tensors. Equation (4.1) may be used to define a fourth-order tensor $\mathbf{C}$ once the dot product $\mathbf{C}.\mathbf{v}_i$ has been defined.

The dot product is defined by writing:

$$(\mathbf{a} \otimes \mathbf{b} \otimes \mathbf{c} \otimes \mathbf{d}).\mathbf{v} = (\mathbf{a} \otimes \mathbf{b} \otimes \mathbf{c})(\mathbf{d}.\mathbf{v}), \qquad \forall \mathbf{v} \ . \tag{4.2}$$

Equation (4.2) defines this dot product and the **tensor product of four vectors** $\mathbf{a} \otimes \mathbf{b} \otimes \mathbf{c} \otimes \mathbf{d}$, which is clearly a fourth-order tensor.

The properties of fourth-order tensors are similar to those of second- and third-order tensors. For example, any fourth-order tensor may be written as the sum of 81 tensor products of four vectors. There is, however, only one such tensor which is of any real interest, and that is the most general fourth-order isotropic tensor.

4.2 ISOTROPIC TENSORS

4.2.1 Definition

It has been stressed in this book that the fundamental property of Cartesian tensors is their invariance under rotations. Although the individual components of a tensor will, in general, depend upon the orientation of the basis used, the tensor itself is an entity which remains invariant. We have, however, come across two examples of tensors with the very special property that their components are themselves invariant under rotations. These are the second-order Unit Tensor (see Problems 2.2 No. 3, 2.5 No. 6) and the third-order Alternate Tensor (see section 3.2.1). Such tensors are termed **isotropic**.

We shall, in the next few pages, seek to establish the forms of the most general isotropic tensors of the 2nd, 3rd and 4th order. We shall first prove, however, that there is no such thing as an isotropic vector.

Theorem 4.2.1 There is no isotropic tensor of the first-order.

Proof. The essential algebra for this proof has already been carried out in section 3.4.1. Let us assume that an isotropic vector **u** exists. Then, for any rotation θ about a line in the direction of the unit vector **n**:

$$u_j = \theta_{ji} u_i \ , \tag{4.3}$$

where θ_{ji} is as given by (3.50). But (4.3) is just (3.56) with $\lambda = 1$ which as we have seen is the only real value of λ for which solutions of (3.56) exist. We have also seen that the solution of (3.56) corresponding to $\lambda = 1$ is $\mathbf{u} = \mathbf{n}$. Thus the only vector whose components are unchanged by a rotation is one which is parallel to the axis of rotation. Therefore there can be no vector whose components remain unchanged by any rotation, i.e. there is no isotropic vector.

4.2.2 Theorem 4.2.2 the most general isotropic tensor of the second-order is the Unit Tensor

Proof. To prove this theorem we make use of the theory of small rotations (see Examples 3.4, No. 3). Suppose T_{ik} is a component of an isotropic second-order tensor.

$$\text{Then } T^*_{ik} = \theta_{ij}\theta_{kl} T_{jl} = T_{ik} \ . \tag{4.4}$$

Let us specify a rotation by a small angle θ about a line **n**. We have (Examples 3.4, No. 3):

$$\theta_{ij} \approx \delta_{ij} + n_p \epsilon_{ijp} \sin\theta \ . \tag{4.5}$$

$$\text{Therefore } T_{ik} = (\delta_{ij} + n_p \epsilon_{ijp} \sin\theta)(\delta_{kl} + n_q \epsilon_{klq} \sin\theta) T_{jl} \ ,$$

$$\text{or } \quad T_{ik} = T_{ik} + (n_p \epsilon_{ijp} T_{jk} + n_q \epsilon_{klq} T_{il}) \sin\theta + O(\theta^2).$$

Hence, by equating the coefficient of θ to zero,

$$n_p \epsilon_{ijp} T_{jk} + n_q \epsilon_{klq} T_{il} = 0 \ , \tag{4.6}$$

$$\text{or } \quad -\mathbf{n} \times \mathbf{T} + \mathbf{T} \times \mathbf{n} = \mathbf{0} \ . \tag{4.7}$$

But we have already seen (Problems 3.3, No. 3) that the only tensor **T** to satisfy (4.7) for any vector **n** is a multiple of the unit tensor. This proves the theorem.

4.2.3 Theorem 4.2.3 The most general isotropic tensor of the third-order is the Alternate Tensor

Proof. We again make use of the theory of small rotations and obtain corresponding to (4.6),

$$n_p \epsilon_{ijp} B_{jkm} + n_q \epsilon_{klq} B_{ilm} + n_r \epsilon_{mnr} B_{ikn} = 0 \ , \tag{4.8}$$

where **B** is an isotropic third-order tensor and **n** is an arbitrary unit vector. Let $\mathbf{n} = \mathbf{e}_t$. (4.8) becomes:

$$\epsilon_{ijt}B_{jkm} + \epsilon_{kjt}B_{ijm} + \epsilon_{mjt}B_{ikj} = 0 \ . \tag{4.9}$$

Let $t = k = m \neq i$.
Then, remembering the properties of ϵ_{ijk} expressed in (3.15),

$$\epsilon_{ijt}B_{jtt} = 0, \quad \text{therefore} \quad B_{jtt} = 0 \, (j \neq t) \ , \tag{4.10}$$

where no summation is implied over the index t. Similarly, by letting $t = i = m \neq k$ and $t = i = k \neq m$ in turn, we conclude that all components of **B** with exactly two equal suffices are zero. Now let $m = k \neq i \neq t$ in (4.9) to obtain:

$$\epsilon_{ikt}B_{kkk} + \epsilon_{kit}B_{iik} + \epsilon_{kit}B_{iki} = 0 \ ,$$

where no summation is implied over the indices i or k.

Hence $\quad B_{kkk} = 0 \ . \qquad (4.11)$

Combining the results of (4.10) and (4.11) we conclude that all the components of **B** with *at least* two equal suffices are all zero. Finally we let $t = i \neq k = m$ to obtain:

$$B_{ijk} = -B_{ikj} \ ,$$

then $t = k \neq m = i$ to obtain:

$$B_{kji} = -B_{ikj} \ , \tag{4.12}$$

and $t = m \neq k = i$ to obtain:

$$B_{jik} = -B_{ijk} \ .$$

The results (4.10)-(4.12) are sufficient to prove that B_{ijk} is a linear multiple of ϵ_{ijk}.

4.2.4 Theorem 4.2.4 The most general isotropic tensor of the fourth-order is the tensor $\mu_1 \delta_{ij}\delta_{kl} + \mu_2 \delta_{ik}\delta_{jl} + \mu_3 \delta_{il}\delta_{jk}$.

Proof. From the theory of small rotations we obtain, corresponding to (4.9):

$$\epsilon_{mip}C_{pjkl} + \epsilon_{mjp}C_{ipkl} + \epsilon_{mkp}C_{ijpl} + \epsilon_{mlp}C_{ijkp} = 0 \ , \tag{4.13}$$

where **C** is an isotropic fourth-order tensor. Let $i = j = k = m \neq l$.

Then $\quad \epsilon_{mlp}C_{iiip} = 0 \ ,$

therefore $\quad C_{iiip} = 0 \ , \qquad (4.14)$

where no summation is carried out over the index i. Similarly, if we let $i = j = l = m \neq k$, we obtain:

$$C_{iipi} = 0 \quad \text{(no summation)}.$$

Proceeding in this way we may readily show that all components with precisely three equal suffices are zero.

We seek next to prove a similar result for all terms with three different suffices. Let $i = j = k \neq l = m$.

Then $$\epsilon_{mip}C_{piim} + \epsilon_{mip}C_{ipim} + \epsilon_{mip}C_{iipm} = 0 ,$$

where no summation is to be carried out over the index i.

Therefore $$C_{ipim} + C_{iipm} + C_{iipm} = 0 . \quad (4.15)$$

Let $i = j \neq k \neq l = m$.

Then $$\epsilon_{mik}C_{piim} + \epsilon_{mik}C_{ikkm} + \epsilon_{mki}C_{iiim} = 0 ,$$

therefore $$C_{kikm} + C_{ikkm} = 0 . \quad (4.16)$$

Equations (4.15) and (4.16) are together sufficient to prove that

$$C_{iipm} = 0 . \quad (4.17)$$

In a similar way we may establish that all components with three different suffices are zero.

We next examine components of the form T_{iijj}. Let $i = j = m \neq k \neq l$.

Then $$\epsilon_{ikl}C_{iill} + \epsilon_{ilk}C_{iikk} = 0 ,$$

therefore $$C_{iill} = C_{iikk} ,$$

where no summation is implied and $k \neq l \neq i$. It is easy to show that all such components are equal.

Let $$A = C_{1122} = C_{1133} = C_{2211} = \text{etc.} \quad (4.18)$$

By letting $i = k = m \neq j \neq l$, we may show that

$$C_{mlml} = C_{mjmj} . \quad (4.19)$$

Again all such components are equal. We therefore let

$$B = C_{1212} = C_{1313} = C_{2121} = \text{etc.} \quad (4.20)$$

Next we let $i \neq j = k = m \neq l$ to obtain:

$$C_{lkkl} = C_{ikki} .$$

These components too are all equal, so we let

$$C = C_{1221} = C_{1331} = C_{2112} = \text{etc.} \quad (4.21)$$

Finally, components with four equal indices are dealt with as follows. Let $i = j = k \neq l \neq m$.

Then $$\epsilon_{mil}C_{liil} + \epsilon_{mil}C_{ilil} + \epsilon_{mil}C_{iill} + \epsilon_{mli}C_{iiii} = 0 \ ,$$

therefore $$C_{iiii} = C_{liil} + C_{ilil} + C_{iill} = A + B + C \ . \tag{4.22}$$

We have now dealt with all the components of the most general isotropic tensor of the fourth-order, and conclude that these elements depend on at most three constants. The tensor

$$C_{ikmp} = \mu_1 \delta_{ik}\delta_{mp} + \mu_2 \delta_{im}\delta_{kp} + \mu_3 \delta_{ip}\delta_{km} \tag{4.23}$$

fulfils these requirements.

This proves the theorem.

4.2.5 Complex Isotropic Vectors

Although, as we have seen, there are no real isotropic vectors, it is possible to find complex eigenvalues λ of the rotation matrix, and corresponding eigenvectors **v** with complex components which satisfy the relation:

$$v_i^* = \lambda v_i \tag{4.24}$$

under the rotation. Such vectors are called **complex isotropic vectors.**

The eigenvalues of the rotation matrix Θ satisfy the relation:

$$|\theta_{ij} - \lambda\delta_{ij}| = 0 \ . \tag{4.25}$$

We know already that $|\theta_{ij} - \delta_{ij}| = 0$ (see Problems 3.4, No. 4), therefore $\lambda = 1$ is an eigenvalue (which leads of course to the solution that **v** is along the axis of rotation.)

To find the other two solutions of (4.25) we expand the determinant to obtain:

$$-\lambda^3 + \theta_{ii}\lambda^2 - \lambda A_{ii} + 1 = 0 \ , \tag{4.26}$$

where A_{ij} is the cofactor of θ_{ij}. Since $\lambda = 1$ is a solution of (4.26) we conclude that $A_{ii} = \theta_{ii}$, and (4.26) becomes:

$$(\lambda - 1)\,[\lambda^2 + (1 - \theta_{ii})\lambda + 1] = 0 \ . \tag{4.27}$$

The solutions of (4.27) are:

$$\lambda = 1, \lambda = \tfrac{1}{2}\,[\theta_{ii} - 1 \pm \sqrt{\{(\theta_{ii} - 1)^2 - 4\}}] \ . \tag{4.28}$$

Now, from (3.55), $\theta_{ii} - 1 = 2\cos\theta$, where θ is the angle of rotation.

Hence
$$\lambda = \cos\theta \pm i \sin\theta \ . \tag{4.29}$$

Examples 4.2

(1) **S** and **T** are two second-order tensors such that **T** = **A**:**S** where **A** is a fourth-order isotropic tensor. Find **S** in terms of **T** when the components of **A** are given by

(i) $A_{ijkl} = \mu\delta_{ik}\delta_{jl}$ (ii) $A_{ijkl} = \mu\delta_{il}\delta_{jk}$.

(i) We have $T_{ij} = \mu\delta_{ik}\delta_{jl}S_{lk} = \mu S_{ji}$.

Hence $\mathbf{S} = (1/\mu)\mathbf{T}'$.

(ii) We have $T_{ij} = \mu\delta_{il}\delta_{jk}S_{lk} = \mu S_{ij}$.

Hence $\mathbf{S} = (1/\mu)\mathbf{T}$.

(2) Consider a rotation of angle θ in the positive sense about the x_3-axis. The rotation matrix is given by:

$$\Theta = \begin{bmatrix} \cos\theta & \sin\theta & 0 \\ -\sin\theta & \cos\theta & 0 \\ 0 & 0 & 1 \end{bmatrix}.$$

The eigenvalues are $\lambda = 1, \lambda = \cos\theta \pm i\sin\theta$.
The corresponding eigenvectors are:

$$\mathbf{v}_1 = (0, 0, 1)\ ,$$

$$\mathbf{v}_2 = (1, i, 0)\ ,$$

$$\mathbf{v}_3 = (1, -i, 0)\ .$$

$\mathbf{v}_2$ and $\mathbf{v}_3$ are the 'complex isotropic vectors'.

(3) The most general isotropic tensor of order r is found as follows:

if r is even the tensor consists of a linear combination of $\dfrac{r!}{(\frac{1}{2}r)!(2!)^{\frac{1}{2}r}}$ products of $(\frac{1}{2}r)\delta_{ij}$'s. If r is odd the tensor consists of a linear combination of

$$\frac{r!}{\left(\frac{r-3}{2}\right)!3!(2!)^{\frac{1}{2}(r-3)}}$$

products of ϵ_{ijk} and $\frac{1}{2}(r-3)\delta_{lm}$'s.

The most general isotropic tensor of order 5 is a linear combination of 10 terms of the form $\epsilon_{ijk}\delta_{lm}$. The most general isotropic tensor of order 6 is a linear combination of 15 terms of the form $\delta_{ij}\delta_{kl}\delta_{mn}$.

Problems 4.2

(1) **S** and **T** are two second-order tensors such that $\mathbf{T} = \mathbf{A}:\mathbf{S}$, where **A** is a fourth-order isotropic tensor. Find **S** in terms of **T** when the components of **A** are given by

(i) $A_{ijkl} = \mu_1\delta_{ij}\delta_{kl} + \mu_2\delta_{ik}\delta_{jl}$;

(ii) $A_{ijkl} = \mu_1\delta_{ij}\delta_{kl} + \mu_2\delta_{ik}\delta_{jl} + \mu_3\delta_{il}\delta_{jk}$.

(2) Find the rotation matrix corresponding to a rotation of $\pi/3$ in the positive sense about the line parallel to the vector (1, 1, 1). Hence determine the 'complex isotropic vectors' corresponding to this rotation.

(3) Find the form of the most general isotropic tensor of order 7.

(4) Show that the eigenvectors **u** of the rotation matrix Θ corresponding to the complex eigenvalues $\lambda = \cos\theta \pm i\sin\theta$ satisfy the relation

$$\bar{\mathbf{u}} \times \mathbf{n} = \pm i\mathbf{u} ,$$

where **n** is a unit vector along the axis of rotation given by (3.52). Hence show that the eigenvectors are given explicitly by:

$$\mathbf{u} = (\mathbf{n} \times \mathbf{m}) \times \mathbf{n} \pm i(\mathbf{n} \times \mathbf{m}) ,$$

where **m** is an arbitrary real vector. Compare this result with those of Example 2, and Problem 2.

CHAPTER 5

The Inertia Tensor

5.1 THE INERTIA TENSOR

5.1.1 Definition

We are now in a position to make use of the algebra developed in the previous chapters, to assist in the treatment of certain three-dimensional mechanical problems involving rigid bodies. Fundamental to the formulation of any rigid body problem is the inertia tensor. The tensor appears in the formulae for angular momentum and kinetic energy, and from it the moment of inertia about a given axis may be deduced.

The value of the inertia tensor varies from point to point throughout the body; it may therefore be regarded as a tensor field. However, once the value of the tensor is known at the centroid, it is easy to deduce its value at any other point.

We shall now see how the inertia tensor arises naturally in the formulae for the angular momentum and kinetic energy of a rotating body.

Consider a rigid body rotating with angular velocity $\boldsymbol{\omega}$ about a fixed point O. At a given instant the angular momentum at O is given by the integral formula:

$$\mathbf{H}(\mathrm{O}) = \int_V \rho(\mathbf{r})\mathbf{r} \times \mathbf{v}\,\mathrm{d}V \ , \tag{5.1}$$

where $\mathbf{r}$ is the position vector of the point whose velocity is $\mathbf{v}$, $\rho(\mathbf{r})$ is the density and V is the volume of the body.

Now $\qquad \mathbf{v} = \boldsymbol{\omega} \times \mathbf{r} \ ,$

therefore $$\mathbf{H}(\mathrm{O}) = \int_V \rho(\mathbf{r})\mathbf{r} \times (\boldsymbol{\omega} \times \mathbf{r})\mathrm{d}V \ . \tag{5.2}$$

Expanding the triple vector product in (5.2), we have:

$$\mathbf{H}(\mathrm{O}) = \int_V \rho(r)[\boldsymbol{\omega}\mathbf{r}^2 - \mathbf{r}(\boldsymbol{\omega}.\mathbf{r})]\,\mathrm{d}V \ . \tag{5.3}$$

We next introduce the tensor

$$\mathbf{I}(\mathrm{O}) = \int_V \rho(\mathbf{r})(\mathbf{r}^2\mathbf{1} - \mathbf{r}\otimes\mathbf{r})\mathrm{d}V \ . \tag{5.4}$$

It is now possible to write the angular velocity as a common factor:

$$\mathbf{H}(\mathrm{O}) = \mathbf{I}(\mathrm{O}).\boldsymbol{\omega} \ . \tag{5.5}$$

The tensor $\mathbf{I}(\mathrm{O})$ is the **inertia tensor**. From the definition (5.4) it is apparent that the tensor is a property of the body and not of its motion. Having evaluated the tensor for a given body the angular momentum, kinetic energy, and moment of inertia about a given axis, may all be derived in a straightforward way.

It is also apparent from the definition that the inertia tensor is symmetric. It may therefore be diagonalized according to the method presented in Chapter 2, and eigenvalues and eigenvectors found. The eigenvectors are termed the **principal axes of inertia**, and the eigenvalues, which turn out to be moments of inertia about the principal axes, are termed the **principal moments of inertia**.

Recalling the method of diagonalization, which is to seek values of λ and $\boldsymbol{\omega}$ which satisfy the equation

$$\mathbf{I}(\mathrm{O}).\boldsymbol{\omega} = \lambda\boldsymbol{\omega} \ ,$$

we observe that, when a body rotates about a principal axis the angular velocity and angular momentum vectors are parallel.

5.1.2 Use of the Inertia Tensor – Kinetic Energy

The formula for angular momentum, involving the inertia tensor, has already been given (5.5). Consider now the kinetic energy of a rigid body rotating with angular velocity $\boldsymbol{\omega}$ about a fixed point O. The kinetic energy is given by:

$$T = \int_V \tfrac{1}{2}\rho(\mathbf{r})\mathbf{v}^2\,\mathrm{d}V \ , \tag{5.6}$$

$$= \int_V \tfrac{1}{2}\rho(\mathbf{r})\,(\boldsymbol{\omega}\times\mathbf{r})^2\mathrm{d}V \ , \tag{5.7}$$

$$= \int_V \tfrac{1}{2}\rho(\mathbf{r})\{(\mathbf{r}^2\,\boldsymbol{\omega}^2 - (\boldsymbol{\omega}.\mathbf{r})^2\}\,\mathrm{d}V \ , \tag{5.8}$$

$$= \int_V \tfrac{1}{2}\rho(\mathbf{r})\,\{\mathbf{r}^2\,\mathbf{1} - \mathbf{r}\otimes\mathbf{r}\} : \boldsymbol{\omega}\otimes\boldsymbol{\omega}\,\mathrm{d}V \ . \tag{5.9}$$

Here we have made use of the expansion (1.98) to reduce the product of 4 vectors in (5.7). In (5.9) we have reintroduced the inertia tensor **I**(O), thus we have:

$$T = \tfrac{1}{2}\mathbf{I}(\mathrm{O}):\boldsymbol{\omega}\otimes\boldsymbol{\omega} \ , \tag{5.10}$$

$$= \tfrac{1}{2}\boldsymbol{\omega}.\mathbf{I}(\mathrm{O}).\boldsymbol{\omega} \ . \tag{5.11}$$

We note also that, in terms of the angular momentum, **H**(O),

$$T = \tfrac{1}{2}\mathbf{H}(\mathrm{O}).\boldsymbol{\omega} \ . \tag{5.12}$$

Suppose now that the point O is no longer fixed but moving with velocity **V**. The velocity of the point whose position vector is **r** with respect to O is then given by:

$$\mathbf{v} = \mathbf{V} + \boldsymbol{\omega}\times\mathbf{r} \ . \tag{5.13}$$

Hence
$$T = \int_V \tfrac{1}{2}\rho(\mathbf{r})(\mathbf{V}+\boldsymbol{\omega}\times\mathbf{r})^2\,\mathrm{d}V \ , \tag{5.14}$$

$$= \int_V \tfrac{1}{2}\rho(\mathbf{r})\mathbf{V}^2\,\mathrm{d}V + \int_V \tfrac{1}{2}\rho(\mathbf{r})(\boldsymbol{\omega}\times\mathbf{r})^2\,\mathrm{d}V + \int_V \rho(\mathbf{r})\mathbf{V}.(\boldsymbol{\omega}\times\mathbf{r})\mathrm{d}V \ . \tag{5.15}$$

Now **V** is unique for the body, and may therefore be taken out of the sign of integration whenever it occurs. Thus we obtain:

$$T = \tfrac{1}{2}M\mathbf{V}^2 + \tfrac{1}{2}\mathbf{I}(\mathrm{O}):\boldsymbol{\omega}\otimes\boldsymbol{\omega} + \mathbf{V}\times\boldsymbol{\omega}.\int_V \rho(\mathbf{r})\mathbf{r}\,\mathrm{d}V \ , \tag{5.16}$$

where $M = \int_V \rho(\mathbf{r})\mathrm{d}V$ is the mass of the body.

The expression (5.16) gives the kinetic energy in the general case of a body with a point O moving with velocity **V**. It is really too general to be of much practical use, and it is usually more convenient to choose the point O to be coincident with the centroid G, an arrangement which leads to a simplified expression.

The position vector of the centroid G is given by:

$$\bar{\mathbf{r}} = \frac{\int_V \rho(\mathbf{r})\mathbf{r}\,\mathrm{d}V}{\int_V \rho(\mathbf{r})\mathrm{d}V} \ . \tag{5.17}$$

Hence, when O is coincident with G,

$$\int_V \rho(\mathbf{r})\mathbf{r}\,\mathrm{d}V = 0 \ ,$$

and the expression (5.16) for kinetic energy reduces to:

$$T = \tfrac{1}{2}MV^2 + \tfrac{1}{2}\mathbf{I}(G): \boldsymbol{\omega} \otimes \boldsymbol{\omega} \ , \tag{5.18}$$

where **V** is now the velocity of the centroid, and **I**(G) is the inertia tensor calculated at the centroid.

Expression (5.18) shows that the kinetic energy of a rigid body in general motion is made up of two parts: (i) the kinetic energy of a single particle of mass equal to that of the body, situated at the centroid, and moving with the velocity of the centroid, and (ii) the kinetic energy due to the rotation of the body about the centroid.

5.1.3 The Moment of Inertia

Let **n** be the unit vector along the instantaneous axis of rotation. Then $\boldsymbol{\omega} = \omega\mathbf{n}$ and (5.10, 5.11) yield:

$$T = \tfrac{1}{2}\mu\omega^2 \ , \tag{5.19}$$

where
$$\mu = \mathbf{I}(O):\mathbf{n} \otimes \mathbf{n} \ , \tag{5.20}$$

$$= \mathbf{n}.\mathbf{I}(O).\mathbf{n} \ . \tag{5.21}$$

Here ω is the scalar angular speed about the axis of rotation and μ is the **moment of inertia** about the same axis.

An integral formula for μ is provided by (5.7):

$$\mu = \int_V \rho(\mathbf{r})(\mathbf{n} \times \mathbf{r})^2 \mathrm{d}V \ ,$$

$$= \int_V \rho(\mathbf{r}) d^2 \mathrm{d}V \ , \tag{5.22}$$

where the point $\mathbf{r} = \mathbf{0}$ is on the axis of rotation, and d is the distance of a typical point **r** from that axis, i.e. $d = |\mathbf{n} \times \mathbf{r}|$.

Suppose now that the body is rotating about a principal axis. The vector **n** is now a unit eigenvector, and $\mathbf{I}(O).\mathbf{n} = \lambda\mathbf{n}$ where λ is the corresponding eigenvalue. From (5.21) μ is now given by

$$\mu = \mathbf{n}.(\lambda\mathbf{n}) = \lambda \ .$$

Hence, as we anticipated at the end of section 5.1.1, the eigenvalues are moments of inertia about the corresponding principal axes.

5.1.4 Radius of Gyration

Another term which is sometimes encountered is the **radius of gyration**. This is best illustrated by an example. We shall see later that the moment of inertia of a uniform solid sphere of mass M and radius a, about a diameter, is $\tfrac{2}{5}Ma^2$. The distance from the axis of a single particle of mass M which has the same moment of inertia about the given axis is termed the radius of gyration k.

Thus $$\mu = Mk^2 = \tfrac{2}{5}Ma^2 ,$$

or $$k = a\sqrt{(\tfrac{2}{5})} .$$

In general the radius of gyration is given by:

$$k = \sqrt{(\mu/M)} ,$$

where μ is the moment of inertia.

5.1.5 Momental Ellipsoid

For a body with inertia tensor **I**(O) the relation

$$\mathbf{I}(\mathrm{O}):\mathbf{r} \otimes \mathbf{r} = 1$$

defines an ellipsoid called the **momental ellipsoid**. Clearly the principal axes of the ellipsoid coincide with those of the body. Referred to these axes the equation becomes:

$$I_{11}x_1^2 + I_{22}x_2^2 + I_{33}x_3^2 = 1 .$$

Examples 5.1

(1) The inertia tensor of a rigid body with respect to an origin O and a given set of axes is

$$\mathbf{I}(\mathrm{O}) = \begin{bmatrix} 2 & -2 & -1 \\ -2 & 4 & 0 \\ -1 & 0 & 6 \end{bmatrix} .$$

The body rotates about the line $2x_1 = x_2 = 3x_3$ with angular speed 7. Calculate the moment of inertia about this line, also the angular momentum and kinetic energy.

Write down the Cartesian equations of all lines such that when the body rotates about any one of them, the angular velocity and angular momentum vectors are parallel. What are the moments of inertia about these lines?

The unit vector along the given line is

$$\mathbf{n} = \frac{(\frac{1}{2}, 1, \frac{1}{3})}{\sqrt{(\frac{1}{4} + 1 + \frac{1}{9})}} = \frac{1}{7}(3, 6, 2) .$$

Hence $$\mathbf{I}(\mathrm{O}).\mathbf{n} = \frac{1}{7}(-8, 18, 9) .$$

The moment of inertia about the given line is given by:

$$\mu = \mathbf{n}.\mathbf{I}(\mathrm{O}).\mathbf{n} = \frac{1}{49}(-24 + 108 + 18) \quad ,$$

i.e. $$\mu = 102/49 \quad .$$

The angular velocity is given by:

$$\boldsymbol{\omega} = 7\mathbf{n} = (3, 6, 2) \quad .$$

Hence $$\mathbf{H}(\mathrm{O}) = \mathbf{I}(\mathrm{O}).\boldsymbol{\omega} = (-8, 18, 9) \quad .$$

The kinetic energy is given by:

$$T = \tfrac{1}{2}\mathbf{H}(\mathrm{O}).\boldsymbol{\omega} = 51 \quad .$$

The second part of the question may equally well have been written 'Diagonalize the matrix representing the inertia tensor, and determine the principal axes and moments of inertia'.

For the angular velocity and angular momentum vectors to be parallel, we must have:

$$\mathbf{I}(\mathrm{O}).\boldsymbol{\omega} = \lambda\boldsymbol{\omega} \quad ,$$

or $$[\mathbf{I}(\mathrm{O}) - \lambda\mathbf{1}].\boldsymbol{\omega} = \mathbf{0} \quad .$$

Values of λ yielding solutions for $\boldsymbol{\omega}$ are found from the characteristic equation:

$$\begin{vmatrix} 2-\lambda & -2 & -1 \\ -2 & 4-\lambda & 0 \\ -1 & 0 & 6-\lambda \end{vmatrix} = 0 \quad ,$$

i.e. $\lambda = 5$ or $\lambda = \frac{1}{2}(7 \pm \sqrt{33})$.

The three values of λ are the principal moments of inertia, i.e. the moments of inertia about the three principal axes. When $\lambda = 5$ we find $\boldsymbol{\omega} \parallel (1, -2, 1)$, and when $\lambda = \frac{1}{2}(7 \pm \sqrt{33})$ we find $\boldsymbol{\omega} \parallel (-8, 1 \pm \sqrt{33}, 10 \pm 2\sqrt{33})$. The three vectors $\boldsymbol{\omega}$ are parallel to the principal axes of inertia.

The Cartesian equations of lines through the origin and parallel to $\mathbf{H}(\mathrm{O})$ and $\boldsymbol{\omega}$ are therefore:

$$2x_1 = -x_2 = 2x_3 \quad ,$$

and $$\frac{-x_1}{8} = \frac{x_2}{1 \pm \sqrt{33}} = \frac{x_3}{10 \pm 2\sqrt{33}} \quad .$$

(2) The inertia tensor of a plane lamina of surface area A and surface density $\sigma(\mathbf{r})$ is given by:

$$\mathbf{I}(\mathrm{O}) = \int_A \sigma(\mathbf{r})(\mathbf{r}^2\mathbf{1} - \mathbf{r}\otimes\mathbf{r})\mathrm{d}A \ .$$

Problems 5.1

(1) The Inertia Tensor of a rigid body with respect to an origin O and a given set of axes is:

$$\mathbf{I}(\mathrm{O}) = \begin{bmatrix} 3 & -1 & 0 \\ -1 & 1 & 0 \\ 0 & 0 & 2 \end{bmatrix} .$$

The body rotates about the line $x_1 = x_2 = 2x_3$ with angular speed 3. Calculate the moment of inertia about this line, also the angular momentum and kinetic energy.

Write down the equations of all lines such that when the body rotates about any one of them, the angular momentum and angular velocity are parallel. What are the moments of inertia about these lines?

(2) Repeat Problem (1) for a body with inertia tensor

$$\mathbf{I}(\mathrm{O}) = \begin{bmatrix} 16 & \sqrt{6} & -\sqrt{2} \\ \sqrt{6} & 15 & \sqrt{3} \\ -\sqrt{2} & \sqrt{3} & 17 \end{bmatrix} ,$$

rotating about the line $7x_1 = 7x_2 = 4x_3$, with angular speed 9.

(3) Show that it is, in general, possible to calculate the components of the inertia tensor of a body from a knowledge of its moments of inertia about six axes through the origin of coordinates. Calculate the components of the inertia tensor given that the moments of inertia are $\mu_1, \mu_2, \ldots, \mu_6$ respectively about axes parallel to (1, 0, 0); (0, 1, 0); (0, 0, 1);(0, 1, 1), (1, 0, 1); (1, 1, 0).

(4) Repeat question (3) for axes parallel to (1, 0, 0); (0, 1, 0); (0, 0, 1); (−1, 1, 1); (1, −1, 1); (1, 1, −1).

(5) Repeat question (3) for axes parallel to (0, 1, 1); (1, 0, 1); (1, 1, 0); (−1, 1, 1); (1, −1, 1); (1, 1, −1).

(6) Show that it is, in general, possible to calculate the components of the inertia tensor of a body from a knowledge of its angular velocity and angular momentum about three axes through the origin of coordinates.

(7) A certain rigid body has mass M and principal moments of inertia I_1, I_2, I_3 at its centre of mass, with $I_1 > I_2 > I_3$. Choose x_1-, x_2-, x_3-axes to lie along

the respective principal directions. Write down in matrix form the inertia tensor at an arbitrary point (a_1, a_2, a_3).

Consider the quadric surface:

$$\frac{x_1^2}{(I_1/M)-\lambda} + \frac{x_2^2}{(I_2/M)-\lambda} + \frac{x_3^2}{(I_3/M)-\lambda} = 1 ,$$

with λ chosen so that the surface passes through the point (a_1, a_2, a_3). Show that the normal to the surface at that point is along a principal axis of the inertia tensor. What are the three principal moments of inertia at (a_1, a_2, a_3)?

5.2 PROPERTIES OF THE INERTIA TENSOR

5.2.1 The Components of the Inertia Tensor

The expression (5.4), defining the inertia tensor, when written in component form becomes:

$$I(\mathrm{O}) = \int_V \rho(\mathbf{r}) \begin{bmatrix} x_2^2 + x_3^2 & -x_1x_2 & -x_1x_3 \\ -x_1x_2 & x_3^2 + x_1^2 & -x_2x_3 \\ -x_1x_3 & -x_2x_3 & x_1^2 + x_2^2 \end{bmatrix} \mathrm{d}V . \tag{5.23}$$

The moment of inertia about the $\mathbf{e}_1$-axis is given by $\mathbf{e}_1.\mathbf{I}(\mathrm{O}).\mathbf{e}_1$.

But $$\mathbf{e}_1.\mathbf{I}(\mathrm{O}).\mathbf{e}_1 = \int_V \rho(\mathbf{r})(x_2^2 + x_3^2)\mathrm{d}V = I_{11} . \tag{5.24}$$

The other diagonal terms, I_{22} and I_{33}, are the moments of inertia about the $\mathbf{e}_2$- and $\mathbf{e}_3$-axes respectively.

The off-diagonal terms:

$$I_{ij} = -\int_V \rho(\mathbf{r})x_ix_j\mathrm{d}V, \qquad i \neq j , \tag{5.25}$$

are called **products of inertia**. They vanish when the inertia tensor is referred to principal axes (see section 5.1.1), and is therefore diagonal.

5.2.2 The Parallel Axis Theorem

Before discussing the calculation of the inertia tensor for bodies of several standard shapes, it is useful to prove a theorem which will enable us to write down the inertia tensor of a body at any point O in terms of the inertia tensor at the centroid of that body, G.

The name of the theorem, the Parallel Axis theorem, is derived from the fact that, as a corollary, it is possible to write down the moment of inertia about an axis through O in terms of the moment of inertia about a *parallel axis* through G. It is this corollary which is generally termed 'the Parallel Axis Theorem' in elementary books on mechanics.

Theorem 5.2.2 Given a rigid body of mass M and centroid G, the inertia tensor at any point O is given by:

$$\mathbf{I}(\mathrm{O}) = \mathbf{I}(\mathrm{G}) + M(\bar{\mathbf{r}}^2 \mathbf{1} - \bar{\mathbf{r}} \otimes \bar{\mathbf{r}}) \, , \tag{5.26}$$

where $\bar{\mathbf{r}}$ is the position vector of G with respect to O.

Proof.

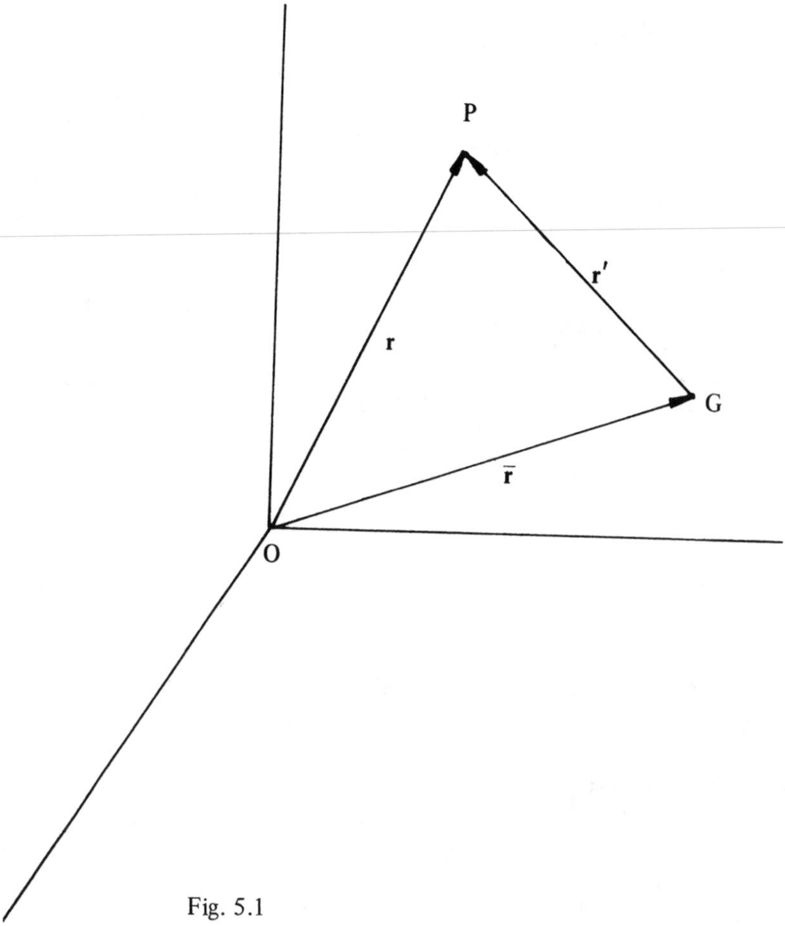

Fig. 5.1

Let P be the point with position vector $\mathbf{r}$ (Fig. 5.1), and let $\mathbf{r}' = \mathrm{GP}$. Then the inertia tensor of the body is given by:

$$\begin{aligned}
\mathbf{I}(\mathrm{O}) &= \int_V \rho(\mathbf{r})(\mathbf{r}^2\mathbf{1} - \mathbf{r}\otimes\mathbf{r})\mathrm{d}V\ ,\\
&= \int_V \rho(\mathbf{r})\ [(\bar{\mathbf{r}}+\mathbf{r}')^2\mathbf{1} - (\bar{\mathbf{r}}+\mathbf{r}')\otimes(\bar{\mathbf{r}}+\mathbf{r}')]\ \mathrm{d}V\ ,\\
&= \int_V \rho(\mathbf{r})\ [(\bar{\mathbf{r}}^2 + 2\bar{\mathbf{r}}.\mathbf{r}' + \mathbf{r}'^2)\mathbf{1} - (\bar{\mathbf{r}}\otimes\bar{\mathbf{r}} + \bar{\mathbf{r}}\otimes\mathbf{r}' + \mathbf{r}'\otimes\bar{\mathbf{r}} + \mathbf{r}'\otimes\mathbf{r}']\ \mathrm{d}V\ ,\\
&= \mathbf{I}(\mathrm{G}) + \int_V \rho(\mathbf{r})\ [(\bar{\mathbf{r}}^2 + 2\bar{\mathbf{r}}.\mathbf{r}')\mathbf{1} - (\bar{\mathbf{r}}\otimes\bar{\mathbf{r}} + \bar{\mathbf{r}}\otimes\bar{\mathbf{r}}, + \mathbf{r}'\otimes\bar{\mathbf{r}})]\ \mathrm{d}V,\\
&= \mathbf{I}(\mathrm{G}) + M(\bar{\mathbf{r}}^2\mathbf{1} - \bar{\mathbf{r}}\otimes\bar{\mathbf{r}}) + \int_V \rho(\mathbf{r})(2\bar{\mathbf{r}}.\mathbf{r}'\mathbf{1} - \bar{\mathbf{r}}\otimes\mathbf{r}' - \mathbf{r}'\otimes\bar{\mathbf{r}})\mathrm{d}V\ .
\end{aligned}$$

The integral term vanishes because $\int_V \rho(\mathbf{r})\mathbf{r}'\,\mathrm{d}V$ is a common factor and is equal to zero (see Section 5.1.2). Thus the desired result is proved.

The result (5.26) provides a simple formula for evaluating the inertia tensor at any point O in terms of the inertia tensor at G. We now deduce, as a corollary, the result which is known as the Parallel Axis Theorem in elementary text books on dynamics.

From (5.26) we have, for any unit vector $\mathbf{n}$:

$$\mathbf{I}(\mathrm{O}).\mathbf{n} = \mathbf{I}(\mathrm{G}).\mathbf{n} + M[\bar{\mathbf{r}}^2\mathbf{n} - \bar{\mathbf{r}}(\bar{\mathbf{r}}.\mathbf{n})]\ ,$$

therefore $$\mathbf{n}.\mathbf{I}(\mathrm{O}).\mathbf{n} = \mathbf{n}.\mathbf{I}(\mathrm{G}).\mathbf{n} + M[\bar{\mathbf{r}}^2 - (\bar{\mathbf{r}}.\mathbf{n})^2]\ .$$

Let $\mu(\mathrm{O}) = \mathbf{n}.\mathbf{I}(\mathrm{O}).\mathbf{n}$ be the moment of inertia about an axis through O parallel to the unit vector $\mathbf{n}$. Similarly let $\mu(\mathrm{G}) = \mathbf{n}.\mathbf{I}(\mathrm{G}).\mathbf{n}$ be the moment of inertia about a parallel axis through G.

Then $$\begin{aligned}\mu(\mathrm{O}) &= \mu(\mathrm{G}) + M[\bar{\mathbf{r}}^2 - (\bar{\mathbf{r}}.\mathbf{n})^2]\ ,\\ &= \mu(\mathrm{G}) + M(\mathbf{n}\times\bar{\mathbf{r}})^2\ . \qquad (5.27)\end{aligned}$$

We observe that $|\mathbf{n}\times\bar{\mathbf{r}}|$ is the perpendicular distance between the two parallel axes. If this distance is d, we have:

$$\mu(\mathrm{O}) = \mu(\mathrm{G}) + Md^2\ . \qquad (5.28)$$

The second term in (5.27) or (5.28) represents the moment of inertia of a single particle of mass M, situated at G, about an axis through O parallel to $\mathbf{n}$ (see (5.43)).

5.2.3 The Perpendicular Axes Theorem for Laminae

The Perpendicular Axes theorem, although its name is similar to that of the theorem discussed in the previous section, is really a much more restricted result in that it applies only to laminae. It is moreover a theorem about moments of inertia rather than about inertia tensors.

Theorem 5.2.3 The moment of inertia of a uniform lamina about an axis perpendicular to its plane is equal to the sum of the moments of inertia about two perpendicular axes within the plane.

Proof.

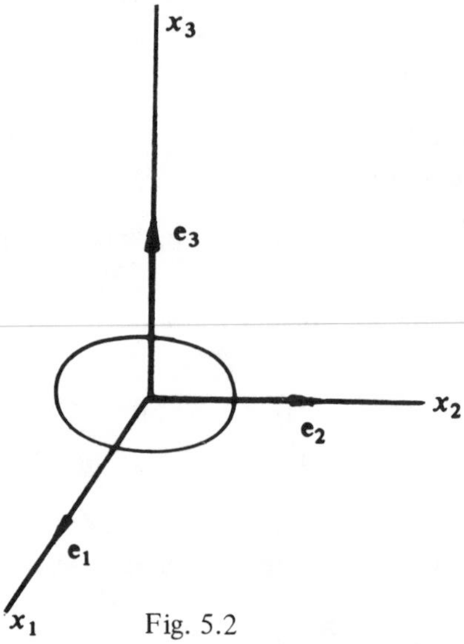

Fig. 5.2

Let axes x_1, x_2 be drawn in the plane and let the x_3-axis be drawn perpendicular to it. Remembering that $x_3 = 0$ on the lamina we have, from the definition of the inertia tensor:

$$\mathbf{I}(O) = \int_A \sigma \begin{bmatrix} x_2^2 & -x_1x_2 & 0 \\ -x_1x_2 & x_1^2 & 0 \\ 0 & 0 & x_1^2 + x_2^2 \end{bmatrix} dA\,, \tag{5.29}$$

where A is the area of the lamina and σ is the surface density, supposed constant.

Clearly $\quad I_{33} = I_{11} + I_{22}\,.$ (5.30)

Thus the Perpendicular Axes Theorem is proved.

Examples 5.2

(1) A given body has mass M and centroid G. The inertia tensor $\mathbf{I}(\mathrm{G})$ is given by:

$$\mathbf{I}(\mathrm{G}) = M\begin{bmatrix} 4 & -1 & -3 \\ -1 & 3 & -2 \\ -3 & -2 & 2 \end{bmatrix},$$

and G is the point (1, 0, 0). Calculate the inertia tensor at the origin.

We have, from the Parallel Axis theorem:

$$\mathbf{I}(\mathrm{O}) = \mathbf{I}(\mathrm{G}) + M(\bar{\mathbf{r}}^2\mathbf{1} - \bar{\mathbf{r}} \otimes \bar{\mathbf{r}}) ,$$

where $\bar{\mathbf{r}} = (1, 0, 0) = \mathbf{e}_1$.

Hence $$\bar{\mathbf{r}}^2\mathbf{1} - \bar{\mathbf{r}} \otimes \bar{\mathbf{r}} = \mathbf{1} - \mathbf{e}_1 \otimes \mathbf{e}_1 = \begin{bmatrix} 0 & 0 & 0 \\ 0 & 1 & 0 \\ 0 & 0 & 1 \end{bmatrix}.$$

Thus $$\mathbf{I}(\mathrm{O}) = M\begin{bmatrix} 4 & -1 & -3 \\ -1 & 4 & -2 \\ -3 & -2 & 3 \end{bmatrix}.$$

(2) The moment of inertia of a uniform circular disc of mass M and radius a about its axis is $\frac{1}{2}Ma^2$. Write down the moment of inertia about a diameter.

From the symmetry properties of the disc the moment of inertia is the same about all diameters. In particular it is the same about two diameters at right angles. Hence from the Perpendicular Axes theorem this moment of inertia is $\frac{1}{4}Ma^2$.

Problems 5.2

(1) The origin of a set of Cartesian axes is chosen to coincide with the centroid of a body of mass M. The inertia tensor of the body at the centroid is:

$$\mathbf{I}(\mathrm{O}) = \mathbf{I}(\mathrm{G}) = M\begin{bmatrix} 3 & -1 & 0 \\ -1 & 1 & 0 \\ 0 & 0 & 2 \end{bmatrix}.$$

Calculate the inertia tensor at the point (1, 1, 3).

(2) The moment of inertia of an elliptical lamina of mass M and bounded by the ellipse

$$(x_1/a)^2 + (x_2/b)^2 = 1, \qquad x_3 = 0$$

about the x_1-axis is $\frac{1}{4}Mb^2$ and about the x_2-axis is $\frac{1}{4}Ma^2$. Write down the moment of inertia of the lamina about the x_3-axis.

(3) The inertia tensor of a body of mass M at the origin is given by:

$$\mathrm{I}(\mathrm{O}) = M \begin{bmatrix} 2 & -1 & -1 \\ -1 & 2 & -2 \\ -1 & -2 & 4 \end{bmatrix}.$$

If the centroid is situated at the point $(1, 1, 0)$ calculate the inertia tensor at the point $(1, 1, 1)$.

5.3 CALCULATION OF THE INERTIA TENSORS OF RECTANGULAR BODIES

5.3.1 Introduction

The formal definition of the inertia tensor, as given in (5.4), will now be used to calculate the tensor for a number of bodies of uniform density and standard geometrical shape. The theorems of parallel and perpendicular axes will be used wherever possible to reduce the work involved.

The bodies which we shall be discussing fall into groups or families. For example the rectangular lamina may be regarded as a special case of the rectangular block, and the straight rod may, in turn, be regarded as a special case of the rectangular lamina. In this sense the block, lamina, and rod form a family. Of course some bodies may belong to more than one family. Thus sphere, disc (= circular lamina), and rod form a family, and right circular cylinder, disc, and rod form another family.

Now there are two general methods of approaching the problem of calculating the inertia tensor of a solid body. The first is to calculate it directly from the formula (5.4) using triple integration. The inertia tensors of the two- and one-dimensional bodies in the same family may then in certain cases be found by setting the appropriate parameter equal to zero. In favour of this method is the fact that the integration needs to be performed only once. On the other hand, except in the case of the rectangular block, the limits are apt to be such that the integration is unnecessarily complicated, at least when Cartesian co-ordinates are employed. This method will be termed the **analytical method**.

The alternative method is to calculate first of all the inertia tensor of the one-dimensional body (rod) and from it build up the required two-dimensional body (rectangular lamina, triangle, disc etc.) by a further integration. A third integration is then required to build up the inertia tensor of the three-dimensional body. This method will be termed the **synthetic method**.

In general the synthetic method is to be preferred because the integrations are all single integrations. The rectangular block has been treated by both methods for purposes of comparison.

5.3.2 The Inertia Tensor of a Uniform Rectangular Block

Consider a uniform solid rectangular block of density ρ and mass M, defined by $|x_1| \leqslant a$, $|x_2| \leqslant b$, $|x_3| \leqslant c$, (Fig. 5.3).

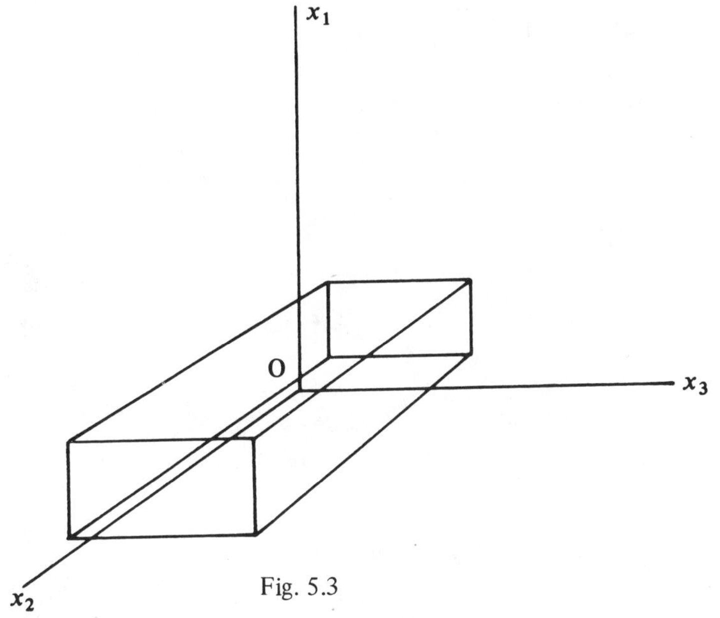

Fig. 5.3

We require to find the inertia tensor at the centroid, which we shall take as the origin of coordinates, with axes parallel to the edges of the block. We have, from (5.23):

$$I_{11} = \int_V \rho(x_2^2 + x_3^2)\mathrm{d}V \quad , \tag{5.31}$$

$$= 2a\rho \int_{-b}^{b} \int_{-c}^{c} (x_2^2 + x_3^2)\mathrm{d}x_2\,\mathrm{d}x_3 \quad ,$$

$$= 8a\rho bc(b^2 + c^2)/3 \; .$$

But $M = 8\rho abc$, therefore

$$I_{11} = \tfrac{1}{3}M(b^2 + c^2) \; . \tag{5.32}$$

The remaining terms on the leading diagonal of $\mathbf{I}(\mathrm{O})$ are found in the same way.

Thus
$$I_{22} = \tfrac{1}{3}M(c^2 + a^2) \; , \tag{5.33}$$

$$I_{33} = \tfrac{1}{3}M(a^2 + b^2) \; . \tag{5.34}$$

The off-diagonal terms, which are the products of inertia, are found as follows:

$$I_{23} = -\int_V \rho x_2 x_3 \mathrm{d}V \ , \tag{5.35}$$

$$= -2a\rho \int_{-b}^{b} \int_{-c}^{c} x_2 x_3 \mathrm{d}x_2 \mathrm{d}x_3 = 0 \ . \tag{5.36}$$

The other off-diagonal terms are also zero, and we obtain finally:

$$\mathbf{I}(\mathrm{O}) = \tfrac{1}{3}M \begin{bmatrix} b^2 + c^2 & 0 & 0 \\ 0 & c^2 + a^2 & 0 \\ 0 & 0 & a^2 + b^2 \end{bmatrix} , \tag{5.37}$$

or $$\mathbf{I}(\mathrm{O}) = \tfrac{1}{3}M[a^2(\mathbf{1} - \mathbf{e}_1 \otimes \mathbf{e}_1) + b^2(\mathbf{1} - \mathbf{e}_2 \otimes \mathbf{e}_2) + c^2(\mathbf{1} - \mathbf{e}_3 \otimes \mathbf{e}_3)] \ . \tag{5.38}$$

Equations (5.37) and (5.38) give the inertia tensor of a rectangular block, calculated at the centroid. The fact that the products of inertia are zero and the inertia tensor is therefore diagonal indicates that we chose at the outset axes coincident with the principal axes of inertia (see section 5.1.1). The moments of inertia I_{11}, I_{22}, I_{33} are thus the principal moments of inertia.

The inertia tensors of a uniform rectangular lamina and of a uniform rod may now be deduced from (5.38) as special cases. In the limit as $c \to 0$ we obtain the inertia tensor of a rectangular lamina in the (x_1, x_2)–plane.

$$\mathbf{I}(\mathrm{O}) = \tfrac{1}{3}M[a^2(\mathbf{1} - \mathbf{e}_1 \otimes \mathbf{e}_1) + b^2(\mathbf{1} - \mathbf{e}_2 \otimes \mathbf{e}_2)] \ . \tag{5.39}$$

When written out in full this tensor becomes:

$$\mathbf{I}(\mathrm{O}) = \tfrac{1}{3}M \begin{bmatrix} b^2 & 0 & 0 \\ 0 & a^2 & 0 \\ 0 & 0 & (a^2 + b^2) \end{bmatrix} , \tag{5.40}$$

from which it may be observed at once that the Perpendicular Axes theorem holds.

If we now allow $b \to 0$ in (5.39) we obtain for the inertia tensor of a uniform rod of length $2a$, referred to its centroid:

$$\mathbf{I}(\mathrm{O}) = \tfrac{1}{3}Ma^2(\mathbf{1} - \mathbf{e}_1 \otimes \mathbf{e}_1) \ . \tag{5.41}$$

If, ultimately, we allow $a \to 0$ while M remains finite we conclude that the null matrix represents the inertia tensor of a particle of mass M situated at the origin. At first sight this result appears trivial. It is, however, possible to apply the parallel axis theorem to determine the inertia tensor for a particle situated at P whose position vector is **r**.

Thus $$\mathbf{I}(\mathrm{P}) = M(\mathbf{r}^2\mathbf{1} - \mathbf{r} \otimes \mathbf{r}) \;. \tag{5.42}$$

The expression (5.42) gives the inertia tensor of a single particle of mass M situated at a point P(**r**). Alternatively (5.42) follows at once from (5.4) when the body reduces to a particle. From it, it is possible in principle to build up the inertia tensor for any configuration of particles and, hence, for any solid body. It also follows from (5,42) that the moment of inertia of a single particle at P about an axis through O parallel to the unit vector **n** is given by

$$\mu = \mathbf{n}.\mathbf{I}(\mathrm{P}).\mathbf{n} = M(\mathbf{r} \times \mathbf{n})^2 = Md^2 \;, \tag{5.43}$$

where d is the distance from P to the axis.

5.3.3 Equimomental Systems

We have seen (5.41) that the inertia tensor of a uniform rod of length $2a$ and mass M, with respect to its centroid, is:

$$\mathbf{I}(\mathrm{G}) = \tfrac{1}{3}Ma^2(\mathbf{1} - \mathbf{e}_1 \otimes \mathbf{e}_1) \;.$$

We have also seen (5.42) that the inertia tensor at the origin of a single particle of equal mass and situated at $(a, 0, 0)$ is:

$$\mathbf{I}(\mathrm{O}) = Ma^2(\mathbf{1} - \mathbf{e}_1 \otimes \mathbf{e}_1) \;. \tag{5.44}$$

Consider now a system of two equal particles each of mass $\frac{1}{6}M$, situated at $(\pm\, a, 0, 0)$. The inertia tensor of this system is given by:

$$\mathbf{I}(\mathrm{O}) = \mathbf{I}(\mathrm{G}) = \tfrac{1}{3}Ma^2(\mathbf{1} - \mathbf{e}_1 \otimes \mathbf{e}_1) \;. \tag{5.45}$$

Thus, compared with the rod, the two-particle system has the same centroid and the same inertia tensor calculated at the centroid. It does not, however, have the same mass. Finally, therefore, we introduce a mass $\frac{2}{3}M$ at O to construct a system which has the same mass, centroid and inertia tensor calculated at the centroid, as the uniform rod. Such systems are said to be **equimomental**. It is a direct consequence of the Parallel Axis theorem that equimomental systems have the same inertia tensor everywhere.

The idea of the equimomental system can, in certain circumstances, be very useful indeed. We shall see later (section 5.4.3) how, by replacing a triangular lamina with an equimomental system, it is possible to arrive at a symmetrical expression for the inertia tensor which would otherwise have been more difficult to obtain.

5.3.4 The Inertia Tensor of a Uniform Rectangular Block – An Alternative Approach

We have, in the previous section, found the inertia tensor of a rectangular block by direct integration, and deduced from it the inertia tensors for a rectangular lamina, a uniform rod and a single particle, by successive suppression of the parameters.

This method has been termed the analytical method (section 5.3.1). It is instructive to derive these results afresh by the alternative, synthetic method, which makes considerable use of the Parallel Axis theorem.

We start from the result that the inertia tensor of a single particle at the origin is the null tensor with respect to the origin (section 5.3.2).

Suppose now we wish to build up a rod of length $2a$ situated along the x_1-axis from $-a$ to $+a$, with line density λ (Fig. 5.4).

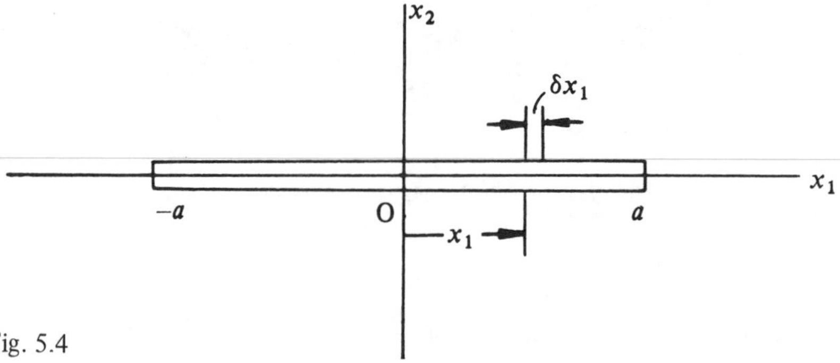

Fig. 5.4

Consider an elementary section of rod situated at $(x_1, 0, 0)$ and of length δx_1. The inertia tensor of this elementary section is, by (5.42):

$$\delta \mathbf{I}(\mathrm{O}) = \lambda(x_1^2 \mathbf{1} - x_1^2 \mathbf{e}_1 \otimes \mathbf{e}_1)\delta x_1 \ .$$

Hence, for the complete rod:

$$\mathbf{I}(\mathrm{O}) = \int_{-a}^{a} \lambda(\mathbf{1} - \mathbf{e}_1 \otimes \mathbf{e}_1)x_1^2 \mathrm{d}x_1 \ ,$$

$$= \lambda(\mathbf{1} - \mathbf{e}_1 \otimes \mathbf{e}_1)\left[\tfrac{1}{3}x_1^3\right]_{-a}^{a} \ ,$$

$$= \tfrac{2}{3}\lambda a^3(\mathbf{1} - \mathbf{e}_1 \otimes \mathbf{e}_1) \ .$$

But $M = 2\lambda a$. Hence

$$\mathbf{I}(\mathrm{O}) = \tfrac{1}{3}Ma^2(\mathbf{1} - \mathbf{e}_1 \otimes \mathbf{e}_1) \ . \qquad (5.46)$$

Thus we have recovered the result, already expressed in (5.41), for the inertia tensor of a uniform rod of length $2a$, lying along the x_1-axis, with its centroid at the origin.

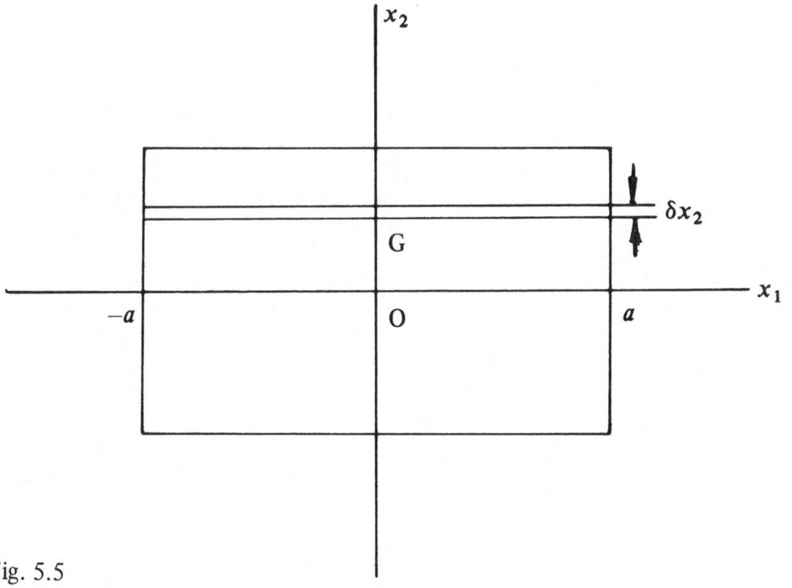

Fig. 5.5

We will now build up a rectangular lamina of area $2a \times 2b$ (Fig. 5.5) and surface density σ, from rods of length $2a$ lying parallel to the x_1-axis.

Consider such an elementary rod, of width δx_2, distant x_2 from the x_1-axis. The inertia tensor of this rod *at its centroid* is given by (5.41).

Thus $\delta \mathbf{I}(G) = \frac{2}{3} a\sigma \delta x_2 a^2 (\mathbf{1} - \mathbf{e}_1 \otimes \mathbf{e}_1)$. We now appeal to the Parallel Axis theorem, given in (5.26), to find $\delta \mathbf{I}(O)$.

Thus $\delta \mathbf{I}(O) = \delta \mathbf{I}(G) + 2a\sigma \delta x_2 x_2^2 (\mathbf{1} - \mathbf{e}_2 \otimes \mathbf{e}_2)$. Hence, for the complete lamina,

$$\mathbf{I}(O) = 2a\sigma \int_{-b}^{b} \left[\tfrac{1}{3}a^2(\mathbf{1} - \mathbf{e}_1 \otimes \mathbf{e}_1) + x_2^2(\mathbf{1} - \mathbf{e}_2 \otimes \mathbf{e}_2)\right] dx_2$$

$$= 2a\sigma \left\{ \tfrac{1}{3}a^2(\mathbf{1} - \mathbf{e}_1 \otimes \mathbf{e}_1)\Big[x_2\Big]_{-b}^{b} + (\mathbf{1} - \mathbf{e}_2 \otimes \mathbf{e}_2)\Big[\tfrac{1}{3}x_2^3\Big]_{-b}^{b} \right\},$$

$$= \tfrac{4}{3}ab\sigma\left[a^2(\mathbf{1} - \mathbf{e}_1 \otimes \mathbf{e}_1) + b^2(\mathbf{1} - \mathbf{e}_2 \otimes \mathbf{e}_2)\right].$$

But $M = 4ab\sigma$.

Hence $$\mathbf{I}(\mathrm{O}) = \tfrac{1}{3}M\left[a^2(\mathbf{1} - \mathbf{e}_1 \otimes \mathbf{e}_1) + b^2(\mathbf{1} - \mathbf{e}_2 \otimes \mathbf{e}_2)\right]. \tag{5.47}$$

We have now recovered the result, expressed in (5.39), for the inertia tensor of a rectangular lamina of dimensions $2a \times 2b$.

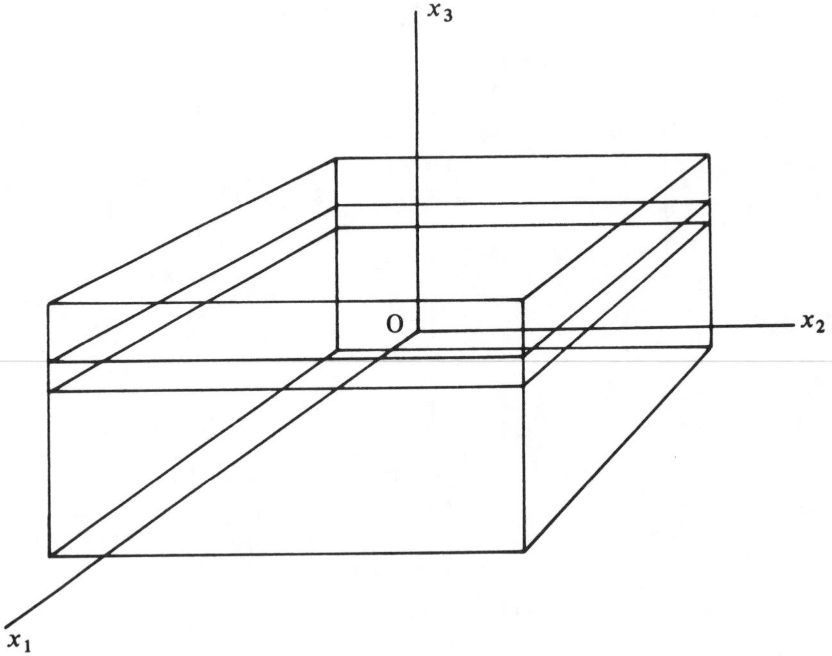

Fig. 5.6

The final stage in the synthetic method is to build up a rectangular block of volume $2a \times 2b \times 2c$ (Fig. 5.6) and density ρ from laminae of area $2a \times 2b$ lying perpendicular to the x_3-axis.

Consider such an elementary lamina, of thickness δx_3, distant x_3 from the (x_1,x_2)-coordinate plane. The inertia tensor of this lamina *at its centroid* is given by (5.39).

Thus $$\delta\mathbf{I}(\mathrm{G}) = \tfrac{4}{3}ab\rho\delta x\{a^2(\mathbf{1} - \mathbf{e}_1 \otimes \mathbf{e}_1) + b^2(\mathbf{1} - \mathbf{e}_2 \otimes \mathbf{e}_2)\} \ .$$

From the Parallel Axis theorem we find the inertia tensor of the lamina, evaluated at the origin:

$$\delta\mathbf{I}(\mathrm{O}) = \delta\mathbf{I}(\mathrm{G}) + 4ab\rho\delta x_3 x_3^2(\mathbf{1} - \mathbf{e}_3 \otimes \mathbf{e}_3) \ .$$

Hence, for the rectangular block,

$$\mathbf{I}(\mathrm{O}) = 4ab\rho\int_{-c}^{c}\left[\tfrac{1}{3}a^2(\mathbf{1}-\mathbf{e}_1\otimes\mathbf{e}_1)+\tfrac{1}{3}b^2(\mathbf{1}-\mathbf{e}_2\otimes\mathbf{e}_2)\right.$$
$$\left. + x_3^2(\mathbf{1}-\mathbf{e}_3\otimes\mathbf{e}_3)\right]\mathrm{d}x_3 ,$$
$$= 4ab\rho\left\{\tfrac{1}{3}a^2(\mathbf{1}-\mathbf{e}_1\otimes\mathbf{e}_1)\Big[x_3\Big]_{-c}^{c} + \tfrac{1}{3}b^2(\mathbf{1}-\mathbf{e}_2\otimes\mathbf{e}_2)\Big[x_3\Big]_{-c}^{c} + (\mathbf{1}-\mathbf{e}_3\otimes\mathbf{e}_3)\Big[\tfrac{1}{3}x_3^3\Big]_{-c}^{c}\right\}.$$

But $M = 8abc\rho$.

Hence $$\mathbf{I}(\mathrm{O}) = \tfrac{1}{3}M\,[a^2(\mathbf{1}-\mathbf{e}_1\otimes\mathbf{e}_1)+b^2(\mathbf{1}-\mathbf{e}_2\otimes\mathbf{e}_2) + c^2(\mathbf{1}-\mathbf{e}_3\otimes\mathbf{e}_3)] . \tag{5.48}$$

We have at last obtained, through a process of repeated integration and application of the Parallel Axis theorem, a result which we obtained by one triple integration at the beginning of section 5.3.2. Although in the case of a rectangular block the synthetic method is much longer than the analytical method, the reverse is true for bodies with curved boundaries. To determine the inertia tensor of a solid sphere, for example, by direct integration is not an easy matter because of the limits involved in the volume integral which arises. Further, having found the inertia tensor of a solid sphere there is no obvious way of reducing it to a disc.

Examples 5.3

(1) The inertia tensor of a cube side $2a$ and centre the origin is $\tfrac{2}{3}Ma^2\mathbf{1}$. This follows from (5.37) with $c = b = a$, and is true independently of the orientation of the edges.

Suppose the edges of the cube are parallel to the coordinate axes (Fig. 5.7). The inertia tensor calculated at the vertex $A(a,a,a)$ is given by:

$$\mathbf{I}(\mathrm{A}) = \tfrac{2}{3}Ma^2\,\mathbf{1} + Ma^2(3\mathbf{1}-\mathbf{r}\otimes\mathbf{r}) , \tag{5.49}$$

where $\mathbf{r} = \mathbf{e}_1 + \mathbf{e}_2 + \mathbf{e}_3$,

i.e. $$\mathbf{I}(\mathrm{A}) = Ma^2\begin{bmatrix} 8/3 & -1 & -1 \\ -1 & 8/3 & -1 \\ -1 & -1 & 8/3 \end{bmatrix} . \tag{5.50}$$

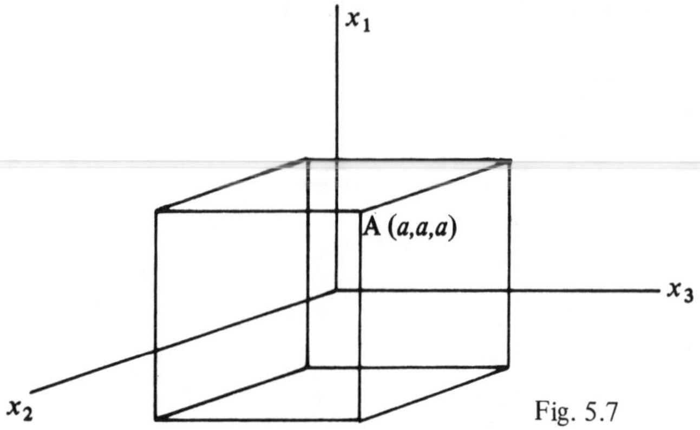

Fig. 5.7

The moment of inertia of the cube about an edge through A parallel to $\mathbf{e}_3$ is:

$$\mu = \mathbf{e}_3.\mathbf{I}(\mathrm{A}).\mathbf{e}_3 \ ,$$

or

$$\mu = 8Ma^3/3 \ . \tag{5.51}$$

(2) Consider equation (5.38) which gives the inertia tensor of a rectangular block. The block is clearly equimomental with a system of six particles each of mass $\frac{1}{6}M$, situated at the centres of the faces.

Problems 5.3

(1) Find the moment of inertia of the cube referred to in Examples 5.3 No. 1 about a line through a corner, lying in a face, and inclined at an angle θ to an edge.

(2) Find the inertia tensor at the centroid of a uniform rectangular block of mass M and sides $2a$, $2a$, $4a$, with respect to principal axes through the centroid. Hence find the inertia tensor at a corner with respect to axes along an edge of length $4a$, a diagonal of a square face, and a mutual perpendicular. Find also the principal axes at the corner.

(3) Find a system of three rods and a particle which is equimomental with the rectangular block whose inertia tensor is given by (5.38). (Note: the 'particle' is allowed to have negative mass).

(4) Find the inertia tensor at a corner of a rectangular lamina of dimensions $2a \times 2b$. Show that two of the principal axes at the corner are in the plane of the lamina and inclined to the edges at an angle θ where:

$$3ab \tan\theta = 2(b^2 - a^2) \pm \sqrt{[4(a^4 + b^4) + a^2b^2]} \ .$$

(5) A massless cube of side $2a$ has masses m, $2m$, $3m$, $4m$, $5m$, $6m$ attached to the centres of the six faces so that the sum of the masses on any pair of opposite faces is $7m$. Calculate the components of the inertia tensor of the system at the

geometric centre of the cube with respect to axes parallel to the edges. Find the position of the centre of mass and show that the principal moments of inertia there are $12\frac{1}{3}ma^2$, $12\frac{1}{3}ma^2$ and $14ma^2$.

(Liverpool University)

(6) A homogeneous rectangular block of total mass M has sides of length $2a$, $2a$ and $4a$. Show that if the x_3-axis be taken upwards along the vertical edge of length $4a$, and the x_2-axis along the diagonal of the lower square face, then the equation of the momental ellipsoid at the corner is:

$$Ma^2(23x_1^2 + 17x_2^2 + 8x_3^2 - 12\sqrt{2}x_2x_3) = \text{const} .$$

Hence obtain the directions of the principal axes at the corner.

(Liverpool University)

5.4 THE INERTIA TENSORS OF NON-RECTANGULAR BODIES

5.4.1 The Inertia Tensor of a Uniform Parallelogram Lamina

Consider a uniform parallelogram lamina of surface density σ, mass M and sides $2a$ and $2b$ inclined at an acute angle θ (Fig. 5.8).

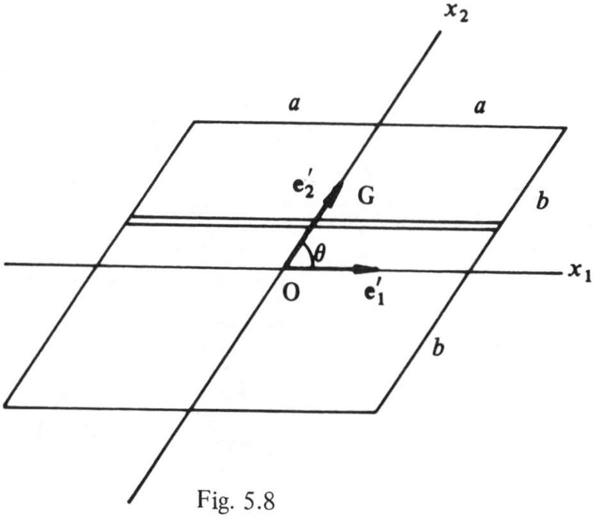

Fig. 5.8

It is convenient to choose axes x_1, x_2 inclined at an angle θ and parallel to the sides of the parallelogram. Let unit vectors $\mathbf{e}_1'$, $\mathbf{e}_2'$ lie along the axes.

Consider an elementary strip of length $2a$, width $\sin\theta\,\delta x_2$, and centroid $G(0, x_2)$. The inertia tensor of this rod, calculated at G, is, by (5.41):

$$\delta\mathbf{I}(G) = \tfrac{2}{3}a^3\sigma \sin\theta\, \delta x_2(\mathbf{1} - \mathbf{e}_1' \otimes \mathbf{e}_1') . \tag{5.52}$$

Hence, by use of the Parallel Axis theorem,

$$\delta\mathbf{I}(\mathrm{O}) = \delta\mathbf{I}(\mathrm{G}) + 2a\sigma \sin\theta\, x_2^2 \delta x_2 (\mathbf{1} - \mathbf{e}_2' \otimes \mathbf{e}_2') .$$

For the lamina we have, therefore,

$$\mathbf{I}(\mathrm{O}) = \int_{-b}^{b} 2a\sigma \sin\theta \left[\tfrac{1}{3}a^2(\mathbf{1} - \mathbf{e}_1' \otimes \mathbf{e}_1') + x_2^2(\mathbf{1} - \mathbf{e}_2' \otimes \mathbf{e}_2')\right] \mathrm{d}x_2 ,$$

$$= (4/3)\sigma ab \sin\theta \left[a^2(\mathbf{1} - \mathbf{e}_1' \otimes \mathbf{e}_1') + b^2(\mathbf{1} - \mathbf{e}_2' \otimes \mathbf{e}_2')\right] .$$

But $M = 4\,\sigma ab \sin\theta$.

Therefore $$\mathbf{I}(\mathrm{O}) = \tfrac{1}{3}M[a^2(\mathbf{1} - \mathbf{e}_1' \otimes \mathbf{e}_1') + b^2(\mathbf{1} - \mathbf{e}_2' \otimes \mathbf{e}_2')] . \quad (5.53)$$

The expression (5.53) gives the inertia tensor of a parallelogram lamina, calculated at its centroid. It is formally the same as expression (5.39) which gives the inertia tensor of a rectangular lamina; and, of course, when $\mathbf{e}_1'$ and $\mathbf{e}_2'$ are perpendicular the expressions are identical. It is very important to remember, however, that the unit tensor in (5.53) must be referred to rectangular axes. We may *not*, therefore, replace (5.53) by expression (5.40).

5.4.2 The Inertia Tensor of a Uniform Solid Parallelepiped

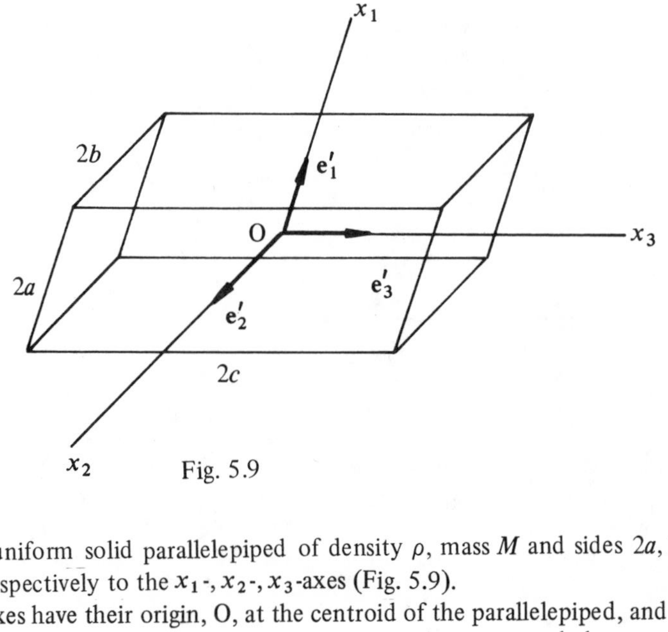

Fig. 5.9

Consider a uniform solid parallelepiped of density ρ, mass M and sides $2a$, $2b$, $2c$ parallel respectively to the x_1-, x_2-, x_3-axes (Fig. 5.9).

Let the axes have their origin, O, at the centroid of the parallelepiped, and let unit vectors $\mathbf{e}_1', \mathbf{e}_2', \mathbf{e}_3'$ lie along the axes. Let ϕ be the acute angle between $\mathbf{e}_1'$

and $\mathbf{e}_2'$ and let θ be the acute angle between $\mathbf{e}_3'$ and the plane of $\mathbf{e}_1'$ and $\mathbf{e}_2'$.

Consider an elementary parallelogram lamina parallel to the plane of $\mathbf{e}_1'$ and $\mathbf{e}_2'$, with sides $2a$, $2b$, thickness $\delta x_3 \sin\theta$ and centroid at $G(0, 0, x_3)$. The inertia tensor of this lamina, calculated at G is, by (5.53):

$$\delta \mathbf{I}(G) = (4/3)ab\rho \sin\theta \sin\phi\, \delta x_3 [a^2(\mathbf{1} - \mathbf{e}_1' \otimes \mathbf{e}_1') + b^2(\mathbf{1} - \mathbf{e}_2' \otimes \mathbf{e}_2')] \quad . \tag{5.54}$$

Hence, by use of the Parallel Axis theorem:

$$\delta \mathbf{I}(O) = \delta \mathbf{I}(G) + 4ab\rho \sin\theta \sin\phi\, \delta x_3 x_3^2 (\mathbf{1} - \mathbf{e}_3' \otimes \mathbf{e}_3') \quad .$$

For the parallelepiped we have, therefore:

$$\mathbf{I}(O) = \int_{-c}^{c} 4ab\rho \sin\theta \sin\phi \Big[\tfrac{1}{3}a^2(\mathbf{1} - \mathbf{e}_1' \otimes \mathbf{e}_1') + \tfrac{1}{3}b^2(\mathbf{1} - \mathbf{e}_2' \otimes \mathbf{e}_2') + x_3^2(\mathbf{1} - \mathbf{e}_3' \otimes \mathbf{e}_3') \Big] \mathrm{d}x_3 \quad .$$

But $M = 8abc\rho \sin\theta \sin\phi$.

Hence
$$\mathbf{I}(O) = \tfrac{1}{3}M\Big[a^2(\mathbf{1} - \mathbf{e}_1' \otimes \mathbf{e}_1') + b^2(\mathbf{1} - \mathbf{e}_2' \otimes \mathbf{e}_2') + c^2(\mathbf{1} - \mathbf{e}_3' \otimes \mathbf{e}_3')\Big] . \tag{5.55}$$

Again, we have an expression which is formally the same as (5.38) the inertia tensor for a rectangular block. It differs fundamentally from (5.38), however, in that it is not diagonal. The principal axes of a parallelepiped do not therefore coincide with any of the axes we have chosen, unless at least one face is rectangular.

5.4.3 The Inertia Tensor of a Uniform Triangular Lamina

Consider a uniform triangular lamina, of surface density σ, mass M, base $2a$ and length of median from base to opposite vertex b (Fig. 5.10). Choose axes x_1 parallel to the base, and x_2 along the median as shown. Let θ be the acute angle between the axes.

Consider an elementary strip parallel to the x_1-axis, centroid $G(0, x_2)$, width $\sin\theta\, \delta x_2$, and length $2ax_2/b$. The inertia tensor of this strip, calculated at G, is, by (5.41):

$$\delta \mathbf{I}(G) = \tfrac{2}{3}(ax_2/b)^3 \sigma \sin\theta\, \delta x_2 (\mathbf{1} - \mathbf{e}_1' \otimes \mathbf{e}_1') \quad . \tag{5.56}$$

Hence, by use of the Parallel Axis theorem:

$$\delta \mathbf{I}(O) = \delta \mathbf{I}(G) + 2a(x_2^3/b)\sigma \sin\theta\, \delta x_2 (\mathbf{1} - \mathbf{e}_2' \otimes \mathbf{e}_2') \quad .$$

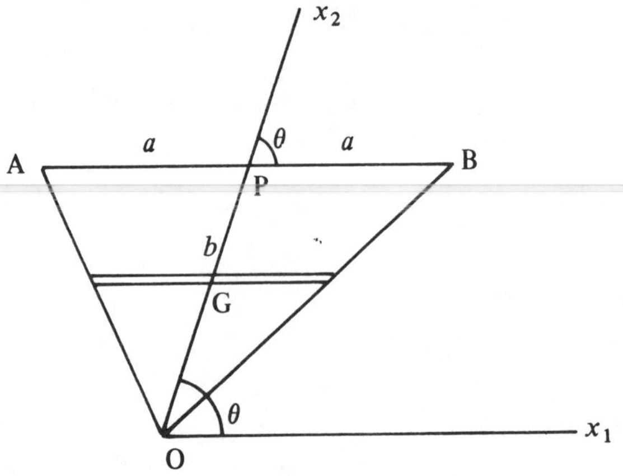

Fig. 5.10

For the triangular lamina we have, therefore,

$$\mathbf{I}(\mathrm{O}) = \int_0^b \frac{2\sigma \sin\theta\, a}{b}\left[\tfrac{1}{3}(a/b)^2 x_2^3(\mathbf{1} - \mathbf{e}_1' \otimes \mathbf{e}_1') + x_2^3(\mathbf{1} - \mathbf{e}_2' \otimes \mathbf{e}_2')\right]\mathrm{d}x_2 ,$$

$$= 2\sigma \sin\theta (a/b)[\tfrac{1}{12}(a/b)^2 b^4(\mathbf{1} - \mathbf{e}_1' \otimes \mathbf{e}_1') + \tfrac{1}{4}b^4(\mathbf{1} - \mathbf{e}_2' \otimes \mathbf{e}_2')] .$$

But $M = ab\sigma \sin\theta$. Therefore

$$\mathbf{I}(\mathrm{O}) = \frac{M}{6}a^2(\mathbf{1} - \mathbf{e}_1' \otimes \mathbf{e}_1') + \frac{M}{2}b^2(\mathbf{1} - \mathbf{e}_2' \otimes \mathbf{e}_2') . \tag{5.57}$$

The expression (5.57) gives the inertia tensor of the triangular lamina, calculated at the vertex. We may use the Parallel Axis theorem to find the inertia tensor at the centroid of the lamina G$'(0, \frac{2}{3}b)$.

Thus $\quad \mathbf{I}(\mathrm{G}') = \mathbf{I}(\mathrm{O}) - M(\mathbf{1}\bar{\mathbf{r}}^2 - \bar{\mathbf{r}} \otimes \bar{\mathbf{r}})$, where $\bar{\mathbf{r}} = \frac{2}{3}b\mathbf{e}_2'$.

Hence
$$\mathbf{I}(\mathrm{G}') = \frac{M}{6}a^2(\mathbf{1} - \mathbf{e}_1' \otimes \mathbf{e}_1') + \frac{M}{18}b^2(\mathbf{1} - \mathbf{e}_2' \otimes \mathbf{e}_2') . \tag{5.58}$$

It is also useful to evaluate $\mathbf{I}(\mathrm{P})$, the inertia tensor at the mid-point of the base.

Thus $\quad \mathbf{I}(\mathrm{P}) = \mathbf{I}(\mathrm{G}') + M(\mathbf{1}r^2 - \mathbf{r} \otimes \mathbf{r})$, where $\mathbf{r} = \frac{1}{3}b\mathbf{e}_2'$.

Hence
$$\mathbf{I}(\mathrm{P}) = \frac{M}{6}a^2(\mathbf{1} - \mathbf{e}_1' \otimes \mathbf{e}_1') + \frac{M}{6}b^2(\mathbf{1} - \mathbf{e}_2' \otimes \mathbf{e}_2') . \tag{5.59}$$

We are now in a position to look for an equimomental system of particles (see section 5.3.3). The first term of (5.59) suggests $M/12$ at A and B. Place another mass $M/12$ at O to preserve the position of the centroid and we observe that a further mass $3M/4$ placed at G$'$ makes up the mass of the lamina, while preserving the position of the centroid. A simple calculation shows that **I**(P) for the system of particles is identical with (5.59). Hence the two systems are equimomental. The equimomental system of particles has the advantage of being symmetrical. It leads directly to a symmetrical form for **I**(G) (see Problems 5.4, No. 2).

5.4.4 The Inertia Tensor of a Uniform Solid Tetrahedron

Consider a uniform solid tetrahedron OABC of density ρ and mass M. The base of the tetrahedron is a triangle ABC with AB $= 2a$ and CP $= b$, where P is the mid-point of AB. A$\hat{\text{P}}$C $= \theta$ and G$'$ is the centroid of the base, thus G$'$P $= \frac{1}{3}b$ (Fig. 5.11). Let OG$'$ $= c$ and let G be the centroid of the tetrahedron. Then G is on OG$'$ and GG$'$ $= \frac{1}{4}c$. Let ϕ be the angle between OG$'$ and the plane of Δ ABC.

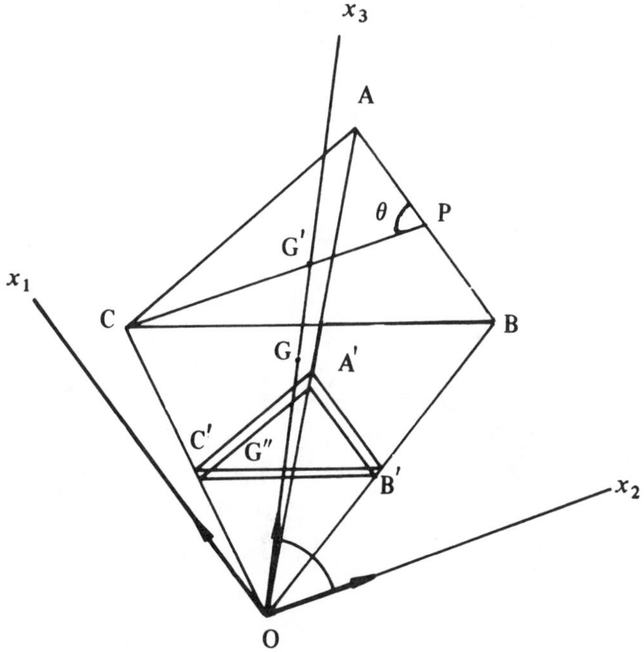

Fig. 5.11

Choose axes x_1 parallel to AB, x_2 parallel to CP and x_3 along OG$'$. Unit vectors $\mathbf{e}_1', \mathbf{e}_2', \mathbf{e}_3'$ lie along x_1, x_2, x_3.

Consider an elementary triangular lamina A$'$B$'$C$'$ whose plane is parallel to that of ABC. Let the centroid of A$'$B$'$C$'$ be G$''$ $(0, 0, x_3)$. The thickness of the lamina is $\delta x_3 \sin\phi$. The inertia tensor of this lamina, calculated at G$''$ is by (5.58):

$$\delta\mathbf{I}(\mathrm{G}'') = \left(\frac{x_3}{c}\right)^4 \left[\frac{a^2}{6}(\mathbf{1} - \mathbf{e}'_1 \otimes \mathbf{e}'_1) + \frac{b^2}{18}(\mathbf{1} - \mathbf{e}'_2 \otimes \mathbf{e}'_2)\right] \rho ab \sin\theta \sin\phi \delta x_3 \ . \tag{5.60}$$

Hence, by the Parallel Axis theorem:

$$\delta\mathbf{I}(\mathrm{O}) = \delta\mathbf{I}(\mathrm{G}'') + \rho ab \sin\theta \sin\phi \delta x_3 \frac{x_3^4}{c^2}(\mathbf{1} - \mathbf{e}'_3 \otimes \mathbf{e}'_3) \ .$$

For the complete body we have, therefore:

$$\mathbf{I}(\mathrm{O}) = \int_0^c \left\{\frac{a^2}{6}(\mathbf{1} - \mathbf{e}'_1 \otimes \mathbf{e}'_1) + \frac{b^2}{18}(\mathbf{1} - \mathbf{e}'_2 \otimes \mathbf{e}'_2) + c^2(\mathbf{1} - \mathbf{e}'_3 \otimes \mathbf{e}'_3)\right\} \rho ab \sin\theta \sin\phi \left(\frac{x_3}{c}\right)^4 \mathrm{d}x_3 \ ,$$

$$= (1/5)\rho abc \sin\theta \sin\phi \left[\frac{a^2}{6}(\mathbf{1} - \mathbf{e}'_1 \otimes \mathbf{e}'_1) + \frac{b^2}{18}(\mathbf{1} - \mathbf{e}'_2 \otimes \mathbf{e}'_2) + c^3(\mathbf{1} - \mathbf{e}'_3 \otimes \mathbf{e}'_3)\right] .$$

But $M = \frac{1}{3}\rho abc \sin\theta \sin\phi$, therefore

$$\mathbf{I}(\mathrm{O}) = \frac{3M}{5}\left[\frac{a^2}{6}(\mathbf{1} - \mathbf{e}'_1 \otimes \mathbf{e}'_1) + \frac{b^2}{18}(\mathbf{1} - \mathbf{e}'_2 \otimes \mathbf{e}'_2) + c^2(\mathbf{1} - \mathbf{e}'_3 \otimes \mathbf{e}'_3)\right] . \tag{5.61}$$

The expression (5.61) gives the inertia tensor of a tetrahedron, calculated at a vertex. Employing again the Parallel Axis theorem we may calculate the inertia tensor at the centroid $(0, 0, \frac{3}{4}c)$.

Thus $$\mathbf{I}(\mathrm{G}) = \mathbf{I}(\mathrm{O}) - \tfrac{9}{16}Mc^2(\mathbf{1} - \mathbf{e}'_3 \otimes \mathbf{e}'_3) \ .$$

$$= \frac{3M}{5}\left[\frac{a^2}{6}(\mathbf{1} - \mathbf{e}'_1 \otimes \mathbf{e}'_1) + \frac{b^2}{18}(\mathbf{1} - \mathbf{e}'_2 \otimes \mathbf{e}'_2) + \frac{c^2}{16}(\mathbf{1} - \mathbf{e}'_3 \otimes \mathbf{e}'_3)\right] . \tag{5.62}$$

Further, the inertia tensor at G′ $(0, 0, c)$, the centroid of the base, is:

$$\mathbf{I}(\mathrm{G}') = \frac{3M}{5}\left[\frac{a^2}{6}(\mathbf{1} - \mathbf{e}'_1 \otimes \mathbf{e}'_1) + \frac{b^2}{18}(\mathbf{1} - \mathbf{e}'_2 \otimes \mathbf{e}'_2) + \frac{c^2}{6}(\mathbf{1} - \mathbf{e}'_3 \otimes \mathbf{e}'_3)\right] . \tag{5.63}$$

A symmetrical form for the inertia tensor at G is given in Problems 5.4, No. 3. Finally we observe that $\mathbf{I}(\mathrm{P})$ is given by:

$$\mathbf{I}(\mathrm{P}) = \mathbf{I}(\mathrm{G}) + M(\mathbf{r}^2\mathbf{1} - \mathbf{r} \otimes \mathbf{r}) \ ,$$

where $\mathbf{r} = \frac{1}{4}c\mathbf{e}'_3 + \frac{1}{3}b\mathbf{e}'_2$. Therefore

$$\mathbf{I}(P) = M\left[\frac{a^2}{10}(\mathbf{1} - \mathbf{e}'_1 \otimes \mathbf{e}'_1) + \frac{13b^2}{90}(\mathbf{1} - \mathbf{e}'_2 \otimes \mathbf{e}'_2)\right.$$
$$\left. + \frac{c^2}{10}(\mathbf{1} - \mathbf{e}'_3 \otimes \mathbf{e}'_3) + \frac{bc}{12}(2\mathbf{e}'_2.\mathbf{e}'_3\mathbf{1} - \mathbf{e}'_2 \otimes \mathbf{e}'_3 - \mathbf{e}'_3 \otimes \mathbf{e}'_2)\right] . \tag{5.64}$$

We may use expression (5.64) to calculate the moment of inertia about an edge (see Problems 5.4, No. 4).

Example 5.4

(1) Write down the components of the inertia tensor of a uniform parallelogram lamina with respect to rectangular Cartesian axes.

In Fig. 5.8 let $\mathbf{e}_1, \mathbf{e}_2, \mathbf{e}_3$ be an orthogonal triad of unit vectors as usual (Fig, 5.12).

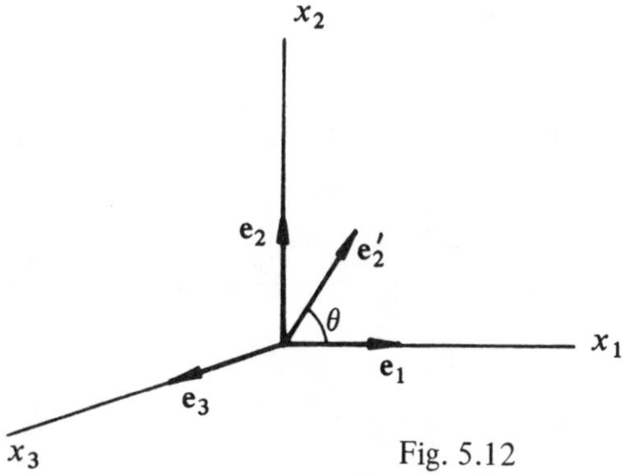

Fig. 5.12

Then, from (5.53), the inertia tensor of the lamina is:

$$\mathbf{I}(O) = \tfrac{1}{3}M[a^2(\mathbf{1} - \mathbf{e}_1 \otimes \mathbf{e}_1) + b^2(\mathbf{1} - \mathbf{e}'_2 \otimes \mathbf{e}'_2))] ,$$

where $\mathbf{e}'_2 = \mathbf{e}_1 \cos\theta + \mathbf{e}_2 \sin\theta$.

Therefore

$$\mathbf{I}(O) = \tfrac{1}{3}M[a^2(\mathbf{e}_2 \otimes \mathbf{e}_2) + a^2(\mathbf{e}_3 \otimes \mathbf{e}_3)$$
$$+ b^2(\mathbf{e}_1 \otimes \mathbf{e}_1 \sin^2\theta + \mathbf{e}_2 \otimes \mathbf{e}_2 \cos^2\theta + \mathbf{e}_3 \otimes \mathbf{e}_3)$$
$$- b^2(\mathbf{e}_1 \otimes \mathbf{e}_2 + \mathbf{e}_2 \otimes \mathbf{e}_1)\sin\theta\cos\theta] ,$$

$$= \tfrac{1}{3}M\begin{bmatrix} b^2\sin^2\theta & -b^2\sin\theta\cos\theta & 0 \\ -b^2\sin\theta\cos\theta & a^2 + b^2\cos^2\theta & 0 \\ 0 & 0 & a^2 + b^2 \end{bmatrix}.$$

Problems 5.4

(1) A uniform rhombus shaped lamina has sides inclined at an angle of 45°. Find the principal axes at the centroid.

(2) A uniform triangular lamina $A_1 A_2 A_3$ has mass M and centroid G which is distant h_i from A_i. If $\mathbf{e}_i'$ is a unit vector along GA_i show that $\mathbf{I}(G)$ is given by:

$$\mathbf{I}(G) = \frac{M}{12}\left[h_1^2(1 - \mathbf{e}_1' \otimes \mathbf{e}_1') + h_2^2(1 - \mathbf{e}_2' \otimes \mathbf{e}_2') + h_3^2(1 - \mathbf{e}_3' \otimes \mathbf{e}_3')\right].$$

[Hint: replace the lamina by an equimomental system of particles].

(3) A uniform solid tetrahedron $A_1 A_2 A_3 A_4$ has mass M and centroid G, which is distant h_i from A_i. Show that the tetrahedron is equimomental with a system of 4 particles each of mass $M/20$ situated at the vertices and a particle of mass $4M/5$ at G. Deduce that $\mathbf{I}(G)$ is given by:

$$\mathbf{I}(G) = \frac{M}{20}\left[h_1^2(1 - \mathbf{e}_1' \otimes \mathbf{e}_1') + h_2^2(1 - \mathbf{e}_2' \otimes \mathbf{e}_2') + h_3^2(1 - \mathbf{e}_3' \otimes \mathbf{e}_3') + h_4^2(1 - \mathbf{e}_4' \otimes \mathbf{e}_4')\right],$$

where $\mathbf{e}_i'$ is a unit vector along GA_i.

(4) Use (5.64) to calculate the moment of inertia of a uniform equilateral tetrahedron of side $2a$ about an edge. Calculate also the moment of inertia of the corresponding equimomental system of particles described in problem (3), about an edge.

5.5 THE INERTIA TENSORS OF BODIES WITH CURVED BOUNDARIES

5.5.1 The Inertia Tensor of a Ring

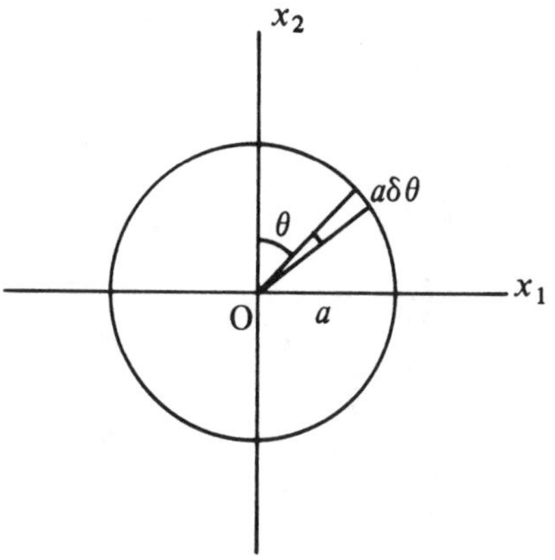

Fig. 5.13

A ring, like a rod, is considered to have mass but no thickness. Let the radius be a and let λ be the line density. Let M be the mass of the ring. Choose rectangular Cartesian axes x_1, x_2, x_3, with the origin at the centre of the ring, and x_3 along the axis of the ring.

Consider an elementary section of the ring, of length $a\delta\theta$, situated at $\mathbf{r} = a(\mathbf{e}_1 \sin\theta + \mathbf{e}_2 \cos\theta)$, where θ is a parameter. The inertia tensor of the elementary section is:

$$\delta\mathbf{I}(\mathrm{O}) = \lambda a\delta\theta(\mathbf{r}^2\mathbf{1} - \mathbf{r}\otimes\mathbf{r}) \ .$$

For the complete ring, therefore:

$$\mathbf{I}(\mathrm{O}) = \int_0^{2\pi} \lambda a^3 \Big[\mathbf{1} - (\mathbf{e}_1 \sin\theta + \mathbf{e}_2\cos\theta)\otimes(\mathbf{e}_1\sin\theta + \mathbf{e}_2\cos\theta)\Big]\mathrm{d}\theta \ ,$$

$$= \lambda a^3 \int_0^{2\pi} \begin{bmatrix} \cos^2\theta & -\sin\theta\cos\theta & 0 \\ -\sin\theta\cos\theta & \sin^2\theta & 0 \\ 0 & 0 & 0 \end{bmatrix} \mathrm{d}\theta \ , \qquad (5.65)$$

$$= \lambda a^3 \begin{bmatrix} \pi & 0 & 0 \\ 0 & \pi & 0 \\ 0 & 0 & 2\pi \end{bmatrix} .$$

But $M = 2\pi a\lambda$.

Hence $$\mathbf{I}(\mathrm{O}) = \tfrac{1}{2}Ma^2 \begin{bmatrix} 1 & 0 & 0 \\ 0 & 1 & 0 \\ 0 & 0 & 2 \end{bmatrix} , \qquad (5.66)$$

$$= \tfrac{1}{2}Ma^2\left[(\mathbf{1} - \mathbf{e}_1\otimes\mathbf{e}_1) + (\mathbf{1} - \mathbf{e}_2\otimes\mathbf{e}_2)\right] \ . \qquad (5.67)$$

We notice from (5.66) that the Perpendicular Axes theorem for laminae is obeyed, and from (5.67) that the ring is equimomental with four particles each of mass $\frac{1}{4}M$ distributed evenly around the ring.

5.5.2 The Inertia Tensor of a Circular Lamina (Disc)

Let a be the radius, M the mass and σ the surface density of a circular lamina in the (x_1, x_2)-plane, with its centre at the origin (Fig. 5.14).

Consider an elementary ring of radius r, width δr, and mass $2\pi r\sigma\delta r$. The inertia tensor of the ring is given by:

$$\delta\mathbf{I}(\mathrm{O}) = \pi\sigma r^3\delta r\left[(\mathbf{1} - \mathbf{e}_1\otimes\mathbf{e}_1) + (\mathbf{1} - \mathbf{e}_2\otimes\mathbf{e}_2)\right] \ .$$

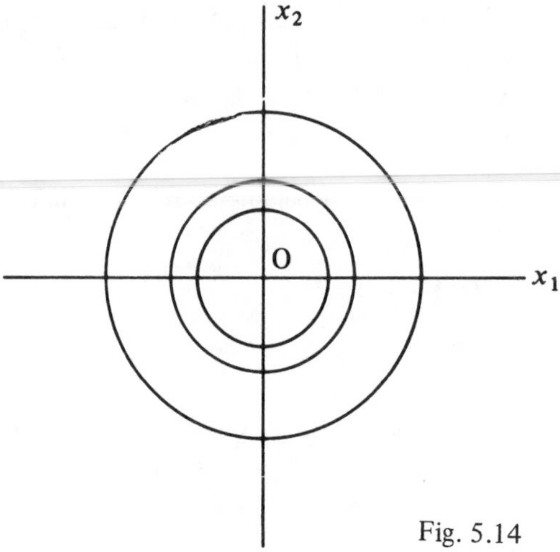

Fig. 5.14

For the circular lamina, therefore:

$$\mathbf{I}(\mathrm{O}) = \pi\sigma \int_0^a r^3 \Big[(\mathbf{1} - \mathbf{e}_1 \otimes \mathbf{e}_1) + (\mathbf{1} - \mathbf{e}_2 \otimes \mathbf{e}_2) \Big] \mathrm{d}r \ ,$$

$$= \pi\sigma \frac{a^4}{4} \Big[(\mathbf{1} - \mathbf{e}_1 \otimes \mathbf{e}_1) + (\mathbf{1} - \mathbf{e}_2 \otimes \mathbf{e}_2) \Big] \ . \qquad (5.68)$$

But $M = \pi a^2 \sigma$.

Hence $$\mathbf{I}(\mathrm{O}) = \tfrac{1}{4} Ma^2 \Big[(\mathbf{1} - \mathbf{e}_1 \otimes \mathbf{e}_1) + (\mathbf{1} - \mathbf{e}_2 \otimes \mathbf{e}_2) \Big] \ . \qquad (5.69)$$

In matrix form (5.69) becomes:

$$\mathbf{I}(\mathrm{O}) = \tfrac{1}{4} Ma^2 \begin{bmatrix} 1 & 0 & 0 \\ 0 & 1 & 0 \\ 0 & 0 & 2 \end{bmatrix} , \qquad (5.70)$$

from which it follows that the moment of inertia of the lamina about its axis is $\frac{1}{2}Ma^2$, and about a diameter is $\frac{1}{4}Ma^2$.

5.5.3 The Inertia Tensor of a Hollow Right Circular Cylinder

Consider a hollow circular cylinder of length $2l$, radius a, surface density σ and mass M.

Choose axes x_2, x_2, x_3 such that O is at the centroid of the cylinder and the x_3-axis coincides with the axis of the cylinder (Fig. 5.15).

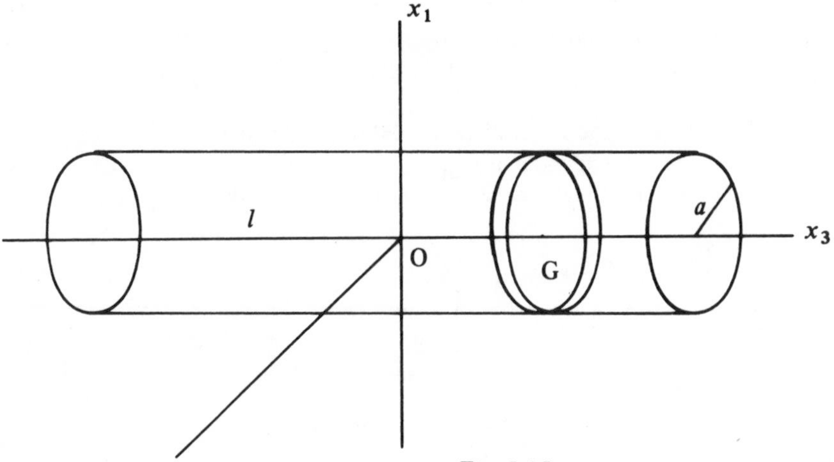

Fig. 5.15

Consider an elementary ring centre $G(0, 0, x_3)$ width δx_3. The inertia tensor of this ring, calculated at its centre is:

$$\delta \mathbf{I}(G) = \pi a \sigma \delta x_3 a^2 \left[(\mathbf{1} - \mathbf{e}_1 \otimes \mathbf{e}_1) + (\mathbf{1} - \mathbf{e}_2 \otimes \mathbf{e}_2) \right] . \tag{5.71}$$

Hence $\delta \mathbf{I}(O) = \delta \mathbf{I}(G) + 2\pi a \sigma x_3^2 \delta x_3 (\mathbf{1} - \mathbf{e}_3 \otimes \mathbf{e}_3)$.

For the complete cylinder, therefore:

$$\mathbf{I}(O) = \int_{-l}^{l} \pi a \sigma \left[a^2 (\mathbf{1} - \mathbf{e}_1 \otimes \mathbf{e}_1) + a^2 (\mathbf{1} - \mathbf{e}_2 \otimes \mathbf{e}_2) + 2x_3^2 (\mathbf{1} - \mathbf{e}_3 \otimes \mathbf{e}_3) \right] \mathrm{d}x_3 ,$$

$$= 2\pi a \sigma l \left[a^2 (\mathbf{1} - \mathbf{e}_1 \otimes \mathbf{e}_1) + a^2 (\mathbf{1} - \mathbf{e}_2 \otimes \mathbf{e}_2) + \tfrac{2}{3} l^2 (\mathbf{1} - \mathbf{e}_3 \otimes \mathbf{e}_3) \right] .$$

But $M = 4\pi \sigma a l$.

Therefore
$$\mathbf{I}(O) = M \left[\frac{a^2}{2} (\mathbf{1} - \mathbf{e}_1 \otimes \mathbf{e}_1) + \frac{a^2}{2} (\mathbf{1} - \mathbf{e}_2 \otimes \mathbf{e}_2) + \frac{l^2}{3} (\mathbf{1} - \mathbf{e}_3 \otimes \mathbf{e}_3) \right] . \tag{5.72}$$

In matrix form this expression becomes

$$\mathbf{I}(O) = M \begin{bmatrix} \frac{1}{2}a^2 + \frac{1}{3}l^2 & 0 & 0 \\ 0 & \frac{1}{2}a^2 + \frac{1}{3}l^2 & 0 \\ 0 & 0 & a^2 \end{bmatrix} . \tag{5.73}$$

Expressions (5.72) and (5.73) give the inertia tensor of a hollow cylinder. If we let $a = 0$ we recover the inertia tensor for a rod (5.41). Alternatively, if $l = 0$ we obtain expression (5.66) for the inertia tensor of a ring.

5.5.4 The Inertia Tensor of a Uniform Solid Right Circular Cylinder

A solid cylinder may be built up from hollow cylinders or circular laminae. We shall employ the first method here and leave the second as an exercise for the student.

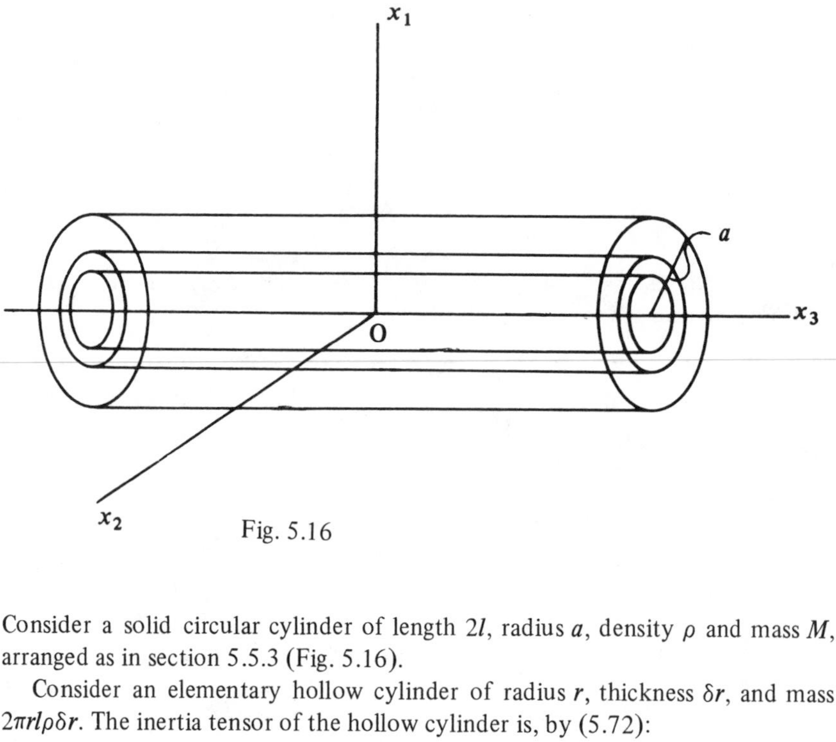

Fig. 5.16

Consider a solid circular cylinder of length $2l$, radius a, density ρ and mass M, arranged as in section 5.5.3 (Fig. 5.16).

Consider an elementary hollow cylinder of radius r, thickness δr, and mass $2\pi r l \rho \delta r$. The inertia tensor of the hollow cylinder is, by (5.72):

$$\delta \mathbf{I}(\mathrm{O}) = 4\pi r l \rho \delta r \left[\frac{r^2}{2}(\mathbf{1} - \mathbf{e}_1 \otimes \mathbf{e}_1) + \frac{r^2}{2}(\mathbf{1} - \mathbf{e}_2 \otimes \mathbf{e}_2) + \frac{l^2}{3}(\mathbf{1} - \mathbf{e}_3 \otimes \mathbf{e}_3) \right] . \tag{5.74}$$

Hence, for the solid cylinder:

$$\mathbf{I}(\mathrm{O}) = 4\pi l \rho \int_0^a \left[\frac{r^3}{2}(\mathbf{1} - \mathbf{e}_1 \otimes \mathbf{e}_1) + \frac{r^3}{2}(\mathbf{1} - \mathbf{e}_2 \otimes \mathbf{e}_2) + \frac{rl^2}{3}(\mathbf{1} - \mathbf{e}_3 \otimes \mathbf{e}_3) \right] dr ,$$

$$= 4\pi l \rho \left[\frac{a^4}{8}(\mathbf{1} - \mathbf{e}_1 \otimes \mathbf{e}_1) + \frac{a^4}{8}(\mathbf{1} - \mathbf{e}_2 \otimes \mathbf{e}_2) + \frac{a^2 l^2}{6}(\mathbf{1} - \mathbf{e}_3 \otimes \mathbf{e}_3) \right] .$$

But $M = 2\pi\rho a^2 l$. Therefore

$$\mathbf{I}(O) = M\left[\frac{a^2}{4}(\mathbf{1} - \mathbf{e}_1 \otimes \mathbf{e}_1) + \frac{a^2}{4}(\mathbf{1} - \mathbf{e}_2 \otimes \mathbf{e}_2) + \frac{l^2}{3}(\mathbf{1} - \mathbf{e}_3 \otimes \mathbf{e}_3)\right]. \tag{5.75}$$

In matrix form this expression becomes:

$$\mathbf{I}(O) = M\begin{bmatrix} \frac{1}{4}a^2 + \frac{1}{3}l^2 & 0 & 0 \\ 0 & \frac{1}{4}a^2 + \frac{1}{3}l^2 & 0 \\ 0 & 0 & \frac{1}{2}a^2 \end{bmatrix}. \tag{5.76}$$

Expressions (5.75) and (5.76) give the inertia tensor for a uniform solid cylinder. We observe that in the limit as $a \to 0$ we recover the expression for the inertia tensor of a rod of length $2l$ (5.41), and in the limit as $l \to 0$ we recover the expressions for the inertia tensor of a circular lamina (5.69), (5.70).

5.5.5 The Inertia Tensor of a Uniform Elliptic Ring

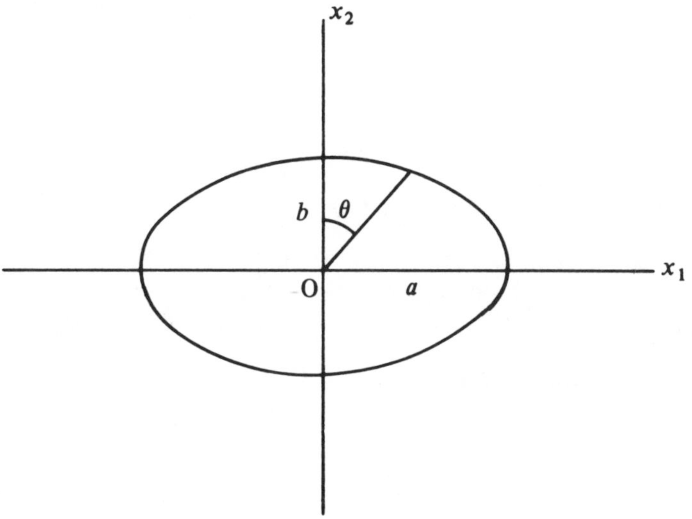

Fig. 5.17

When considering the elliptic ring we meet for the first time an inertia tensor the components of which cannot be expressed in terms of elementary functions. This is because the arc length of an ellipse can only be written down as an elliptic integral. Such integrals are tabulated or are expressible as an infinite series.

The ring is considered to have mass but no thickness. Let λ be the line density,

M the mass, and a, b the semi-major and semi-minor axes respectively, of the ring.

Choose rectangular Cartesian axes x_1, x_2, x_3, with origin at the centre of the ring and x_1, x_2 along the semi-major and semi-minor axes respectively (Fig. 5.17). The parametric equations of the ring are thus:

$$x_1 = a \sin\theta, \qquad x_2 = b\cos\theta \ . \tag{5.77}$$

Consider an elementary section of the ring of length $\sqrt{(a^2 \cos^2\theta + b^2 \sin^2\theta)}\,\delta\theta$ situated at $\mathbf{r} = a\sin\theta\mathbf{e}_1 + b\cos\theta\mathbf{e}_2$. The inertia tensor of this section of ring is:

$$\delta\mathbf{I}(\mathrm{O}) = \delta M(\mathbf{r}^2\mathbf{1} - \mathbf{r}\otimes\mathbf{r}) \ ,$$

where $\qquad \delta M = \lambda\sqrt{(a^2 \cos^2\theta + b^2 \sin^2\theta)}\delta\theta \quad .$

For the complete ring, therefore:

$$\mathbf{I}(\mathrm{O}) = \int_0^{2\pi} \lambda\sqrt{(a^2 \cos^2\theta + b^2\sin^2\theta)}\Big[(a^2\sin^2\theta + b^2\cos^2\theta)\mathbf{1}$$

$$- (a\sin\theta\,\mathbf{e}_1 + b\cos\theta\,\mathbf{e}_2)\otimes(a\sin\theta\,\mathbf{e}_1 + b\cos\theta\,\mathbf{e}_2)\Big]\mathrm{d}\theta \ , \tag{5.78}$$

$$= \int_0^{2\pi} \begin{bmatrix} b^2\cos^2\theta & 0 & 0 \\ 0 & a^2\sin^2\theta & -ab\sin\theta\cos\theta \\ 0 & -ab\sin\theta\cos\theta & a^2\sin^2\theta + b^2\cos^2\theta \end{bmatrix} f(\theta)\,\mathrm{d}\theta \tag{5.79}$$

where $f(\theta) = \lambda\sqrt{(a^2\cos^2\theta + b^2\sin^2\theta)}$.

The evaluation of the integrals in (5.79) presents some difficulty. It is easy to see that all the off-diagonal terms are zero – and, hence, that we have again chosen principal axes – but the terms on the leading diagonal may only be expressed in terms of elliptic integrals. We therefore introduce the **complete elliptic integral of the first kind**:

$$K = \int_0^{\pi/2} \frac{\mathrm{d}\theta}{\sqrt{(1 - e^2\sin^2\theta)}} \ , \tag{5.80}$$

and the **complete elliptic integral of the second kind**:

$$E = \int_0^{\pi/2} \sqrt{(1 - e^2\sin^2\theta)}\mathrm{d}\theta \ . \tag{5.81}$$

It is now a straightforward matter to show that the integral expressing the mass of the elliptic ring, namely

$$M = \lambda\int_0^{2\pi} \sqrt{(a^2\cos^2\theta + b^2\sin^2\theta)}\mathrm{d}\theta \ , \tag{5.82}$$

is given by $M = 4aE$, (5.83)

where e in (5.80) and (5.81) is now the eccentricity of the ellipse given by:

$$e^2 = 1 - (b^2/a^2) . \tag{5.84}$$

The integral expressing the first diagonal term I_{11} is:

$$I_{11} = 4ab^2\lambda \int_0^{\pi/2} \cos^2\theta \sqrt{(1 - e^2 \sin^2\theta)} \mathrm{d}\theta . \tag{5.85}$$

Similarly, for the second diagonal term,

$$I_{22} = 4a^3\lambda \int_0^{\pi/2} \sin^2\theta \sqrt{(1 - e^2 \sin^2\theta)} \mathrm{d}\theta , \tag{5.86}$$

while $I_{33} = I_{11} + I_{22}$, as we would expect from the Perpendicular Axes theorem.

Let us consider the integral

$$I = \int_0^{\pi/2} \sin^2\theta \sqrt{(1 - e^2 \sin^2\theta)} \mathrm{d}\theta . \tag{5.87}$$

Our object is to express I in terms of K and E.

Clearly $$I = \int_0^{\pi/2} \frac{(1 - \cos 2\theta)}{2} \sqrt{[1 - \tfrac{1}{2}e^2(1 - \cos 2\theta)]} \, \mathrm{d}\theta ,$$

$$= \tfrac{1}{2}E - \tfrac{1}{2}\int_0^{\pi/2} \cos 2\theta \sqrt{[\alpha + \beta \cos 2\theta]} \, \mathrm{d}\theta , \tag{5.88}$$

where $\alpha = 1 - \tfrac{1}{2}e^2, \beta = \tfrac{1}{2}e^2.$ (5.89)

The integral in (5.88) may be integrated by parts to give:

$$I = \tfrac{1}{2}E - (\beta/4)\int_0^{\pi/2} \frac{\sin^2 2\theta \, \mathrm{d}\theta}{\sqrt{(\alpha + \beta \cos 2\theta)}} ,$$

which, after some manipulation, leads to:

$$I = \tfrac{1}{2}E - (\beta/4)K + \tfrac{1}{4}(E - 2I) - \frac{\alpha}{4\beta} E + \frac{\alpha^2}{4\beta} K .$$

Recalling the definitions of α and β given in (5.89) we have, finally:

$$I = [(2e^2 - 1)E + (1 - e^2)K] / 3e^2 . \tag{5.90}$$

Hence, introducing the expression for M given in (5.83), we have, from (5.86):

$$I_{22} = Ma^2 [2e^2 - 1 + (1 - e^2)K/E] / 3e^2 . \tag{5.91}$$

We now note that the integral

$$J = \int_0^{\pi/2} \cos^2\theta\sqrt{(1-e^2\sin^2\theta)}\mathrm{d}\theta \ , \tag{5.92}$$

is given by $J = E - I$,

i.e. $$J = [(e^2+1)E + (e^2-1)K]/3e^2 \ , \tag{5.93}$$

therefore $$I_{11} = Mb^2[e^2+1+(e^2-1)K/E]/3e^2 \ , \tag{5.94}$$

$$= Ma^2[1-e^4-(1-e^2)^2K/E]/3e^2 \ . \tag{5.95}$$

Finally $$I_{33} = I_{11} + I_{22} = \tfrac{1}{3}Ma^2[2-e^2+(1-e^2)K/E]. \tag{5.96}$$

We have now evaluated the principal moments of inertia of an elliptic ring in terms of the complete elliptic integrals K and E. For small values of the eccentricity e, expansions may be used to evaluate K and E.

Thus $$K = \frac{\pi}{2}\left(1 + \frac{1^2}{2^2}e^2 + \frac{1^2.3^2}{2^2.4^2}e^4 + \frac{1^2.3^2.5^2}{2^2.4^2.6^2}e^6 + \ldots\right), \tag{5.97}$$

$$E = \frac{\pi}{2}\left(1 - \frac{1}{2^2}e^2 - \frac{1^2.3}{2^2.4^2}e^4 - \frac{1^2.3^2.5}{2^2.4^2.6^2}e^6 - \ldots\right), \tag{5.98}$$

$$\frac{1}{E} = \frac{2}{\pi}\left(1 + \frac{1}{2^2}e^2 + \frac{7}{2^2.4^2}e^4 + \frac{135}{2^2.4^2.6^2}e^6 + \ldots\right). \tag{5.99}$$

By taking the product of (5.97) and (5.99) we find that:

$$\frac{K}{E} = 1 + \tfrac{1}{2}e^2 + \frac{5}{16}e^4 + \frac{7}{32}e^6 + \frac{337}{2048}e^8 + \ldots\ . \tag{5.100}$$

Hence, from (5.95), (5.91) and (5.96), we have:

$$I_{11} = Ma^2\left(\frac{1}{2} - \frac{7}{16}e^2 - \frac{1}{32}e^4 - \frac{27}{2048}e^6 - \ldots\right), \tag{5.101}$$

$$I_{22} = Ma^2\left(\frac{1}{2} - \frac{1}{16}e^2 - \frac{1}{32}e^4 - \frac{37}{2048}e^6 - \ldots\right), \tag{5.102}$$

$$I_{33} = Ma^2\left(1 - \frac{1}{2}e^2 - \frac{1}{16}e^4 - \frac{1}{32}e^6 - \frac{37}{2048}e^8 - \ldots\right). \tag{5.103}$$

Expansions (5.101) - (5.103) give approximations to the principal moments of inertia for small values of the eccentricity. In the limit as $e \to 0$ we recover expression (5.66) which gives the inertia tensor of a circular ring. Reference to the literature on elliptic integrals will provide tabulated values or alternative expansions which converge more quickly.

5.5.6 The Inertia Tensor of an Elliptic Lamina

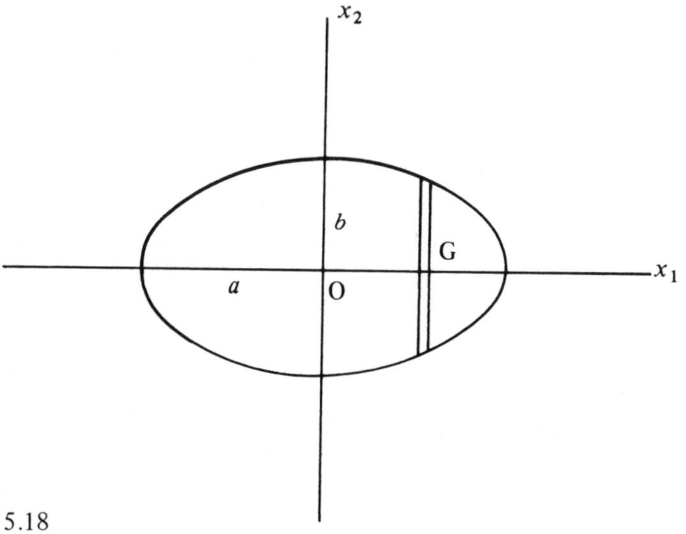

Fig. 5.18

We have seen, in the last section, that it is not possible to write down the inertia tensor of an elliptic ring in terms of elementary functions. For this reason we shall not attempt to build up our elliptic lamina from elliptic rings but from elementary strips.

Consider an elliptic lamina of semi-axes a and b, surface density σ and mass M. Choose rectangular Cartesian axes x_1, x_2 along the semi-axes, with origin at the centroid of the lamina (Fig. 5.18).

Consider an elementary rod parallel to the x_2-axis, of length $2x_2$, width δx_1 and situated a distance x_1 from the x_2-axis. The mass of the rod is $2\sigma x_2 \delta x_1$. The inertia tensor of the rod, calculated at its centroid G, is:

$$\delta \mathbf{I}(\mathrm{G}) = \tfrac{2}{3}\sigma x_2 \delta x_1 x_2^2(\mathbf{1} - \mathbf{e}_2 \otimes \mathbf{e}_2) \; . \tag{5.104}$$

The inertia tensor at O, therefore, is by the Parallel Axis theorem:

$$\delta \mathbf{I}(\mathrm{O}) = \delta \mathbf{I}(\mathrm{G}) + 2\sigma x_2 \delta x_1 x_1^2(\mathbf{1} - \mathbf{e}_1 \otimes \mathbf{e}_1) \; .$$

For the elliptic lamina, therefore,

$$\mathbf{I}(\mathrm{O}) = \int \tfrac{2}{3}\sigma(\mathbf{1} - \mathbf{e}_2 \otimes \mathbf{e}_2)x_2^3 \mathrm{d}x_1 + \int 2\sigma(\mathbf{1} - \mathbf{e}_1 \otimes \mathbf{e}_1)x_1^2 x_2 \mathrm{d}x_1 \; ,$$

where the integration is carried out over the ellipse. In order to perform this integration we resort to the parametric equations of the ellipse:

$$x_1 = a \cos\theta, \qquad x_2 = b \sin\theta \; .$$

Hence $$\mathbf{I}(\mathrm{O}) = \int_0^\pi \frac{2\sigma}{3}(\mathbf{1} - \mathbf{e}_2 \otimes \mathbf{e}_2)ab^3 \sin^4\theta \, \mathrm{d}\theta$$

$$+ \int_0^\pi 2\sigma(\mathbf{1} - \mathbf{e}_1 \otimes \mathbf{e}_1)a^3 b \sin^2\theta \cos^2\theta \, \mathrm{d}\theta \; .$$

Now $$\int_0^\pi \sin^4\theta \, \mathrm{d}\theta = 3\pi/8 \quad \text{and} \quad \int_0^\pi \sin^2\theta \cos^2\theta \, \mathrm{d}\theta = \pi/8.$$

We note also that $M = \pi ab\sigma$. Therefore

$$\mathbf{I}(\mathrm{O}) = \frac{M}{4}\left[a^2(\mathbf{1} - \mathbf{e}_1 \otimes \mathbf{e}_1) + b^2(\mathbf{1} - \mathbf{e}_2 \otimes \mathbf{e}_2)\right] . \tag{5.105}$$

5.5.7 The Inertia Tensor of a Hollow Torus

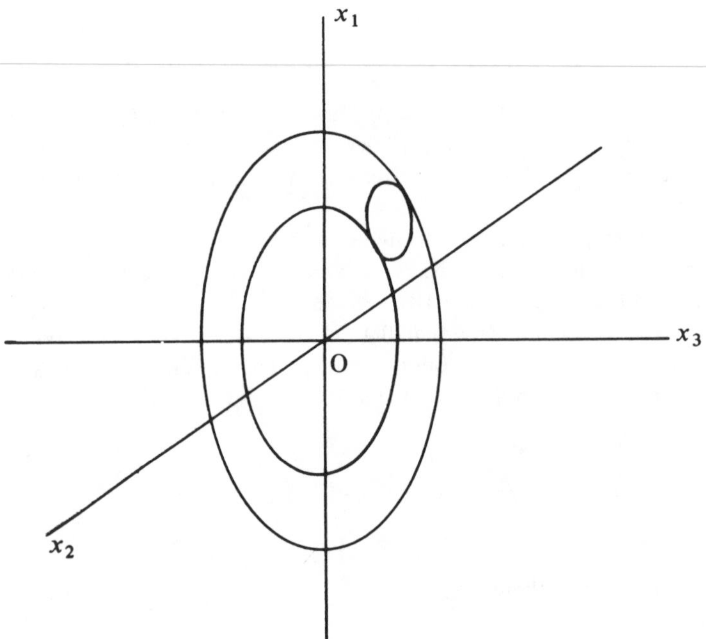

Fig. 5.19

Consider a hollow torus or anchor ring consisting of a hollow cylinder of radius r_2 bent round into a circle of mean radius r_1 (Figs. 5.19, 5.20).

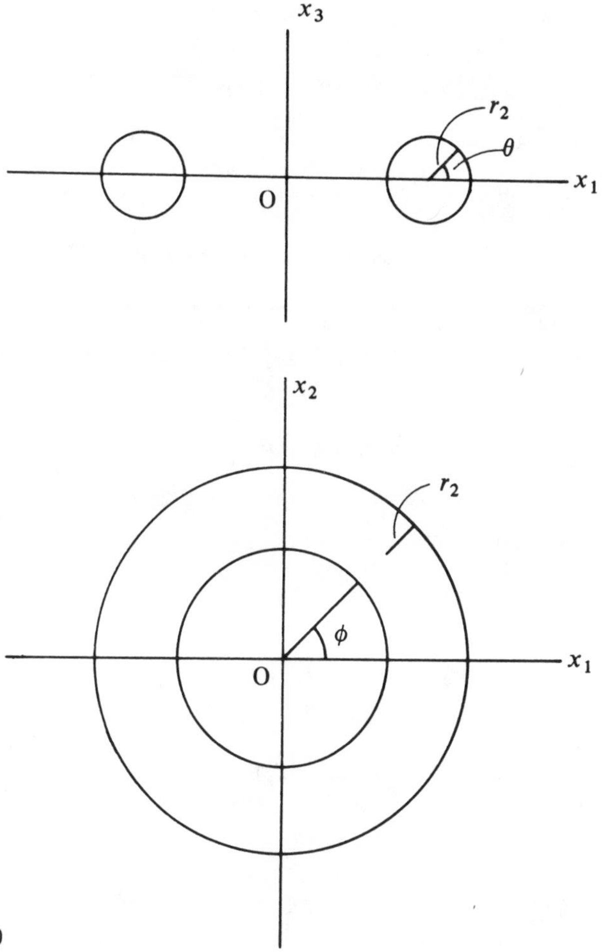

Fig. 5.20

Choose rectangular Cartesian axes x_1, x_2, x_3, with x_3 along the axis of the ring and x_1, x_2 in a plane bisecting the ring as shown. The parametric equations of the torus are thus:

$$\begin{aligned} x_1 &= (r_1 + r_2 \cos\theta)\cos\phi \ , \\ x_2 &= (r_1 + r_2 \cos\theta)\sin\phi \ , \\ x_3 &= r_2 \sin\theta \ , \end{aligned} \tag{5.106}$$

where θ and ϕ are parameters.

The surface area of the torus is given by

$$S = \int_0^{2\pi} 2\pi r_2 (r_1 + r_2 \cos\theta) d\theta = 4\pi^2 r_1 r_2 , \tag{5.107}$$

a result also predicted by the First Theorem of Pappus†.

Hence $\quad M = 4\pi^2 \sigma r_1 r_2 ,$ (5.108)

where M is the mass and σ the surface density of the hollow torus.

Consider an elementary ring of radius $(r_1 + r_2 \cos\theta)$, and width $r_2 \delta\theta$. The inertia tensor of this elementary ring calculated at its centre, G, is, by (5.67):

$$\delta \mathbf{I}(G) = \pi\sigma r_2 \delta\theta (r_1 + r_2 \cos\theta)^3 [(\mathbf{1} - \mathbf{e}_1 \otimes \mathbf{e}_1) + (\mathbf{1} - \mathbf{e}_2 \otimes \mathbf{e}_2)] .$$

At O the inertia tensor is, by the Parallel Axis theorem:

$$\delta \mathbf{I}(O) = \delta \mathbf{I}(G) + 2\pi\sigma r_2 \delta\theta (r_1 + r_2 \cos\theta)(r_2 \sin\theta)^2 (\mathbf{1} - \mathbf{e}_3 \otimes \mathbf{e}_3) .$$

Hence for the complete hollow torus:

$$\begin{aligned} \mathbf{I}(O) = \int_0^{2\pi} & [\pi\sigma r_2 (r_1 + r_2 \cos\theta)^3 \{(\mathbf{1} - \mathbf{e}_1 \otimes \mathbf{e}_1) + (\mathbf{1} - \mathbf{e}_2 \otimes \mathbf{e}_2)\} \\ & + 2\pi\sigma r_2^3 (r_1 + r_2 \cos\theta) \sin^2\theta \{\mathbf{1} - \mathbf{e}_1 \otimes \mathbf{e}_3\}] d\theta , \qquad (5.109) \\ = \pi\sigma r_2 & (2\pi r_1^3 + 3\pi r_1 r_2^2) [(\mathbf{1} - \mathbf{e}_1 \otimes \mathbf{e}_1) + (\mathbf{1} - \mathbf{e}_2 \otimes \mathbf{e}_2)] \\ & + 2\pi^2 \sigma r_2^3 r_1 (\mathbf{1} - \mathbf{e}_3 \otimes \mathbf{e}_3) . \end{aligned}$$

But $M = 4\pi^2 \sigma r_1 r_2$.

Thus

$$\begin{aligned} \mathbf{I}(O) = & \frac{M}{4} (2r_1^2 + 3r_2^2) [(\mathbf{1} - \mathbf{e}_1 \otimes \mathbf{e}_1) + (\mathbf{1} - \mathbf{e}_2 \otimes \mathbf{e}_2)] \\ & + \frac{M}{2} r_2^2 (\mathbf{1} - \mathbf{e}_3 \otimes \mathbf{e}_3) . \end{aligned} \tag{5.110}$$

Expression (5.110) gives the inertia tensor of a torus calculated at its centroid. In the limit as $r_2 \to 0$ we recover the expression for the inertia tensor of a circular ring (5.67).

5.5.8 The Inertia Tensor of a Solid Torus

Suppose now that the torus shown in Fig. 5.19 is solid. Consider an elementary

†The First Theorem of Pappus states: If an arc of a plane curve is rotated about an axis in its plane, which does not intersect it, the surface generated is equal to the length of the arc multiplied by the path length of the centroid of the arc.

hollow torus of radii r_1, r_2 and thickness δr_2. Its mass is $4\pi^2 \rho r_1 r_2 \delta r_2$ where ρ is the density. The inertia tensor of the elementary hollow torus is, by (5.110):

$$\delta \mathbf{I}(O) = \pi^2 \rho r_1 r_2 \delta r_2 (2r_1^2 + 3r_2^2)\,[(\mathbf{1} - \mathbf{e}_1 \otimes \mathbf{e}_1) + (\mathbf{1} - \mathbf{e}_2 \otimes \mathbf{e}_2)] + 2\pi^2 \rho r_1 r_2^3 \delta r_2 (\mathbf{1} - \mathbf{e}_3 \otimes \mathbf{e}_3)\ .$$

Hence, for the complete torus:

$$\begin{aligned} \mathbf{I}(O) &= \pi^2 \rho \int_0^{r_2} \Big[r_1 r_2 (2r_1^2 + 3r_2^2) \{(\mathbf{1} - \mathbf{e}_1 \otimes \mathbf{e}_1) + (\mathbf{1} - \mathbf{e}_2 \otimes \mathbf{e}_2)\} \\ &\quad + 2r_1 r_2^3 (\mathbf{1} - \mathbf{e}_3 \otimes \mathbf{e}_3) \Big] dr_2\ , \\ &= \pi^2 \rho r_1 (r_1^2 r_2^2 + \tfrac{3}{4} r_2^4) \{(\mathbf{1} - \mathbf{e}_1 \otimes \mathbf{e}_1) + (\mathbf{1} - \mathbf{e}_2 \otimes \mathbf{e}_2)\} \\ &\quad + \tfrac{1}{2}\pi^2 \rho r_1\, r_2^4 (\mathbf{1} - \mathbf{e}_3 \otimes \mathbf{e}_3)\ . \end{aligned} \tag{5.111}$$

$$\text{but}\quad M = \int_0^{r_2} 4\pi^2 \rho r_1 r_2\, dr_2 = 2\pi^2 \rho r_1 r_2^2\ , \tag{5.112}$$

a result predicted by the Second Theorem of Pappus†. Therefore

$$\mathbf{I}(O) = M(\tfrac{1}{2} r_1^2 + \tfrac{3}{8} r_2^2)\,[(\mathbf{1} - \mathbf{e}_1 \otimes \mathbf{e}_1) + (\mathbf{1} - \mathbf{e}_2 \otimes \mathbf{e}_2)] + \frac{M}{4} {r_2}^2 (\mathbf{1} - \mathbf{e}_3 \otimes \mathbf{e}_3)\ . \tag{5.113}$$

Expression (5.113) gives the inertia tensor of a solid torus. Again as $r_2 \to 0$ we recover (5.67) the inertia tensor for a ring. An alternative method of arriving at this result is to divide the torus into annulus shaped slices parallel to the $(x_1 x_2)$-plane (see Problems 5.5 No. 12).

5.5.9 The Inertia Tensor of a Spherical Shell

Consider a spherical shell of radius a, surface density σ and mass M. Choose rectangular Cartesian axes centered on the centre of the sphere.

Consider an elementary ring of radius $a \sin\theta$ and width $a\delta\theta$. The inertia tensor of this ring calculated at its centroid G is, from (5.67):

$$\delta \mathbf{I}(G) = \pi\sigma (a \sin\theta)^3 a\delta\theta\, [(\mathbf{1} - \mathbf{e}_1 \otimes \mathbf{e}_1) + (\mathbf{1} - \mathbf{e}_2 \otimes \mathbf{e}_2)]\ . \tag{5.114}$$

The inertia tensor of the elementary ring at the origin O is, by the Parallel Axis theorem:

$$\delta \mathbf{I}(O) = \delta \mathbf{I}(G) + 2\pi\sigma a^4 \sin\theta\,\delta\theta\, \cos^2\theta\ (\mathbf{1} - \mathbf{e}_3 \otimes \mathbf{e}_3)\ . \tag{5.115}$$

†The Second Theorem of Pappus states: If a plane area is rotated about an axis in its plane which does not intersect it, the volume generated is equal to the area multiplied by the length of the path of the centroid of the area.

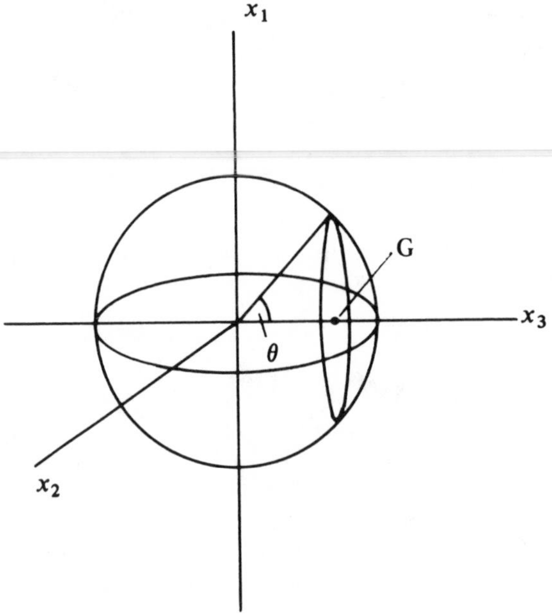

Fig 5.21

Now $\int_0^{\pi} \sin^3\theta \, d\theta = \frac{4}{3}$, and $\int_0^{\pi} \sin\theta\cos^2\theta \, d\theta = \frac{2}{3}$.

Hence, for the spherical shell:

$$\mathbf{I}(O) = \tfrac{4}{3}\pi\sigma a^4 \,(3\mathbf{1} - \mathbf{e}_1 \otimes \mathbf{e}_1 - \mathbf{e}_2 \otimes \mathbf{e}_2 - \mathbf{e}_3 \otimes \mathbf{e}_3).$$

But $M = 4\pi\sigma a^2$, therefore

$$\mathbf{I}(O) = \tfrac{2}{3}Ma^2\mathbf{1} \quad . \tag{5.116}$$

5.5.10 The Inertia Tensor of a Solid Sphere

Suppose the sphere shown in Fig. 5.21 is a solid sphere of density ρ. Consider an elementary circular lamina of radius $a \sin\theta$, thickness $a \sin\theta\delta\theta$, situated a distance $a\cos\theta$ from the (x_1x_2) - plane and parallel to it. The inertia tensor of this lamina at its centroid G is, from (5.69):

$$\delta\mathbf{I}(G) = \tfrac{1}{4}\pi\rho(a\sin\theta)^5\,\delta\theta[(\mathbf{1} - \mathbf{e}_1 \otimes \mathbf{e}_1) + (\mathbf{1} - \mathbf{e}_2 \otimes \mathbf{e}_2)] \ . \tag{5.117}$$

The inertia tensor of the elementary lamina, calculated at the origin O, is by the Parallel Axis theorem:

$$\delta\mathbf{I}(O) = \delta\mathbf{I}(G) = \pi\rho(a\sin\theta)^3\,(a\cos\theta)^2\delta\theta(\mathbf{1} - \mathbf{e}_3 \otimes \mathbf{e}_3) \ . \tag{5.118}$$

Now $\int_0^{\pi} \sin^5\theta \, d\theta = \frac{16}{5}$; $\int_0^{\pi} \sin^3\theta\cos^2\theta \, d\theta = \frac{4}{15}$.

Hence, for the solid sphere:

$$\mathbf{I}(\mathrm{O}) = \tfrac{4}{15}\pi\rho a^5 (3\mathbf{1} - \mathbf{e}_1 \otimes \mathbf{e}_1 - \mathbf{e}_2 \otimes \mathbf{e}_2 - \mathbf{e}_3 \otimes \mathbf{e}_3) .$$

But $M = \frac{4}{3}\pi\rho a^3$, therefore

$$\mathbf{I}(\mathrm{O}) = \tfrac{2}{5}Ma^2 \mathbf{1} . \tag{5.119}$$

Examples 5.5

(1)

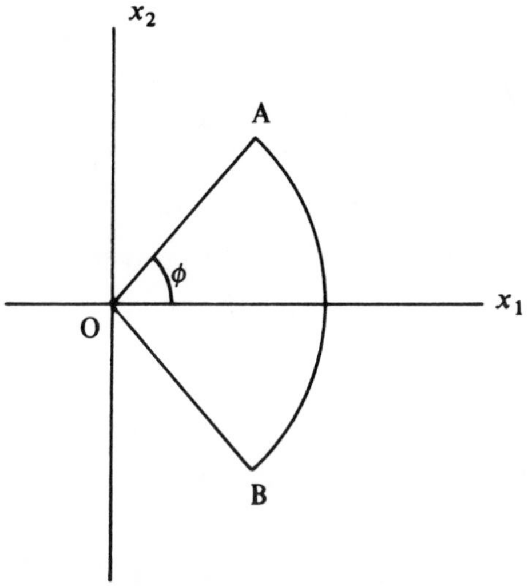

Fig. 5.22

By taking the complement of $\angle\theta$ and changing the limits of integration in (5.65) we may determine the inertia tensor of a circular arc, subtending an angle 2ϕ at the centre (Fig. 5.22).

Thus
$$\mathbf{I}(\mathrm{O}) = \int_{-\phi}^{\phi} \lambda a^3 \begin{bmatrix} \sin^2\theta & -\sin\theta\cos\theta & 0 \\ -\sin\theta\cos\theta & \cos^2\theta & 0 \\ 0 & 0 & 1 \end{bmatrix} \mathrm{d}\theta .$$

Now
$$\int_{-\phi}^{\phi} \sin^2\theta \, \mathrm{d}\theta = \phi - \tfrac{1}{2}\sin 2\phi ,$$

and
$$\int_{-\phi}^{\phi} \cos^2\theta \, \mathrm{d}\theta = \phi + \tfrac{1}{2}\sin 2\phi ,$$

also
$$\int_{-\phi}^{\phi} \sin\theta\cos\theta \, \mathrm{d}\theta = 0, \text{ therefore}$$

$$\mathbf{I}(O) = \lambda a^3 \begin{bmatrix} \phi - \frac{1}{2}\sin 2\phi & 0 & 0 \\ 0 & \phi + \frac{1}{2}\sin 2\phi & 0 \\ 0 & 0 & 2\phi \end{bmatrix} .$$

But $M = 2a\lambda\phi$.

Hence $$\mathbf{I}(O) = \frac{Ma^2}{2}\Big[(\mathbf{1} - \mathbf{e}_1 \otimes \mathbf{e}_1) + (\mathbf{1} - \mathbf{e}_2 \otimes \mathbf{e}_2)\Big]$$
$$+ \frac{Ma^2}{2}\frac{\sin 2\phi}{2\phi}\Big[(\mathbf{1} - \mathbf{e}_1 \otimes \mathbf{e}_1) - (\mathbf{1} - \mathbf{e}_2 \otimes \mathbf{e}_2)\Big] .$$

(2) From (5.68) we may deduce the inertia tensor of an annulus of inner radius b and outer radius a:

$$\mathbf{I}(O) = \tfrac{1}{4}\pi\sigma(a^4 - b^4)\,[(\mathbf{1} - \mathbf{e}_1 \otimes \mathbf{e}_1) + (\mathbf{1} - \mathbf{e}_2 \otimes \mathbf{e}_2)] .$$

But $M = \pi\sigma(a^2 - b^2)$, therefore

$$\mathbf{I}(O) = \frac{M}{4}(a^2 + b^2)\,[(\mathbf{1} - \mathbf{e}_1 \otimes \mathbf{e}_1) + (\mathbf{1} - \mathbf{e}_2 \otimes \mathbf{e}_2)] .$$

(3) The method of section 5.5.9 may be used to determine the inertia tensor of a hollow spherical cap. Suppose the rim of the cap subtends an angle ϕ at the centre, we note that

$$\int_0^\phi \sin^3\theta\, d\theta = \tfrac{2}{3} - \cos\phi + \tfrac{1}{3}\cos^3\phi ,$$

and $$\int_0^\phi \sin\theta \cos^2\theta\, d\theta = \tfrac{1}{3} - \tfrac{1}{3}\cos^3\phi .$$

Hence $$\mathbf{I}(O) = \pi a^4\sigma[(\mathbf{1} - \mathbf{e}_1 \otimes \mathbf{e}_1) + (\mathbf{1} - \mathbf{e}_2 \otimes \mathbf{e}_2)](\tfrac{2}{3} - \cos\phi + \tfrac{1}{3}\cos^3\phi)$$
$$+ 2\pi a^4\sigma(\mathbf{1} - \mathbf{e}_3 \otimes \mathbf{e}_3)(\tfrac{1}{3} - \tfrac{1}{3}\cos^3\phi) .$$

But the mass of the cap is given by:

$$M = 2\pi a^2\sigma(1 - \cos\phi) .$$

Hence $$\mathbf{I}(O) = \tfrac{1}{2}Ma^2[(\mathbf{1} - \mathbf{e}_1 \otimes \mathbf{e}_1) + (\mathbf{1} - \mathbf{e}_2 \otimes \mathbf{e}_2)]$$
$$+ \tfrac{1}{6}Ma^2\,[2(\mathbf{1} - \mathbf{e}_3 \otimes \mathbf{e}_3) - (\mathbf{1} - \mathbf{e}_1 \otimes \mathbf{e}_1)$$
$$- (\mathbf{1} - \mathbf{e}_2 \otimes \mathbf{e}_2)](1 + \cos\phi + \cos^3\phi) .$$

(4) We have seen that the inertia tensors of the spherical shell and solid sphere are multiples of the unit tensor – a fact which could have been anticipated beforehand from considerations of symmetry. By assuming this form it is possible to arrive at the results with very little labour.

Consider the case of the spherical shell. We have:

$$\mathbf{I}(\mathrm{O}) = \int_S \sigma(r^2\mathbf{1} - \mathbf{r}\otimes\mathbf{r})\mathrm{d}S, \text{ therefore}$$

$$I_{11} = \int_S \sigma(r^2 - r_1{}^2)\mathrm{d}S \ ,$$

Thus $$I_{11} + I_{22} + I_{33} = 2\int_S \sigma r^2\,\mathrm{d}S = 2Ma^2 \ .$$

If we assume that $\mathbf{I}(\mathrm{O})$ is a multiple of $\mathbf{1}$, it follows that

$$I_{11} = I_{22} = I_{33} = \tfrac{2}{3}Ma^2, \text{ and } I_{12} \text{ etc.} = 0.$$

Hence $\mathbf{I}(\mathrm{O}) = \tfrac{2}{3}Ma^2\mathbf{1}$.

Problems 5.5

(1) A hollow circular cone has its vertex at the origin and its axis along the x_3-axis. The height is h, the base radius a, and the mass M. Calculate the inertia tensor $\mathbf{I}(\mathrm{O})$, and find the moment of inertia about a diameter through the base.

(2) A solid elliptical cylinder has mass M, length $2l$ and semi-axes a and b. The x_1- and x_2-axes are along the semi-axes a and b respectively, and the x_3 axis coincides with the axis of the cylinder. Calculate the inertia tensor $\mathbf{I}(\mathrm{O})$ where O is the centroid of the cylinder.

(3) A uniform solid right circular cylinder of mass M is bounded by the planes $x_3 = \pm l$ and the cylinder $x_1^2 + x_2^2 = a^2$. Calculate the inertia tensor at O, and show that the moment of inertia about the line

$$\mathbf{r} = \lambda(\cos\alpha, \cos\beta, \cos\gamma), \text{ where } \cos^2\alpha + \cos^2\beta + \cos^2\gamma = 1, \text{ is}$$

$$\mu = M[\tfrac{1}{3}l^2\sin^2\gamma + \tfrac{1}{4}a^2(1 + \cos^2\gamma)] \ .$$

(4) A homogenous right circular cone of mass M, semi-vertical angle α, and slant height l, is placed with its axis of symmetry along the x_3-axis, and its vertex at the origin O. Find the inertia tensor at O and deduce the moment of inertia about a generator.

(5) Find the inertia tensor of a solid sphere by building it up from elementary hollow spheres. Compare your answer with (5.119).

(6) Use the method of section 5.5.10 to determine the inertia tensor of a solid ellipsoid of semi-axes a, b and c.

(7) Show that the inertia tensor of a solid uniform hemisphere calculated at the centre O is:

$$\mathbf{I}(\mathrm{O}) = \tfrac{2}{5}Ma^2\mathbf{1} \ .$$

What is the inertia tensor at the centre of mass?

(8) Deduce from Examples 5.5 No. 1 the inertia tensor at O of the sector shaped lamina OAB (Fig. 5.22), and find the moment of inertia about the line OA.

(9) Let I_{11}, I_{22}, I_{33} be the principal moments of inertia of the elliptic ring discussed in section 5.5.5. Prove that

$$a^2 I_{11} + b^2 I_{22} = Ma^2 b^2 \ .$$

(10) Use the method of Example 5.5 No. 3 to determine the inertia tensor of a solid spherical sector.

(11) Show that the inertia tensor of a hollow sphere of external radius a and internal radius b is:

$$\mathbf{I}(\mathrm{O}) = (2/5)M\frac{(a^5 - b^5)}{(a^3 - b^3)}\mathbf{1} \ .$$

(12) Obtain the inertia tensor of the solid torus shown in Fig. 5.19 by dividing it into annulus shaped laminae parallel to the (x_1, x_2)-plane.

(13) Apply the argument of Examples 5.5 No. 4 to establish the inertia tensor of a solid sphere.

(14) A rigid structure is formed by cementing together three identical uniform spheres (each of mass M and radius a) in such a way that their centres lie on the vertices of an equilateral triangle. Find the principal axes of inertia at the centre of one of the spheres. Find also the corresponding moments of inertia. Suppose the structure is constrained to rotate about an axis joining any two centres. What is the angle between this axis and the angular momentum vector at the centre of one of the spheres?

(Liverpool University)

(15) For the structure defined in question (14), use the Parallel Axis theorem to calculate the components of the inertia tensor at (i) the centroid and (ii) the point when two spheres touch.

(16) If at a point O of a rigid body the equation of the momental ellipsoid relative to axes $\mathrm{O}x_1x_2x_3$ is known, show how the principal moments of inertia and the directions of the principal axes at that point can be calculated.

The principal moments of inertia of a cube, calculated at one corner, are A, C, C. Show that the equation of the momental ellipsoid with respect to axes parallel to the edges of the cube is:

$$\tfrac{1}{3}(A + 2C)(x_1^2 + x_2^2 + x_3^2) + \tfrac{2}{3}(A - C)(x_2x_3 + x_3x_1 + x_1x_2) = 1.$$

The region between a sphere of radius a and a cube of edge $2b$, concentric with the sphere, $b > a$, contains solid homogeneous material of density ρ. Find at a corner of the cube, the principal moments of inertia, the directions of the principal axes, and the equation of the momental ellipsoid referred to axes parallel to the edges of the cube.

5.6 EULER'S EQUATIONS OF MOTION

We shall now establish the equations of motion of a rigid body about its centre of mass, under the influence of an applied couple.

We have, from section 3.3.2, for any tensor **T**, in a frame of reference rotating with angular velocity $\boldsymbol{\omega}$:

$$\frac{d\mathbf{T}}{dt} = \frac{\partial \mathbf{T}}{\partial t} + \boldsymbol{\omega} \times \mathbf{T} - \mathbf{T} \times \boldsymbol{\omega} , \tag{5.120}$$

where $\partial\mathbf{T}/\partial t$ is the rate of change of **T** within the rotating frame, and the cross-product of a vector and a tensor was defined in section 3.3.1.

Now the equation of motion of a body rotating about its centroid G is:

$$\frac{d\mathbf{H}(G)}{dt} = \boldsymbol{\Gamma}(G) , \tag{5.121}$$

where $\boldsymbol{\Gamma}(G)$ is the moment of the applied forces about G.

But $\mathbf{H}(G) = \mathbf{I}(G).\boldsymbol{\omega}$, therefore

$$\boldsymbol{\Gamma}(G) = \frac{d}{dt}[\mathbf{I}(G).\boldsymbol{\omega}] . \tag{5.122}$$

Hence, by use of (1.138), and remembering that the rotating axes are fixed relative to the body so that $\partial\mathbf{I}(G)/\partial t = 0$,

$$\boldsymbol{\Gamma}(G) = \mathbf{I}(G).\frac{d\boldsymbol{\omega}}{dt} + \boldsymbol{\omega} \times [\mathbf{I}(G).\boldsymbol{\omega}] . \tag{5.123}$$

Suppose the inertia tensor is referred to principal axes. Equation (5.123) leads to the three scalar equations:

$$\left.\begin{aligned} \Gamma_1 &= I_{11}\frac{d\omega_1}{dt} + (I_{33} - I_{22})\,\omega_2\omega_3 , \\ \Gamma_2 &= I_{22}\frac{d\omega_2}{dt} + (I_{11} - I_{33})\,\omega_3\omega_1 , \\ \Gamma_3 &= I_{33}\frac{d\omega_3}{dt} + (I_{22} - I_{11})\,\omega_1\omega_2 . \end{aligned}\right\} \tag{5.124}$$

Equations (5.124) are known as **Euler's Equations of Motion**, and give the components of the couple necessary to maintain the angular velocity $\boldsymbol{\omega}$ about the centroid. They are also valid for motion about a fixed point not coinciding with the centroid. It must be remembered, however, that they are true not with respect to any fixed set of axes but with respect to axes which coincide instantaneously with the principal axes of the body.

The special case of zero external couple $[\boldsymbol{\Gamma}(\mathrm{G}) = \mathbf{0}]$ is of interest, as the equations of motion may then be integrated. From (5.122) we have:

$$\frac{\mathrm{d}}{\mathrm{d}t}[\mathbf{I}(\mathrm{G}).\boldsymbol{\omega}] = \mathbf{0} ,$$

i.e.
$$\frac{\mathrm{d}\mathbf{H}(\mathrm{G})}{\mathrm{d}t} = \mathbf{0} . \tag{5.125}$$

The angular momentum $\mathbf{H}(\mathrm{G})$ is thus constant. A second integral may be found by putting $\boldsymbol{\Gamma}(\mathrm{G}) = \mathbf{0}$ in (5.123) and taking the scalar multiple with $\boldsymbol{\omega}$.

Thus
$$\boldsymbol{\omega}.\mathbf{I}(\mathrm{G}).\frac{\mathrm{d}\boldsymbol{\omega}}{\mathrm{d}t} + \boldsymbol{\omega}.[\boldsymbol{\omega} \times \mathbf{I}(\mathrm{G}).\boldsymbol{\omega}] = 0 ,$$

i.e.
$$\boldsymbol{\omega}.\mathbf{I}(\mathrm{G}).\frac{\mathrm{d}\boldsymbol{\omega}}{\mathrm{d}t} = 0 ,$$

or
$$\boldsymbol{\omega}.\mathbf{I}(\mathrm{G}).\boldsymbol{\omega} = \text{constant} . \tag{5.126}$$

The constant in (5.126) is equal to twice the kinetic energy, which quantity is thus also conserved in the case of zero external couple.

A further special case which is of interest is that of a body having axial symmetry. Such a body has two equal principal moments of inertia about axes perpendicular to the axis of symmetry. The inertia tensor is therefore of the form:

$$\mathbf{I}(\mathrm{G}) = \begin{bmatrix} A & 0 & 0 \\ 0 & A & 0 \\ 0 & 0 & C \end{bmatrix} , \tag{5.127}$$

where the axis of symmetry is the x_3-axis. The angular momentum is given by:

$$\mathbf{H}(\mathrm{G}) = \mathbf{I}(\mathrm{G}).\boldsymbol{\omega} = (A\omega_1, A\omega_2, C\omega_3) . \tag{5.128}$$

Now for any vector $\boldsymbol{\omega}$:

$$\boldsymbol{\omega} \equiv (\boldsymbol{\omega}.\mathbf{e}_3)\mathbf{e}_3 - (\boldsymbol{\omega} \times \mathbf{e}_3) \times \mathbf{e}_3 .$$

The component of $\boldsymbol{\omega}$ along the axis of symmetry is called the **spin**, which is usually denoted by n.

Thus $\boldsymbol{\omega} = n\mathbf{e}_3 - (\boldsymbol{\omega} \times \mathbf{e}_3) \times \mathbf{e}_3$. (5.129)

Hence, from (5.128):

$$\mathbf{H}(G) = Cn\mathbf{e}_3 + A\mathbf{e}_3 \times (\boldsymbol{\omega} \times \mathbf{e}_3) \ .$$

Now $\mathbf{e}_3$ is an axis fixed in the body.

Thus $\boldsymbol{\omega} \times \mathbf{e}_3 = \mathrm{d}\mathbf{e}_3/\mathrm{d}t$,

therefore $\mathbf{H}(G) = Cn\mathbf{e}_3 + A\mathbf{e}_3 \times \dfrac{\mathrm{d}\mathbf{e}_3}{\mathrm{d}t}$. (5.130)

Finally we obtain by differentiation of (5.130):

$$\boldsymbol{\Gamma}(G) = C\frac{\mathrm{d}n}{\mathrm{d}t}\mathbf{e}_3 + Cn\frac{\mathrm{d}\mathbf{e}_3}{\mathrm{d}t} + A\mathbf{e}_3 \times \frac{\mathrm{d}^2\mathbf{e}_3}{\mathrm{d}t^2} \ . \qquad (5.131)$$

Examples 5.6

(1) A sphere of mass M and radius a rolls without slipping, down a slope inclined at an angle α (Fig. 5.23). Determine the angular acceleration of the sphere.

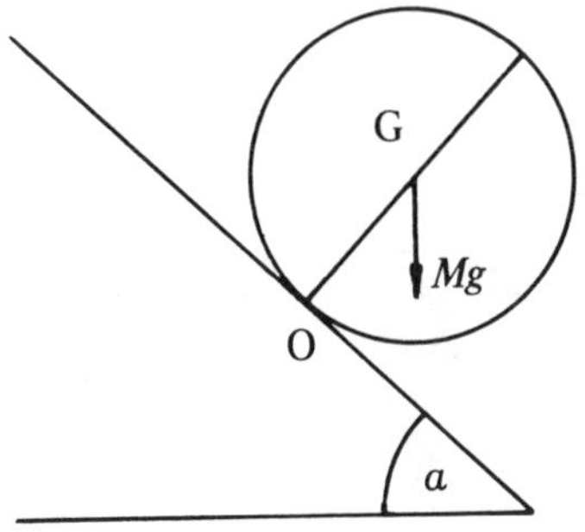

Fig. 5.23

The problem is essentially one-dimensional. It is sufficient therefore to use the first Euler equation with $\omega_1 = \omega, \omega_2 = \omega_3 = 0$.

Thus $Mga \sin\alpha = [\mu(G) + Ma^2]\,\dot{\omega}$. (5.132)

The left-hand side represents the couple about the instantaneous point of contact, O, while $[\mu(G) + Ma^2]$ is the moment of inertia about an axis through O, calculated using the Parallel Axis theorem.

From (5.132) we have:

$$\dot{\omega} = \frac{Mga \sin\alpha}{\mu(G) + Ma^2} \ .$$

(2) Consider a rectangular block of dimensions $2a \times 2b \times 2c$ ($a \neq b \neq c$). It is found that rotation about the longest and shortest principal axes is stable,

whereas rotation about the intermediate principal axes is not (a fact which is easily verified by throwing the block into the air when rotating about the three axes in turn). The explanation is as follows.

Suppose the block rotates approximately about the $2a$ principal axis with angular speed ω_1 in such a way that the product $\omega_2\omega_3$ may be neglected. Euler's equations, with $\mathbf{\Gamma} = \mathbf{0}$, become:

$$\mathrm{d}\omega_1/\mathrm{d}t = 0 \ ,$$

$$I_{22}(\mathrm{d}\omega_2/\mathrm{d}t) = (I_{33} - I_{11})\omega_1\omega_3 \ ,$$

$$I_{33}(\mathrm{d}\omega_3/\mathrm{d}t) = (I_{11} - I_{22})\omega_1\omega_2 \ .$$

Thus, noting that ω_1 is constant, and eliminating ω_3 between the other two equations, we obtain:

$$I_{22}I_{33}\frac{\mathrm{d}^2\omega_2}{\mathrm{d}t^2} + (I_{11} - I_{22})(I_{11} - I_{33})\omega_1^2\omega_2 = 0 \ .$$

This equation leads to an oscillatory solution for ω_2 (which is therefore stable) provided $(I_{11} - I_{22})(I_{11} - I_{33}) > 0$.
This condition is fulfilled *unless*

$$I_{22} < I_{11} < I_{33} \ , \quad \text{or} \quad I_{33} < I_{11} < I_{22} \ .$$

The result follows at once.

(3) A body is constrained to rotate at constant angular velocity $\boldsymbol{\omega}$ about an arbitrary axis through its centroid. Find the couple necessary to maintain the motion.

Euler's equations with $\boldsymbol{\omega}$ = constant are:

$$\mathbf{\Gamma}(\mathrm{G}) = \boldsymbol{\omega} \times [\mathbf{I}(\mathrm{G}).\boldsymbol{\omega}] \ ,$$

or

$$\left.\begin{aligned} \Gamma_1 &= (I_{33} - I_{22})\omega_2\omega_3 \ , \\ \Gamma_2 &= (I_{11} - I_{33})\omega_3\omega_1 \ , \\ \Gamma_3 &= (I_{22} - I_{11})\omega_1\omega_2 \ . \end{aligned}\right\}$$

These equations give the components of the couple necessary to maintain the motion, with respect to the principal axes of the body.

(4) Equation (5.131), having been derived from Euler's equations, applies to motion about a fixed point O. A body which rotates about a fixed point situated on the axis of symmetry is called a **top** (Fig. 5.24).

Consider such a top with centroid G distant l from O and mass m. Let $\mathbf{i}$ and $\mathbf{k}$ be unit vectors along OG and in a vertical direction respectively. Then, from (5.131):

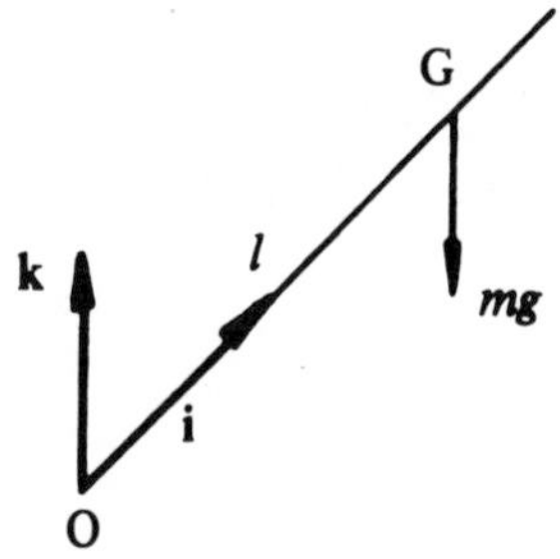

Fig. 5.24

$$-mgl\mathbf{i} \times \mathbf{k} = C\frac{dn}{dt}\mathbf{i} + Cn\frac{d\mathbf{i}}{dt} + A\mathbf{i} \times \frac{d^2\mathbf{i}}{dt^2} \;. \tag{5.133}$$

Clearly from (5.133) $dn/dt = 0$, i.e. the spin remains constant. This is always the case if $\mathbf{\Gamma}$ has no component about the axis of symmetry.

Therefore $$A\mathbf{i} \times \frac{d^2\mathbf{i}}{dt^2} + Cn\frac{d\mathbf{i}}{dt} + mgl\mathbf{i} \times \mathbf{k} = \mathbf{0} \;. \tag{5.134}$$

Problems 5.6

(1) You are provided with two spheres of the same weight and diameter, and are told that one is solid all through, and the other is hollow. Describe an experiment to determine which is which.

(2) A flywheel consisting of a circular disc of mass M and radius a is mounted on an axle which is inclined at a small angle α to its geometrical axis. Find the magnitude of the couple exerted on the bearings.

(3) A toy gyroscope spins with its fixed point placed in a small hollow on top of a pyramid. Show that a solution of the equation of motion exists such that the axis of the gyroscope rotates in a horizontal plane. What is the angular speed of rotation of the axis of symmetry within this plane?

(4) A rigid body has principal axes Oxzy at its centre of mass O, and respective principal moments A, B, C. State Euler's equations for the rotation of the body under a torque $\mathbf{\Gamma}$ acting at O. Show that if the body rotates about a fixed axis with direction cosines $\cos\alpha$, $\cos\beta$, $\cos\gamma$ and with constant angular speed ω, the magnitude of the torque necessary to sustain the motion is:

$$|\mathbf{\Gamma}| = \omega^2 [(B - C)^2 \cos^2\beta \cos^2\gamma + (C - A)^2 \cos^2\gamma \cos^2\alpha + (A - B)^2 \cos^2\alpha \cos^2\beta]^{1/2} \;.$$

Suggest an experimental method of finding the orientations of the principal axes of the inertia at the centre of mass of a given rigid body.

CHAPTER 6

The Application of Cartesian Tensors to Fluid Mechanics

6.1 THE STRESS TENSOR

6.1.1 Introduction

When considering the rotation of a rigid body about a fixed point we were led to the conclusion that it would be a convenience to introduce a tensor, which we termed the inertia tensor, in order to write the angular velocity as a common factor in our equation of motion (5.3).

We now propose to consider the equation of motion of a fluid body, and we shall see that if we are to write it in vector form we are virtually forced to introduce a tensor quantity from the outset.

The derivation of the equation of motion is simple and depends upon equating the rate of change of momentum with the applied forces, according to Newton's Second Law of Motion.

Consider an arbitrary material body of fluid of volume τ. Let the fluid velocity be $\mathbf{u}$ and the density ρ. The linear momentum of this body of fluid is given by the integral:

$$\int_\tau \rho \mathbf{u} \mathrm{d}\tau \ . \tag{6.1}$$

The rate of change of momentum is therefore given by:

$$\int_\tau \rho \frac{\mathrm{D}\mathbf{u}}{\mathrm{D}t} \mathrm{d}\tau \tag{6.2}$$

where $\mathrm{D}\mathbf{u}/\mathrm{D}t$ is the material derivative or derivative following the flow given by:

$$\frac{\mathrm{D}\mathbf{u}}{\mathrm{D}t} = \frac{\partial \mathbf{u}}{\partial t} + \mathbf{u}.\nabla\mathbf{u} \ , \tag{6.3}$$

a formula proved in the opening pages of all standard textbooks on fluid mechanics.

The total force on the body of fluid may be divided conveniently into two parts. In the first place there is the body force which in the case of a fluid at rest is simply the force of gravity. The body force is given by the integral:

$$\int_T \rho \mathbf{F} \, d\tau \ , \tag{6.4}$$

where **F** is the force per unit mass acting on the fluid.

Finally we have to consider the surface force, namely the force exerted by the surrounding fluid on the surface of the body of fluid under consideration.

The total surface force is given by the integral:

$$\int_S \mathbf{\Sigma} \, dS \ ,$$

where S is the surface area of the body of fluid and $\mathbf{\Sigma}$ is the force per unit area on the surface. $\mathbf{\Sigma}$ is called the **stress**.

In general the stress depends on the orientation of the surface, i.e. $\mathbf{\Sigma} = \mathbf{\Sigma}(\mathbf{n})$ where **n** is the unit normal vector outward from the surface. The simplest example of this rule is the non-viscous or inviscid fluid, for which the stress ($\mathbf{\Sigma} = -p\mathbf{n}$, where p is the fluid pressure) is always normal to the fluid surface on which it acts. Such fluids, although they do not occur in nature, form the basis of a highly developed mathematical theory which predicts accurately many of the properties of fluids of low viscosity, such as air and water. Here, however, we shall be concerned with viscous fluids for which $\mathbf{\Sigma}$ is a function of **n**. It is this dependence on another vector which recalls the definition of a second order Cartesian tensor given at the beginning of Chapter 2.

We will now temporarily abandon our equation of motion, leaving it in the form:

$$\int_T \rho \frac{D\mathbf{u}}{Dt} \, d\tau = \int_T \rho \mathbf{F} d\tau + \int_S \mathbf{\Sigma} dS \ , \tag{6.5}$$

while we investigate further the relation between the stress on a surface and the orientation of that surface.

6.1.2 The Stress Tensor

Consider an elementary tetrahedron of fluid OABC (Fig. 6.1). Let δA_i be the area of the face perpendicular to $\mathbf{e}_i$, and let δA be the area of the inclined face, ABC. We shall suppose that the total body forces on the tetrahedron are negligible compared with the surface forces†.

†For a justification of this assumption, see, for example *An Introduction to Fluid Mechanics* by G. K. Batchelor, section 1.3.

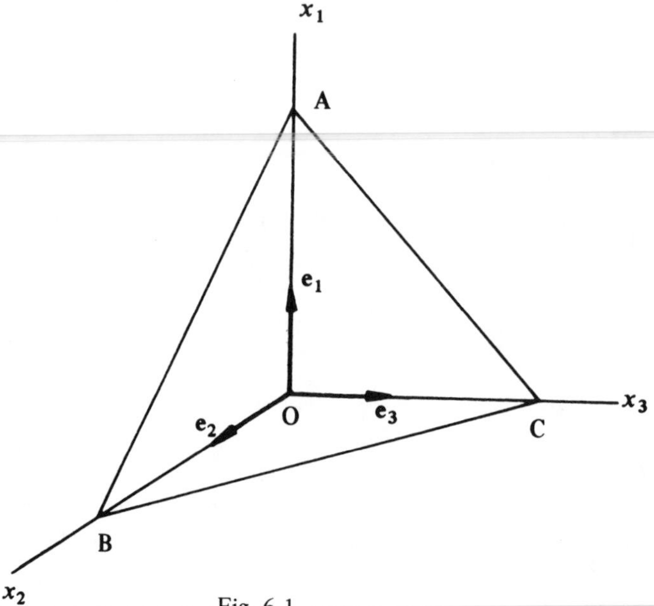

Fig. 6.1

The sum of the surface forces on the tetrahedron is given by:

$$\mathbf{f}_s = \boldsymbol{\Sigma}(\mathbf{n})\delta A + \boldsymbol{\Sigma}(-\mathbf{e}_i)\delta A_i \ , \tag{6.6}$$

where $\mathbf{n}$ is the unit outward normal to the inclined face and $\boldsymbol{\Sigma}(\mathbf{n})$ is the surface force per unit area acting on that face. A summation is implied over the repeated suffix i.

The force $\boldsymbol{\Sigma}$ will depend on position as well as the direction of the normal. We shall assume therefore that the tetrahedron is sufficiently small for that dependence to be ignored.

From the geometry of the figure it is apparent that δA_i is given by:

$$\delta A_i = \mathbf{e}_i.\mathbf{n}\delta A, \qquad i = 1, 2, 3 \ .$$

Therefore $\mathbf{f}_s = \{\boldsymbol{\Sigma}(\mathbf{n}) + \boldsymbol{\Sigma}(-\mathbf{e}_i)\mathbf{e}_i.\mathbf{n}\}\delta A$.

We now invoke Newton's Third Law of Motion: 'Action and reaction are equal and opposite'. Applying this to the forces on either side of the faces of the tetrahedron we have:

$$\boldsymbol{\Sigma}(-\mathbf{e}_i) = -\boldsymbol{\Sigma}(\mathbf{e}_i), \qquad i = 1, 2, 3 \ .$$

Hence $\mathbf{f}_s = \{\boldsymbol{\Sigma}(\mathbf{n}) - \boldsymbol{\Sigma}(\mathbf{e}_i)\mathbf{e}_i.\mathbf{n}\}\delta A$.

We now apply Newton's Second Law of Motion to obtain the relation:

Rate of change of momentum = Resultant body force + resultant surface force. (6.7)

The term on the left-hand side of (6.7) and the first term on the right-hand side are both proportional to the fluid mass, and hence both approach zero as δV when the volume of the tetrahedron tends to zero. The second term on the right-hand side of (6.7), however, only approaches zero as δA. Hence if the equation is to be satisfied the coefficient of δA must vanish identically.

Thus $$\mathbf{\Sigma}(\mathbf{n}) - \mathbf{\Sigma}(\mathbf{e}_i)\mathbf{e}_i.\mathbf{n} = \mathbf{0} \ , \tag{6.8}$$

or $$\mathbf{\Sigma}(\mathbf{n}) - n_i\mathbf{\Sigma}(\mathbf{e}_i) = \mathbf{0} \ , \tag{6.9}$$

therefore $$\Sigma_k(\mathbf{n}) = n_i\Sigma_k(\mathbf{e}_i) = \sigma_{ki}n_i \ , \tag{6.10}$$

where $$\sigma_{ki} = \Sigma_k(\mathbf{e}_i) \ . \tag{6.11}$$

We have introduced the quantity σ_{ki} which is the k - component of the force acting on an element of surface perpendicular to $\mathbf{e}_i$. From (6.9) we observe that $\mathbf{\Sigma}(\mathbf{n})$ is a linear vector function of the vector $\mathbf{n}$. Thus the quantity σ_{ki} in (6.10) is, by definition, a component of a tensor, $\boldsymbol{\sigma}$, which has the property

$$\mathbf{\Sigma}(\mathbf{n}) = \boldsymbol{\sigma}.\mathbf{n} \ . \tag{6.12}$$

The tensor $\boldsymbol{\sigma}$ is known as the **stress tensor**.

6.1.3 Properties of the Stress Tensor

We have seen in the last section that the stress tensor has the property that $\boldsymbol{\sigma}.\mathbf{n}$ is the actual stress vector on a surface element whose unit outward normal is in the direction of the vector **n**.

Consider the fluid forces on a surface whose outward normal is in the direction of $\mathbf{e}_1$ (Fig. 6.2). The stress on this surface is given by:

$$\mathbf{\Sigma} = \boldsymbol{\sigma}.\mathbf{e}_1 \ , \tag{6.13}$$

or, written out in full,

$$\mathbf{\Sigma} = \begin{bmatrix} \sigma_{11} & \sigma_{12} & \sigma_{13} \\ \sigma_{21} & \sigma_{22} & \sigma_{23} \\ \sigma_{31} & \sigma_{32} & \sigma_{33} \end{bmatrix} \begin{bmatrix} 1 \\ 0 \\ 0 \end{bmatrix} \tag{6.14}$$

$$= (\sigma_{11}, \sigma_{21}, \sigma_{31}) \ . \tag{6.15}$$

The component of $\mathbf{\Sigma}$ normal to the surface, i.e. σ_{11}, is called the **normal stress.** The components of $\mathbf{\Sigma}$ within the surface, i.e. σ_{21}, σ_{31} are called the components of **shearing or tangential stress.**

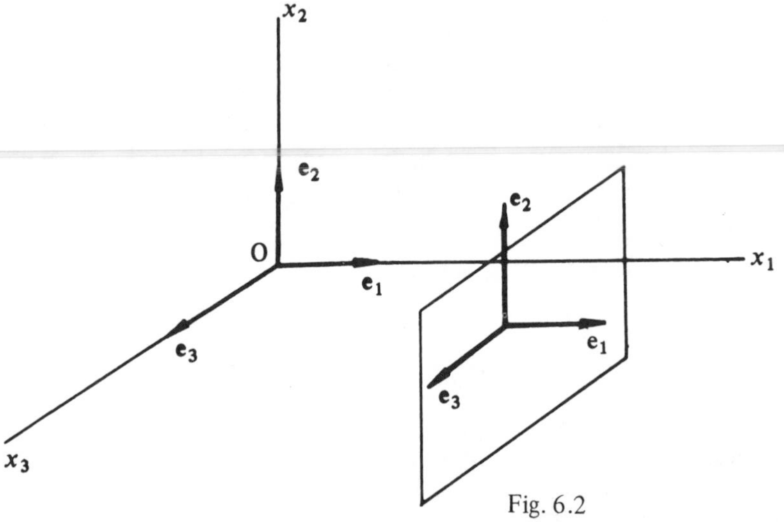

Fig. 6.2

In general, since the stress vector is $\boldsymbol{\sigma}.\mathbf{n}$, the normal stress is given by:

$$\sigma_n = \mathbf{n}.\boldsymbol{\sigma}.\mathbf{n} \ , \tag{6.16}$$

while the shearing stress is given by:

$$\sigma_t = |\boldsymbol{\sigma}.\mathbf{n} - (\mathbf{n}.\boldsymbol{\sigma}.\mathbf{n})\mathbf{n}| \ . \tag{6.17}$$

The shearing stress vector in the example we have been considering is:

$$\boldsymbol{\sigma}.\mathbf{n} - (\mathbf{n}.\boldsymbol{\sigma}.\mathbf{n})\mathbf{n} = \sigma_{21}\mathbf{e}_2 + \sigma_{31}\mathbf{e}_3 \ . \tag{6.18}$$

We have set out the properties of $\boldsymbol{\sigma}$ relative to a particular surface, but the tensor itself may be evaluated at any point without reference to a surface, just as the inertia tensor of any solid body may be evaluated without reference to a rotation. Unlike the inertia tensor, however, the stress tensor cannot be expressed at one point in terms of its value at another point by a theorem analagous to the Parallel Axis Theorem.

One important property which the stress tensor does share with the inertia tensor is symmetry. This result has far-reaching consequences, and we shall seek to establish it in section 6.1.5. First, however, it is necessary to incorporate the stress tensor into the equation of motion.

6.1.4 The Appearance of the Stress Tensor in the Equation of Fluid Motion

At the end of section 6.1.1 we left the equation of fluid motion in the form (6.5). By writing $\boldsymbol{\sigma}.\mathbf{n}$ for $\boldsymbol{\Sigma}$ as in (6.12), we may replace (6.5) by:

$$\int_\tau \rho \frac{\mathrm{D}\mathbf{u}}{\mathrm{D}t} \mathrm{d}\tau = \int_\tau \rho \mathbf{F} \mathrm{d}\tau + \int_S \boldsymbol{\sigma}.\mathrm{d}\mathbf{S} \ . \tag{6.19}$$

We now invoke the special form of Green's Theorem given by (3.74) and remember that $\boldsymbol{\sigma}.\mathrm{d}\mathbf{S} = \mathrm{d}\mathbf{S}.\boldsymbol{\sigma}'$, to convert the surface integral in (6.19) to a volume integral.

Thus
$$\int_\tau \rho \frac{\mathrm{D}\mathbf{u}}{\mathrm{D}t}\,\mathrm{d}\tau = \int_\tau \rho \mathbf{F}\,\mathrm{d}\tau + \int_\tau \operatorname{div} \boldsymbol{\sigma}'\,\mathrm{d}\tau \ . \tag{6.20}$$

Now (6.20) is true for any arbitrary material volume of fluid τ. We now therefore let $\tau \to 0$ to obtain:

$$\frac{\mathrm{D}\mathbf{u}}{\mathrm{D}t} = \mathbf{F} + \frac{1}{\rho}\operatorname{div} \boldsymbol{\sigma}' \ . \tag{6.21}$$

6.1.5 The Symmetry of the Stress Tensor

We are now in a position to show that $\boldsymbol{\sigma}$ is a symmetric tensor. We consider, as in section 6.1.1, an arbitrary material volume of fluid. The linear momentum of this body of fluid is given by (6.1). Therefore the angular momentum (which is the moment of the linear momentum) about the origin O is given by:

$$\int_\tau \rho \mathbf{r} \times \mathbf{u}\,\mathrm{d}\tau \ .$$

Hence the rate of change of angular momentum is given by:

$$\int_\tau \rho \mathbf{r} \times \frac{\mathrm{D}\mathbf{u}}{\mathrm{D}t}\,\mathrm{d}\tau \ . \tag{6.22}$$

Now the rate of change of angular momentum is equal to the applied couple. This couple may be divided into two parts, the moment of the applied forces and a pure couple, $\boldsymbol{\Gamma}(\mathrm{O})$ per unit mass. The moment of the applied forces is given by:

$$\int_\tau \rho \mathbf{r} \times \mathbf{F}\,\mathrm{d}\tau + \int_S \mathbf{r} \times \boldsymbol{\sigma}.\mathrm{d}\mathbf{S} \ , \tag{6.23}$$

and the pure couple by:

$$\int_\tau \rho \boldsymbol{\Gamma}(\mathrm{O})\,\mathrm{d}\tau \ . \tag{6.24}$$

Equating (6.22) with the sum of (6.23) and (6.24), at the same time converting the surface integral in (6.23) to a volume integral, using the special case of Green's theorem contained in Problems 3.5 No. 1 (iii), we see that:

$$\int_\tau \left[\rho \mathbf{r} \times \frac{\mathrm{D}\mathbf{u}}{\mathrm{D}t} - \rho \mathbf{r} \times \mathbf{F} - \mathbf{r} \times \operatorname{div} \boldsymbol{\sigma}' + 2\,\mathrm{vec}\,\boldsymbol{\sigma} - \rho \boldsymbol{\Gamma}(\mathrm{O})\right]\mathrm{d}\tau = \mathbf{0} \ . \tag{6.25}$$

Hence, by virtue of (6.21):

$$\int_\tau [2\,\mathrm{vec}\,\boldsymbol{\sigma} - \rho\boldsymbol{\Gamma}(\mathrm{O})]\,\mathrm{d}\tau = \mathbf{0} \,. \tag{6.26}$$

If we now assume that there is no pure couple $[\boldsymbol{\Gamma}(\mathrm{O}) = \mathbf{0}]$, and let $\tau \to 0$, (6.26) reduces to:

$$\mathrm{vec}\,\boldsymbol{\sigma} = \mathbf{0} \,. \tag{6.27}$$

This shows that in the absence of pure couple the stress tensor is symmetric (see Examples 3.3, No. 2). We may therefore rewrite the equation of motion (6.21) as:

$$\frac{\mathrm{D}\mathbf{u}}{\mathrm{D}t} = \mathbf{F} + \frac{1}{\rho}\,\mathrm{div}\,\boldsymbol{\sigma} \,. \tag{6.28}$$

We note further that it is possible to diagonalize the stress tensor according to the theory of section 2.12.2. The elements of the diagonal tensor are called the **principal stresses**, and act normally on surfaces parallel to the coordinate planes.

6.1.6 Decomposition of the Stress Tensor

In a fluid at rest there are no shear stresses. The only force acting on a material element of fluid is the pressure, which acts normally at every point on the surface of that element. The stress tensor is therefore given by:

$$\boldsymbol{\sigma} = -p\mathbf{1} \,, \tag{6.29}$$

or

$$\sigma_{ik} = -p\delta_{ik} \,, \tag{6.30}$$

and the stress vector by:

$$\boldsymbol{\sigma}.\mathbf{n} = -p\mathbf{n} \,. \tag{6.31}$$

From (6.30) we see that p is given by

$$p = -\tfrac{1}{3}\sigma_{ii} \,. \tag{6.32}$$

For a fluid in motion, therefore, it is logical to define the pressure as minus the mean normal stress. Thus we write:

$$\sigma_{ik} = -p\delta_{ik} + d_{ik} \,, \tag{6.33}$$

or

$$\boldsymbol{\sigma} = -p\mathbf{1} + \mathbf{d} \,, \tag{6.34}$$

where

$$d_{ik} = \sigma_{ik} - \tfrac{1}{3}\sigma_{jj}\delta_{ik} \,. \tag{6.35}$$

The tensor **d**, whose components are given by (6.35), is called the **deviatoric stress tensor**. It follows at once from (6.35) that **d** has zero trace, i.e. $$d_{ii} = 0 \ . \tag{6.36}$$

We have thus split the stress tensor into two parts, an isotropic part $(-p\mathbf{1})$ and the deviatoric part $\mathbf{d}$ which has zero trace. The isotropic tensor corresponds to a change of volume without change of shape. Consider, for example, a spherical droplet. For a positive value of p (negative σ_{ii}) the droplet will be compressed until an equilibrium state is achieved in which the internal forces resisting compression balance the external pressure. The spherical shape will be maintained. Similarly a positive value of σ_{ii} will cause the sphere to expand.

The deviatoric stress tensor, which is entirely due to fluid motion, corresponds to a change of shape with no change of volume.

Examples 6.1

(1) The components σ_{ik} of the stress tensor obey the transformation law for tensors.

We have, from (6.11):

$$\sigma_{ik} = \Sigma_i(\mathbf{e}_k) \ ,$$

therefore
$$\begin{aligned} \sigma^*_{jl} &= \Sigma_j^*(\mathbf{e}^*_l) \ , \\ &= \theta_{ji}\Sigma_i(\theta_{lk}\mathbf{e}_k) \ , \\ &= \theta_{ji}\theta_{lk}\Sigma_i(\mathbf{e}_k) \ , \end{aligned}$$

since $\boldsymbol{\Sigma}_i(\mathbf{n})$ is a linear function of $\mathbf{n}$.

Hence $$\sigma^*_{jl} = \theta_{ji}\theta_{lk}\sigma_{ik} \ .$$

But this is the transformation law for tensors.

(2) The stress tensor at a certain point in a fluid is:

$$\boldsymbol{\sigma} = \begin{bmatrix} 7 & -1 & -2 \\ -1 & 7 & 2 \\ -2 & 2 & 4 \end{bmatrix} .$$

Find the principal stresses.

This is just the eigenvalue problem (see section 2.9.1). The characteristic equation is:

$$\begin{vmatrix} 7-\lambda & -1 & -2 \\ -1 & 7-\lambda & 2 \\ -2 & 2 & 4-\lambda \end{vmatrix} = 0 \ ,$$

i.e. $(\lambda - 6)(\lambda^2 - 12\lambda + 24) = 0$.

$$\text{Hence} \quad \boldsymbol{\sigma}^* = \begin{bmatrix} 6 + 2\sqrt{3} & 0 & 0 \\ 0 & 6 - 2\sqrt{3} & 0 \\ 0 & 0 & 6 \end{bmatrix} .$$

These are the principal stresses.

(3) In many problems it is required to find the orientation of the surface on which the normal stress is a maximum or a minimum. The problem of finding a stationary value of the normal stress may be reduced to the eigenvalue problem as we shall now show.

Consider a surface characterized by its unit outward normal **n**. The normal component of stress on this surface is given by (6.16):

$$\sigma_{\mathrm{n}} = \mathbf{n}.\boldsymbol{\sigma}.\mathbf{n} .$$

In component form this becomes:

$$\sigma_{\mathrm{n}} = n_i \sigma_{ij} n_j . \tag{6.37}$$

We wish to find stationary values of σ_{n} with respect to variations of n_1, n_2, n_3, subject to the restriction:

$$n_1^2 + n_2^2 + n_3^2 = 1, \text{ or } n_i n_i = 1 . \tag{6.38}$$

This is the classic Lagrange Multiplier problem. We write:

$$\sigma_{\mathrm{n}}^* = \sigma_{ij} n_i n_j - \lambda(n_i n_i - 1) . \tag{6.39}$$

$$\text{Therefore} \quad \frac{\partial \sigma_{\mathrm{n}}^*}{\partial n_k} = \sigma_{ij} \frac{\partial}{\partial n_k}(n_i n_j) - \lambda \frac{\partial}{\partial n_k}(n_i n_i) ,$$

$$= n_i \sigma_{ik} + n_j \sigma_{kj} - 2\lambda n_k ,$$

$$= 2(\boldsymbol{\sigma}.\mathbf{n} - \lambda \mathbf{n})_k . \tag{6.40}$$

To find stationary values of σ_{n}^* we put:

$$\frac{\partial \sigma_{\mathrm{n}}^*}{\partial n_k} = 0, k = 1, 2, 3, \tag{6.41}$$

$$\text{i.e.} \quad \boldsymbol{\sigma}.\mathbf{n} - \lambda \mathbf{n} = \mathbf{0} . \tag{6.42}$$

But this is just the eigenvalue problem.

As an example consider a fluid for which the stress tensor at the origin is given by:

$$\boldsymbol{\sigma} = \begin{bmatrix} 1 & 2 & 0 \\ 2 & 1 & 0 \\ 0 & 0 & -2 \end{bmatrix} .$$

When diagonalized (see section 2.12.2) the stress tensor becomes:

$$\boldsymbol{\sigma}^* = \begin{bmatrix} 3 & 0 & 0 \\ 0 & -1 & 0 \\ 0 & 0 & -2 \end{bmatrix} .$$

Thus the principal stresses are 3, −1 and −2, and the maximum normal stress is 3. The outward normals to the surfaces on which the principal stresses act are the eigenvectors of the stress tensor, namely (1, 1, 0); (1, −1, 0) and (0, 0, 1).

(4) The problem of finding stationary values of the shearing stress is slightly more difficult than the corresponding problem for normal stress. This is because the shearing stress vector can be anywhere within the surface on which it acts. We shall suppose that the stress tensor is referred to its principal axes:

$$\boldsymbol{\sigma} = \begin{bmatrix} \sigma_{11} & 0 & 0 \\ 0 & \sigma_{22} & 0 \\ 0 & 0 & \sigma_{33} \end{bmatrix} . \tag{6.43}$$

The stress at any point on a surface whose outward normal is **n** is given by:

$$\boldsymbol{\Sigma} = \boldsymbol{\sigma}.\mathbf{n} = (\sigma_{11}n_1, \sigma_{22}n_2, \sigma_{33}n_3) . \tag{6.44}$$

The normal stress vector is therefore given by:

$$\begin{aligned} \boldsymbol{\Sigma}_n &= (\mathbf{n}.\boldsymbol{\sigma}.\mathbf{n})\mathbf{n} \\ &= (\sigma_{11}n_1^2 + \sigma_{22}n_2^2 + \sigma_{33}n_3^2)\mathbf{n} \end{aligned} \tag{6.45}$$

The difference between these two must be the shearing stress:

$$\boldsymbol{\Sigma}_t = \boldsymbol{\Sigma} - \boldsymbol{\Sigma}_n \tag{6.46}$$

Let $\sigma_t = |\boldsymbol{\Sigma}_t|, \quad \sigma_n = |\boldsymbol{\Sigma}_n|$.

Then $$\sigma_t^2 = \boldsymbol{\Sigma}^2 + \boldsymbol{\Sigma}_n{}^2 - 2\boldsymbol{\Sigma}.\boldsymbol{\Sigma}_n$$

$$= \sigma_{11}^2 n_1^2 + \sigma_{22}^2 n_2^2 + \sigma_{33}^2 n_3^2 - (\sigma_{11} n_1^2 + \sigma_{22} n_2^2 + \sigma_{33} n_3^2)^2. \tag{6.47}$$

We will now look for stationary values of σ_t^2 by the method of Lagrange multipliers.

Let $$(\sigma_t^2)^* = \sigma_t^2 + \lambda(n_1^2 + n_2^2 + n_3^2 - 1)\,, \tag{6.48}$$

then $$\frac{\partial(\sigma_t^2)^*}{\partial n_1} = 2\sigma_{11}^2 n_1 - 2(\sigma_{11} n_1^2 + \sigma_{22} n_2^2 + \sigma_{33} n_3^2) 2\sigma_{11} n_1 + 2\lambda n_1 \tag{6.49}$$

and two similar results. From the condition:

$$\frac{\partial(\sigma_t^2)^*}{\partial n_1} = \frac{\partial(\sigma_t^2)^*}{\partial n_2} = \frac{\partial(\sigma_t^2)^*}{\partial n_3} = 0\,, \tag{6.50}$$

we obtain:

$$\left.\begin{aligned} \lambda + \sigma_{11}^2 &= 2\sigma_{11}(\sigma_{11} n_1^2 + \sigma_{22} n_2^2 + \sigma_{33} n_3^2) \text{ or } n_1 = 0\,, \\ \lambda + \sigma_{22}^2 &= 2\sigma_{22}(\sigma_{11} n_1^2 + \sigma_{22} n_2^2 + \sigma_{33} n_3^2) \text{ or } n_2 = 0\,, \\ \lambda + \sigma_{33}^2 &= 2\sigma_{33}(\sigma_{11} n_1^2 + \sigma_{22} n_2^2 + \sigma_{33} n_3^2) \text{ or } n_3 = 0\,, \end{aligned}\right\} \tag{6.51}$$

while (6.38) has again to be satisfied. Solutions of (6.51) which satisfy (6.38) are:

$$\left.\begin{aligned} \mathbf{n} &= (0, 1/\sqrt{2}, \pm 1/\sqrt{2}); & \sigma_t{}^2 &= \tfrac{1}{4}(\sigma_{22} - \sigma_{33})^2\,, \\ \mathbf{n} &= (1/\sqrt{2}, 0, \pm 1/\sqrt{2}); & \sigma_t{}^2 &= \tfrac{1}{4}(\sigma_{33} - \sigma_{11})^2\,, \\ \mathbf{n} &= (1/\sqrt{2}, \pm 1/\sqrt{2}, 0); & \sigma_t{}^2 &= \tfrac{1}{4}(\sigma_{11} - \sigma_{22})^2\,. \end{aligned}\right\} \tag{6.52}$$

The other set of solutions, namely $\mathbf{n} = (1, 0, 0)$ etc., together with $\sigma_t^2 = 0$, are trivial and follow at once from the fact that we have chosen to refer $\boldsymbol{\sigma}$ to principal axes.

Let us consider again the stress tensor of the previous example, which when diagonalized became:

$$\boldsymbol{\sigma} = \begin{bmatrix} 3 & 0 & 0 \\ 0 & -1 & 0 \\ 0 & 0 & -2 \end{bmatrix}.$$

The stationary values of σ_t^2 together with the normals to the surfaces on which the shear stresses act are:

$$\left.\begin{aligned} \mathbf{n} &= (0, 1/\sqrt{2}, \pm 1/\sqrt{2}); & \sigma_t^2 &= 1/4\,, \\ \mathbf{n} &= (1/\sqrt{2}, 0, \pm 1/\sqrt{2}); & \sigma_t^2 &= 25/4\,, \\ \mathbf{n} &= (1/\sqrt{2}, \pm 1/\sqrt{2}, 0); & \sigma_t^2 &= 16/4\,. \end{aligned}\right\}$$

Hence the maximum shear stress has a magnitude 5/2.

(5) A small cube has one corner at the origin and its faces parallel to the co-ordinate planes (Fig. 6.3). The stress tensor, which is assumed to be constant in the vicinity of the cube, is:

$$\boldsymbol{\sigma} = \begin{bmatrix} 2 & 0 & 0 \\ 0 & 1 & 0 \\ 0 & 0 & -1 \end{bmatrix}.$$

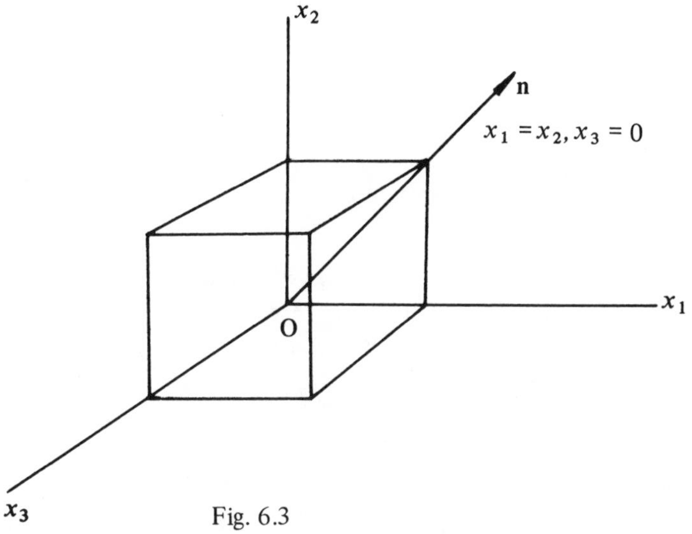

Fig. 6.3

The cube is now rotated through an angle of 90°_0 about the line $x_1 = x_2$, $x_3 = 0$ in the negative sense (i.e. the cube finishes up 'above' the plane $x_1 = x_2$). Find the stresses on the faces of the cube after it has been rotated.

We need to give the coordinate axes the same rotation as that given to the cube. From section 3.4.1 the components of the rotation matrix are given:

$$\theta_{ij} = n_i n_j (1 - \cos\theta) + \cos\theta\,\delta_{ij} + \epsilon_{ijk} n_k \sin\theta\ ,$$

where $\cos\theta = 0, \qquad \sin\theta = -1, \mathbf{n} = (1/\sqrt{2}, 1/\sqrt{2}, 0),$

i.e. $\theta_{ij} = n_i n_j - \epsilon_{ijk} n_k$.

Hence the rotation matrix Θ is given by:

$$\Theta = \begin{bmatrix} 1/2 & 1/2 & 1/\sqrt{2} \\ 1/2 & 1/2 & -1/\sqrt{2} \\ -1/\sqrt{2} & 1/\sqrt{2} & 0 \end{bmatrix}.$$

The components of the stress tensor with respect to the rotated set of axes are given by:

$$\boldsymbol{\sigma}^* = \Theta\boldsymbol{\sigma}\Theta' ,$$

or

$$\boldsymbol{\sigma}^* = \frac{1}{4}\begin{bmatrix} 1 & 1 & \sqrt{2} \\ 1 & 1 & -\sqrt{2} \\ -\sqrt{2} & \sqrt{2} & 0 \end{bmatrix}\begin{bmatrix} 2 & 0 & 0 \\ 0 & 1 & 0 \\ 0 & 0 & -1 \end{bmatrix}\begin{bmatrix} 1 & 1 & -\sqrt{2} \\ 1 & 1 & \sqrt{2} \\ \sqrt{2} & -\sqrt{2} & 0 \end{bmatrix},$$

i.e.

$$\boldsymbol{\sigma}^* = \begin{bmatrix} 1/4 & 5/4 & -\sqrt{2}/4 \\ 5/4 & 1/4 & -\sqrt{2}/4 \\ -\sqrt{2}/4 & -\sqrt{2}/4 & 3/2 \end{bmatrix}.$$

This matrix is to be interpreted as follows. The normal stress on the (x_2, x_3)-face, whose unit normal is $\mathbf{e}_1^*$ after it has been rotated, is $\frac{1}{4}$. The shearing stress on that face after rotation is $(5/4)\mathbf{e}_2^* - (\sqrt{2}/4)\mathbf{e}_3^*$.

(6) Plane Stress. When the stress tensor is given by

$$\boldsymbol{\sigma} = \begin{bmatrix} \sigma_{11} & \sigma_{12} & 0 \\ \sigma_{12} & \sigma_{22} & 0 \\ 0 & 0 & 0 \end{bmatrix},$$

the distribution of stress is said to be plane. Let us suppose, for convenience, that $\sigma_{11} \geqslant \sigma_{22}$. There is no loss of generality here. If $\sigma_{22} > \sigma_{11}$ we may re-label the x_1- and x_2- axes. Let the axes suffer a rotation by an angle θ in the positive sense about the x_3-axis. We may refer to equations (3.65) to write down the new normal and shearing stresses:

$$\sigma_{11}^* = \tfrac{1}{2}(\sigma_{11} + \sigma_{22}) + \tfrac{1}{2}(\sigma_{11} - \sigma_{22})\cos 2\theta + \sigma_{12}\sin 2\theta ,$$

$$\sigma_{12}^* = -\tfrac{1}{2}(\sigma_{11} - \sigma_{22})\sin 2\theta + \sigma_{12}\cos 2\theta .$$

Now $\quad \partial\sigma_{11}^* / \partial\theta = -(\sigma_{11} - \sigma_{22})\sin 2\theta + 2\sigma_{12}\cos 2\theta = 2\sigma_{12}^* \quad .$

Hence the maximum and minimum values of the normal stress σ_{11}^* occur when the shearing stress σ_{12}^* vanishes. They are given by:

$$\sigma_{11}^*(\max) = \tfrac{1}{2}(\sigma_{11} + \sigma_{22}) + \sqrt{[(\sigma_{11} - \sigma_{22})^2 + (2\sigma_{12})^2]}$$
$$\sigma_{11}^*(\min) = \tfrac{1}{2}(\sigma_{11} + \sigma_{22}) - \sqrt{[(\sigma_{11} - \sigma_{22})^2 + (2\sigma_{12})^2]} \quad .$$

However, armed with the information that σ_{11}^* is stationary when σ_{12}^* is zero, it is more convenient to start again, referring the stress tensor to its principal axes. We may now use equations (3.65) to write down the effect of the rotation:

$$\sigma_{11}^* = \tfrac{1}{2}(\sigma_{11} + \sigma_{22}) + \tfrac{1}{2}(\sigma_{11} - \sigma_{22})\cos 2\theta \quad ,$$
$$\sigma_{12}^* = -\tfrac{1}{2}(\sigma_{11} - \sigma_{22})\sin 2\theta \quad .$$

It has already been remarked that equations (3.65) are the parametric equations of a circle with respect to Cartesian axes which are in this case in the directions of σ_{11}^* and σ_{12}^* increasing. This circle is known as **Mohr's circle of stress.** It is a circle centre $[\tfrac{1}{2}(\sigma_{11} + \sigma_{22}), 0]$, and radius $\tfrac{1}{2}(\sigma_{11} - \sigma_{22})$. Clearly the maximum value of σ_{11}^* is σ_{11}, which occurs when $\cos 2\theta = 1$. The minimum value of σ_{11}^* is σ_{22}, which occurs when $\cos 2\theta = -1$. In both cases $\sigma_{12}^* = 0$.

The maximum value of σ_{12}^* occurs when $\sin 2\theta = -1$ and is equal to $\tfrac{1}{2}(\sigma_{11} - \sigma_{22})$. Hence we have the result that the maximum shearing stress is equal to half the difference between the maximum and minimum normal stresses. This result may be extended to the general case (see Problem (5)).

Problems 6.1

(1) The stress tensor at the origin in a certain fluid is given by:

$$\boldsymbol{\sigma} = \begin{bmatrix} 1 & 2 & 0 \\ 2 & 1 & 0 \\ 0 & 0 & 2 \end{bmatrix} .$$

Find the normal stress on the plane $2x_1 + 3x_2 + 4x_3 = 0$. Find also the equation of the plane through the origin on which the normal stress is a maximum.

(2) A cube is centred on the origin with its faces perpendicular to the coordinate axes. The cube is subjected to normal stresses $\sigma_{11} = 6, \sigma_{22} = 4$ and $\sigma_{33} = 1$. Write down the stress tensor and determine the normal and shearing stresses on the cube after it has been rotated by an angle of $\pi/3$, in a positive sense about the line $x_1 = x_2 = x_3$.

(3) The stress tensor at the origin in a fluid is:

$$\boldsymbol{\sigma} = \begin{bmatrix} \frac{1}{4} & 0 & \frac{3\sqrt{3}}{4} \\ 0 & -2 & 0 \\ \frac{3\sqrt{3}}{4} & 0 & \frac{-5}{4} \end{bmatrix}.$$

Consider the stress on a small surface element through O with unit normal **n**. Show that those vectors **n** for which the stress is parallel to **n** satisfy one of the following two relations:

$$\text{(i) } \mathbf{n} \times \mathbf{n}_0 = \mathbf{0} \, ,$$

$$\text{(ii) } \mathbf{n}.\mathbf{n}_0 = 0 \, ,$$

where $\mathbf{n}_0$ is a fixed unit vector (which should be found).

(Liverpool University, June 1974)

(4) Prove that the normal stress has a stationary value when the shear stress is zero.

(5) Prove that the maximum shearing stress is equal to half the difference between the maximum and minimum normal stress.

(6) The stress tensor at the origin in a fluid is given by:

$$\boldsymbol{\sigma} = \begin{bmatrix} 1 & 0 & 3\sqrt{3} \\ 0 & -8 & 0 \\ 3\sqrt{3} & 0 & -5 \end{bmatrix}.$$

Let the unit vector **n** have components:

$$\mathbf{n} = (\sin\theta\cos\phi, \sin\theta\sin\phi, \cos\theta) \, .$$

Show that the normal component of stress on a surface whose outward normal is **n** is given by:

$$\sigma_n = (3\sin\theta\cos\phi + \sqrt{3}\cos\theta)^2 - 8 \, .$$

Hence find the direction of the normal to a surface such that the normal stress shall be a maximum. Show also that the normal stress may take its minimum value provided that $|\sin\theta| > \frac{1}{2}$. Compare these results with those obtained by diagonalizing the matrix representing the stress tensor.

(7) A tetrahedron of fluid has its vertices at $A(\delta, 0, 0)$; $B(0, \delta, 0)$; $C(0, 0, \delta)$; $D(\frac{1}{2}\delta, \frac{1}{2}\delta, \frac{1}{2}\delta)$ respectively, where δ is so small that the stress tensor can be assumed constant throughout the tetrahedron. Show that the unit normal outward from the face BCD has components $(0, \sqrt{2}/2, \sqrt{2}/2)$.

The stresses on the faces BCD, ADC, DAB have the components

$$(1/\sqrt{2})(2, 4, -9),\ (1/\sqrt{2})(-1, 1, -8),\ (1/\sqrt{2})(3, 9, -3)$$

respectively. Find the components of the stress tensor, and hence show that the stress on the face ABC has components $(1/\sqrt{3})(-2, -7, 10)$.

Verify that the total force on the tetrahedron due to stresses is zero.

6.2 ANALYSIS OF FLUID MOTION

6.2.1 The Rate of Strain Tensor

In this section we propose to analyse the motion in the neighbourhood of a typical point within a fluid. We shall see that tensors are useful in carrying out the analysis. We shall then be in a position to put forward a hypothesis which relates one of these tensors to the stress tensor, a hypothesis which will enable us to express the equation of fluid motion in a usable form.

For the time being we shall confine ourselves to kinematics and suppose that a fluid is in motion, without taking any account of the forces responsible for that motion.

Suppose the velocity at the point $\mathbf{r}$ in a fluid, at time t, is $\mathbf{v} = \mathbf{v}(\mathbf{r}, t)$. At a neighbouring point, $\mathbf{r} + \delta\mathbf{r}$, the velocity is given by $\mathbf{v} + \delta\mathbf{v}$ where

$$\mathbf{v} + \delta\mathbf{v} = \mathbf{v}(\mathbf{r} + \delta\mathbf{r}, t) \ . \tag{6.53}$$

Now $$\mathbf{v}(\mathbf{r} + \delta\mathbf{r}, t) = \mathbf{v}(\mathbf{r}) + \delta\mathbf{r}.\nabla\mathbf{v} + \frac{1}{2!}(\delta\mathbf{r}.\nabla)^2\mathbf{v} + \ldots \tag{6.54}$$

therefore $$\delta\mathbf{v} = \delta\mathbf{r}.\nabla\mathbf{v} \ , \tag{6.55}$$

where (6.55) is correct to the 1st order in $|\delta\mathbf{r}|$. Now $\delta\mathbf{r}.\nabla\mathbf{v}$ may be thought of either as the operator $\delta\mathbf{r}.\nabla$ operating on the vector $\mathbf{v}$ or as the inner product of the vector $\delta\mathbf{r}$ and the tensor $\nabla\mathbf{v}$ (defined in section 2.15.2), in which case we may write:

$$\delta\mathbf{r}.\nabla\mathbf{v} = (\nabla\mathbf{v})'.\delta\mathbf{r} \ . \tag{6.56}$$

The vector $\delta\mathbf{v}$ represents the velocity of the fluid at $\mathbf{r} + \delta\mathbf{r}$ relative to the velocity at $\mathbf{r}$.

We now split the transposed tensor $(\nabla\mathbf{v})'$ into its symmetric and antisymmetric parts, $\mathbf{e}$ and $\boldsymbol{\xi}$ respectively.

Thus $$(\nabla\mathbf{v})' = \mathbf{e} + \boldsymbol{\xi} \ , \tag{6.57}$$

where $$\mathbf{e} = \tfrac{1}{2}[(\nabla\mathbf{v})' + \nabla\mathbf{v}] \ , \tag{6.58}$$

and $$\boldsymbol{\xi} = \tfrac{1}{2}[(\nabla\mathbf{v})' - \nabla\mathbf{v}] \ . \tag{6.59}$$

The symmetric tensor $\mathbf{e}$ is known as the **rate of strain tensor.** It makes, as we shall see in the following sections, a contribution to the relative velocity which is basically different from that made by the antisymmetric tensor $\boldsymbol{\xi}$.

6.2.2 The Contribution of the Rate of Strain Tensor to the Relative Velocity

Since by (6.55) and (6.56) we may write:

$$\delta\mathbf{v} = (\nabla\mathbf{v})'.\delta\mathbf{r} \ , \tag{6.60}$$

we may split $\delta\mathbf{v}$ into a 'symmetric' part $\delta\mathbf{v}^{(s)}$, and an 'antisymmetric' part $\delta\mathbf{v}^{(a)}$, where

$$\delta\mathbf{v}^{(s)} = \mathbf{e}.\delta\mathbf{r} \ , \tag{6.61}$$

and

$$\delta\mathbf{v}^{(a)} = \boldsymbol{\xi}.\delta\mathbf{r} \ . \tag{6.62}$$

We shall concentrate first on the 'symmetric' part. Let us suppose that the rate of strain tensor $\mathbf{e}$ is referred to principal axes.

Then

$$\mathbf{e} = \begin{bmatrix} \partial v_1/\partial x_1 & 0 & 0 \\ 0 & \partial v_2/\partial x_2 & 0 \\ 0 & 0 & \partial v_3/\partial x_3 \end{bmatrix} , \tag{6.63}$$

and

$$\delta\mathbf{v}^{(s)} = \left(\frac{\partial v_1}{\partial x_1}\,\delta x_1 , \frac{\partial v_2}{\partial x_2}\,\delta x_2 , \frac{\partial v_3}{\partial x_3}\,\delta x_3\right) . \tag{6.64}$$

The kind of motion represented by (6.64), which is a special case of the motion

$$\delta\mathbf{v}^{(s)} = (\alpha_1\delta x_1 , \alpha_2\delta x_2 , \alpha_3\delta x_3) \ , \tag{6.65}$$

is known as **pure straining motion**. In such a motion a line element which was originally parallel to the x_1-axis would be stretched parallel to that axis. A spherical droplet for example would be distorted into an ellipsoid whose principal axes coincided with those of the rate of strain tensor.

6.2.3 The Rate of Strain

The rate of strain of a line element parallel to the unit vector **n** is defined to be **e.n**. The component of the rate of strain in the direction of the line element is therefore given by **n.e.n**. We notice at once the similarity between these formulae and the formulae for stress and normal stress (6.12) and (6.16). Now **e** is symmetric and given by:

$$\mathbf{e} = \tfrac{1}{2}[(\nabla\mathbf{v})' + \nabla\mathbf{v}] \ .$$

Hence

$$\mathbf{n}.\mathbf{e}.\mathbf{n} = \mathbf{n}.\nabla\mathbf{v}.\mathbf{n} \ . \tag{6.66}$$

Consider the rate of strain of an elementary line element AB (Fig. 6.4) of length $l = |\delta\mathbf{r}|$.

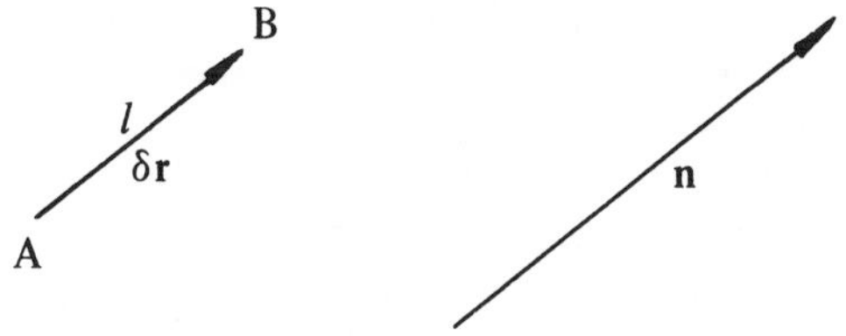

Fig. 6.4

If **v** is the velocity field, $\dot{l} = \delta\mathbf{v}.\mathbf{n}$, where **n** is a unit vector in the direction of AB.

Hence from (6.55) $\dot{l} = (\delta\mathbf{r}.\nabla\mathbf{v}).\mathbf{n}$, (6.67)

therefore $$\frac{1}{l}\frac{\mathrm{d}l}{\mathrm{d}t} = \mathbf{n}.\nabla\mathbf{v}.\mathbf{n} \ . \qquad (6.68)$$

We have recovered in (6.68) the elementary definition of rate of strain for an infinitesimal line element of length l.

6.2.4 The Contribution of the Antisymmetric Tensor ξ to the Relative Velocity

We now consider the 'antisymmetric' part of $\delta\mathbf{v}$ given by (6.62). The components of $\boldsymbol{\xi}$ are given by:

$$2\boldsymbol{\xi} = \begin{bmatrix} 0 & \dfrac{\partial v_1}{\partial x_2} - \dfrac{\partial v_2}{\partial x_1} & \dfrac{\partial v_1}{\partial x_3} - \dfrac{\partial v_3}{\partial x_1} \\ \dfrac{\partial v_2}{\partial x_1} - \dfrac{\partial v_1}{\partial x_2} & 0 & \dfrac{\partial v_2}{\partial x_3} - \dfrac{\partial v_3}{\partial x_2} \\ \dfrac{\partial v_3}{\partial x_1} - \dfrac{\partial v_1}{\partial x_3} & \dfrac{\partial v_3}{\partial x_2} - \dfrac{\partial v_2}{\partial x_3} & 0 \end{bmatrix} . \qquad (6.69)$$

Let us define a vector $\boldsymbol{\omega}$ by:

$$\boldsymbol{\omega} = -2 \operatorname{vec} \boldsymbol{\xi} = \mathbf{E}:\boldsymbol{\xi} \ . \qquad (6.70)$$

Then, from the result of Problems 3.3, No. 5,

$$\boldsymbol{\xi} = -\tfrac{1}{2} \operatorname{tens} \boldsymbol{\omega} = -\tfrac{1}{2}\mathbf{E}.\boldsymbol{\omega} \ . \qquad (6.71)$$

Hence, from (6.62):

$$\delta\mathbf{v}^{(a)} = -\tfrac{1}{2}(\mathbf{E}.\boldsymbol{\omega}).\delta\mathbf{r} \ , \qquad (6.72)$$

or $$\delta v_i^{(a)} = -\tfrac{1}{2}\epsilon_{ijk}\omega_k\delta x_j \ . \qquad (6.73)$$

i.e. $$\delta\mathbf{v}^{(a)} = \tfrac{1}{2}\boldsymbol{\omega}\times\delta\mathbf{r} \ . \qquad (6.74)$$

$\delta \mathbf{v}^{(a)}$ is thus the velocity at $\delta \mathbf{r}$ relative to $\mathbf{r}$ about which there is a rigid body rotation $\frac{1}{2}\boldsymbol{\omega}$. Hence $\boldsymbol{\omega}$ is twice the effective local angular velocity. For this reason, in view of the relation (6.71), the tensor $\boldsymbol{\xi}$ has been called the **Vorticity tensor, Angular velocity tensor** or **Spin tensor**.

In this and the previous section we have shown that the relative motion in the neighbourhood of a point within a fluid may be split into two distinct parts, namely a pure strain and a rigid body rotation, and these two parts are associated respectively with the symmetric and antisymmetric parts of the tensor $(\nabla \mathbf{v})'$.

Examples 6.2

(1) The vector $\boldsymbol{\omega}$ defined in section 6.2.4 is the **vorticity vector**, curl **v**.

We have
$$\omega_i = \epsilon_{ijk}\xi_{kj}$$
$$= \frac{1}{2}\epsilon_{ijk}\left(\frac{\partial v_k}{\partial x_j} - \frac{\partial v_j}{\partial x_k}\right),$$
$$= \epsilon_{ijk}\frac{\partial v_k}{\partial x_j} .$$

Hence
$$\boldsymbol{\omega} = \text{curl } \mathbf{v} . \tag{6.75}$$

(2) With any tensor there is associated a quadric surface $\mathbf{T}:\mathbf{r} \otimes \mathbf{r} = 1$, or $T_{ij}x_ix_j = 1$. The quadric surface $\mathbf{e}:\mathbf{r} \otimes \mathbf{r} = 1$ is called the **rate of strain quadric.** With respect to principal axes the equation of the quadric becomes:

$$e_{11}x_1^2 + e_{22}x_2^2 + e_{33}x_3^2 = 1 . \tag{6.76}$$

Writing this equation in the form $\phi(x_1, x_2, x_3) = 0$, it is clear that the direction of the normal is given by:

$$\nabla\phi = (e_{11}x_1, e_{22}x_2, e_{33}x_3) . \tag{6.77}$$

Now, from (6.63) and (6.64):

$$\delta \mathbf{v}^{s} = (e_{11}\delta x_1, e_{22}\delta x_2, e_{33}\delta x_3) .$$

Hence we have the result that $\delta \mathbf{v}^{(s)}$ is in the direction of the normal to the local rate of strain quadric.

(3) The components of the spin tensor written in terms of the vorticity vector $\boldsymbol{\omega}$ are:

$$\boldsymbol{\xi} = \frac{1}{2}\begin{bmatrix} 0 & -\omega_3 & \omega_2 \\ \omega_3 & 0 & -\omega_1 \\ -\omega_2 & \omega_1 & 0 \end{bmatrix} . \tag{6.78}$$

This follows at once from (6.69) and (6.75).

(4) Discuss the fluid motion represented by:

(i) $\mathbf{v} = kx_2\mathbf{e}_1$; (ii) $\mathbf{v} = k(x_2\mathbf{e}_1 + x_1\mathbf{e}_2)$; (iii) $\mathbf{v} = k(x_2\mathbf{e}_1 - x_1\mathbf{e}_2)$.

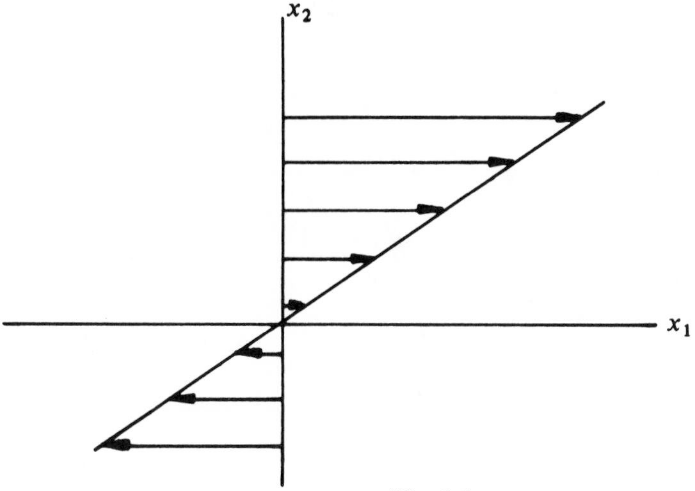

Fig. 6.5

(i) $\mathbf{v} = kx_2\mathbf{e}_1$ (Fig. 6.5).

There is only one non-zero component of the tensor $\nabla\mathbf{v}$, namely $(\nabla\mathbf{v})_{21} = \partial v_1/\partial x_2 = k$.

Thus
$$(\nabla\mathbf{v})' = \begin{bmatrix} 0 & k & 0 \\ 0 & 0 & 0 \\ 0 & 0 & 0 \end{bmatrix}, \quad \mathbf{e} = \frac{1}{2}\begin{bmatrix} 0 & k & 0 \\ k & 0 & 0 \\ 0 & 0 & 0 \end{bmatrix},$$

$$\boldsymbol{\xi} = \frac{1}{2}\begin{bmatrix} 0 & k & 0 \\ -k & 0 & 0 \\ 0 & 0 & 0 \end{bmatrix} \text{ from (6.58), (6.59)} \quad .$$

Hence $\boldsymbol{\omega} = (0, 0, k)$ from Example (1) above. This kind of motion is called **plane shear** or **simple shear**. The flow is everywhere parallel to the x_1-axis, with a velocity gradient perpendicular to that axis. A rectangle would be transformed into a parallelogram of the same area.

(ii) $\mathbf{v} = kx_2\mathbf{e}_1 + kx_1\mathbf{e}_2$. (Fig. 6.6).

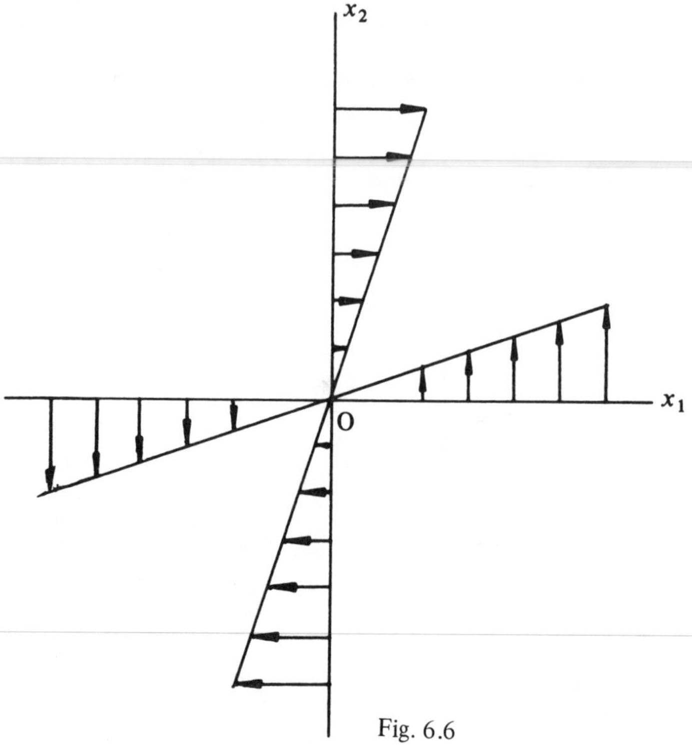

Fig. 6.6

Thus $$\mathbf{e} = \nabla\mathbf{v} = \begin{bmatrix} 0 & k & 0 \\ k & 0 & 0 \\ 0 & 0 & 0 \end{bmatrix}, \quad \boldsymbol{\xi} = \mathbf{0}, \; \boldsymbol{\omega} = \mathbf{0} \; .$$

This motion, with $\mathbf{e} \neq \mathbf{0}$, $\boldsymbol{\xi} = \mathbf{0}$ is pure strain (see section 6.2.2). It is instructive to refer $\mathbf{e}$ and $\mathbf{v}$ to the principal axes of $\mathbf{e}$ (which are inclined at 45° to the given axes) to obtain:

$$\mathbf{e}^* = \begin{bmatrix} k & 0 & 0 \\ 0 & -k & 0 \\ 0 & 0 & 0 \end{bmatrix}, \quad \mathbf{v}^* = k(x_1, -x_2, 0) \; .$$

The path lines of the fluid particles are shown in Fig. 6.7

(iii) $\mathbf{v} = kx_2\mathbf{e}_1 - kx_1\mathbf{e}_2$. (Fig. 6.8).

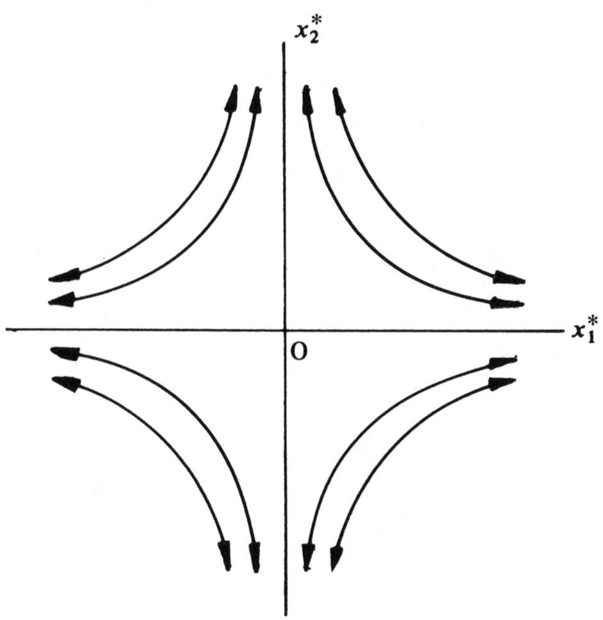

Fig. 6.7

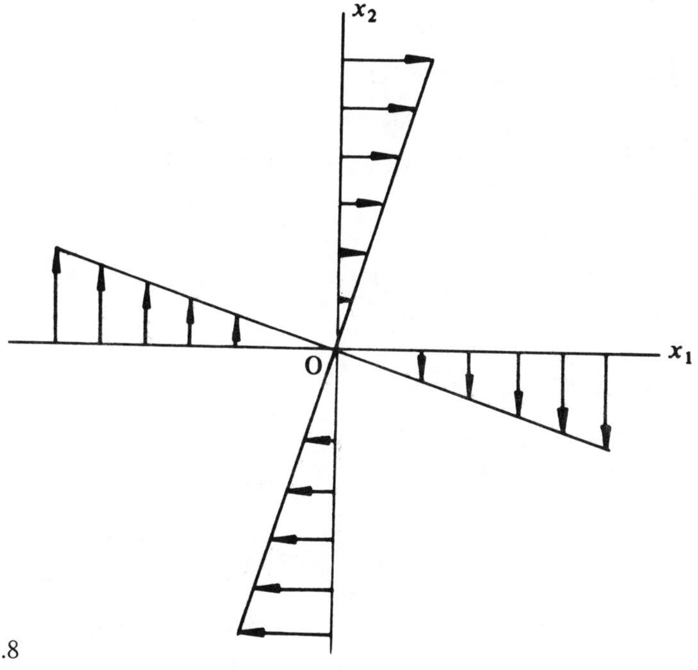

Fig. 6.8

Thus $\nabla \mathbf{v} = \begin{bmatrix} 0 & -k & 0 \\ k & 0 & 0 \\ 0 & 0 & 0 \end{bmatrix}$, $\xi = (\nabla \mathbf{v})' = \begin{bmatrix} 0 & k & 0 \\ -k & 0 & 0 \\ 0 & 0 & 0 \end{bmatrix}$, and $\mathbf{e} = \mathbf{0}$.

Hence $\boldsymbol{\omega} = (0, 0, -2k)$.

The vector $\mathbf{v}$ is everywhere perpendicular to the radius vector from the origin. Let us compare this motion with that of a rigid body rotating about the x_3-axis with angular velocity $-k\mathbf{e}_3$. The velocity at any point is given by:

$$\mathbf{v} = -k\mathbf{e}_3 \times (x_1\mathbf{e}_1 + x_2\mathbf{e}_2 + x_3\mathbf{e}_3)$$

$$= k(x_2\mathbf{e}_1 - x_1\mathbf{e}_2) \ .$$

This is identical to the given velocity, which therefore represents rigid body rotation with angular velocity $\boldsymbol{\omega} = -k\mathbf{e}_3$.

Problems 6.2

(1) Let X, X_1 be neighbouring material points in a continuous body. Let $\mathbf{v}$ be the velocity field of the body. Show that the angular velocity of the material line element XX_1 is, to a first approximation:

$$\boldsymbol{\Omega} = \mathbf{n} \times [(\mathbf{n}.\nabla)\mathbf{v}] \ ,$$

where $\mathbf{n}$ is the unit vector parallel to XX_1. Show also that the rate of strain of the line element XX_1, defined as $\frac{1}{l}\frac{\mathrm{d}l}{\mathrm{d}t}$ where $l(t)$ is the length of XX_1, is equal to $\mathbf{n}.\nabla\mathbf{v}.\mathbf{n}$. Hence show that the motion is rigid if and only if the rate of strain tensor $\mathbf{e}$ vanishes.

(Liverpool University, June 1972).

(2) The vector $\mathbf{u}(\mathbf{r})$ is axisymmetric about an axis through the origin, parallel to the constant vector $\mathbf{n}$. Show that:

(i) $(\mathbf{n} \times \mathbf{r}).\nabla\mathbf{u} = \mathbf{n} \times \mathbf{u}$,
(ii) $[\mathbf{r}^2\mathbf{n} - (\mathbf{n}.\mathbf{r})\mathbf{r}].(\nabla \times \mathbf{u}) = \mathbf{r}.\nabla(\mathbf{n} \times \mathbf{r}.\mathbf{u})$.

(Liverpool University, June 1970).

(3) The velocity field in a fluid is given by:

$$\mathbf{v} = \frac{U}{a^3}\left[z(x^2 - y^2), z(x^2 + y^2), -z^2(x + y)\right] ,$$

where (x, y, z) are Cartesian coordinates, and U and a are constants. Show that div $\mathbf{v} = 0$.

Find the rate of strain tensor and the vorticity, and evaluate them at the point $(a, a, 0)$.

Determine how a small sphere centred on $(a, a, 0)$ is being instantaneously deformed.

(Liverpool University, June 1976).

6.3 THE NAVIER - STOKES EQUATIONS OF FLUID MOTION

In this section we return yet again to the equation of fluid motion which we left in the form (6.28):

$$\frac{D\mathbf{v}}{Dt} = \mathbf{F} + \frac{1}{\rho} \operatorname{div} \boldsymbol{\sigma} \ . \tag{6.79}$$

We have examined the properties of the stress tensor, $\boldsymbol{\sigma}$, and seen how to divide it into an isotropic part and a non-isotropic part with zero trace, called the derivatoric stress.

It is now necessary to determine how the kinematics of fluid motion respond to the forces imposed by the stress. This can only be done by hypothesis supported by experimental observation. For solid bodies the simplest assumption to make is that the stress depends linearly on the strain (Hooke's law in the one-dimensional case). Fluids, unlike solids, exhibit no strain. Put more crudely: a fluid cannot resist any attempt to change its shape. Fluids do, however, exhibit a resistance which depends on the rate at which the change in shape is being made. It is the deviatoric part of the stress tensor which is responsible for change in shape, a fact which leads us to the hypothesis which we must make in order to progress. The hypothesis is that the deviatoric stress is an isotropic function of the rate of strain,

i.e. $$\mathbf{d} = \mathbf{f}(\mathbf{e}) \ . \tag{6.80}$$

It follows therefore from the Representation Theorem (see section 2.14) for an isotropic tensor function that:

$$\mathbf{d} = \alpha \mathbf{1} + \beta \mathbf{e} + \gamma \mathbf{e}.\mathbf{e} \ , \tag{6.81}$$

or $$d_{ik} = \alpha \delta_{ik} + \beta e_{ik} + \gamma e_{im} e_{mk} \ , \tag{6.82}$$

where α, β, γ are functions of the scalar invariants of $\mathbf{e}$, the density and temperature. Fluids for which (6.82) is true are known as **Reiner - Rivlin** fluids.

The special case of a **Newtonian fluid** is derived from (6.82) by putting $\gamma = 0, \beta = \text{const.} = 2\mu$. Then, since by definition,

$$d_{ii} = 0 \ ,$$

whence $3\alpha + 2\mu e_{ii} = 0, \quad \text{or } \alpha = -\tfrac{2}{3}\mu e_{ii} \ .$

Therefore $$d_{ik} = 2\mu(e_{ik} - \tfrac{1}{3}\delta_{ik} e_{mm}) \ . \tag{6.83}$$

We are now in a position to substitute the expression for the derivatoric stress tensor into the equation of motion of (6.79). In component form (6.79) becomes:

$$\frac{Dv_i}{Dt} = F_i + \frac{1}{\rho}\frac{\partial \sigma_{ki}}{\partial x_k} \, .$$

But

$$\begin{aligned}\sigma_{ki} &= -p\delta_{ki} + d_{ki} \, , \\ &= -p\delta_{ki} + 2\mu(e_{ki} - \tfrac{1}{3}\delta_{ik}e_{mm}) \, , \qquad (6.84) \\ &= -p\delta_{ki} + 2\mu\left\{\tfrac{1}{2}\left(\frac{\partial v_i}{\partial x_k} + \frac{\partial v_k}{\partial x_i}\right) - \tfrac{1}{3}\delta_{ki}\frac{\partial v_m}{\partial x_m}\right\} .\end{aligned}$$

Therefore

$$\begin{aligned}\frac{\partial \sigma_{ki}}{\partial x_k} &= -\frac{\partial p}{\partial x_i} + \mu\left(\frac{\partial^2 v_i}{\partial x_k \partial x_k} + \frac{\partial^2 v_k}{\partial x_i \partial x_k}\right) - \frac{2\mu}{3}\frac{\partial^2 v_m}{\partial x_i \partial x_m} \, , \\ &= -\frac{\partial p}{\partial x_i} + \tfrac{1}{3}\mu\frac{\partial}{\partial x_i}(\text{div } \mathbf{v}) + \mu\frac{\partial^2 v_i}{\partial x_k \partial x_k} \, .\end{aligned}$$

Thus

$$\frac{Dv_1}{Dt} = F_i - \frac{1}{\rho}\frac{\partial p}{\partial x_i} + \frac{1}{\rho}\frac{\partial}{\partial x_k}\left[2\mu(e_{ik} - \tfrac{1}{3}\delta_{ik}e_{mm})\right] , \qquad (6.85)$$

or

$$\frac{D\mathbf{v}}{Dt} = \mathbf{F} - \frac{1}{\rho}\nabla p + \nu(\nabla^2\mathbf{v} + \tfrac{1}{3}\text{ grad div }\mathbf{v}) \, , \qquad (6.86)$$

where p is the pressure, $\nu = \mu/\rho$ is the kinematic viscosity, and div $\mathbf{v} = e_{mm}$ is the rate of dilation.

Equation (6.85) or (6.86) is known as the Navier - Stokes equation. A fluid for which div $\mathbf{v} = 0$ is said to be **incompressible**, i.e. a liquid. For such fluids the Navier - Stokes equation reduces to

$$\frac{D\mathbf{v}}{Dt} = \mathbf{F} - \frac{1}{\rho}\nabla p + \nu\nabla^2\mathbf{v} \, . \qquad (6.87)$$

The Navier - Stokes equation is fundamental to Newtonian fluid mechanics. To make further progress it is necessary to look for solutions of the equation subject to appropriate boundary conditions, a search rendered more difficult by the fact that the left-hand side of (6.85) contains a non-linear term [see equation (6.3)]. This work is considered to be beyond the scope of the present book.

CHAPTER 7

The Application of Cartesian Tensors to Elasticity

7.1 STRAIN

7.1.1 The Strain Tensor - Infinitesimal Strain

We now consider deformable solid bodies in which the distance between two particles is in general a function of time. Unlike fluids, solids are able to maintain their shape in the absence of external forces. This shape will be termed the **equilibrium configuration** or **reference configuration**. Any departure from the equilibrium configuration will be specified by a tensor known as the **strain tensor**, a precise definition of which will be given below. The system of forces set up by the departure from equilibrium may be specified by means of the stress tensor already defined in section 6.1.2. Fundamental to the study of elasticity is the relation between the stress and strain tensors. This relation for the case of infinitesimal strain will be postulated in section 7.2.2.

We observed in section 6.3 that the stress in a fluid in motion was dependent upon the *rate of strain*, a dependence which was formalized in equation (6.83), (the strain, of course, being zero throughout). The dependence on rate of strain is no less true for solids, but we shall, for the most part, concern ourselves with cases in which the rate of strain is effectively zero.

Suppose that the point whose position vector is $\mathbf{x} = (x_1, x_2, x_3)$ in a deformable solid is displaced to the point $\mathbf{z} = \mathbf{x} + \mathbf{r}$ when the solid is deformed. In the undeformed state (equilibrium configuration) the element of length ds_0 along some curve C_0 is given by:

$$ds_0^2 = dx_j dx_j \ . \tag{7.1}$$

Suppose the curve deforms into a curve C. The element of length is now given by:

$$ds^2 = dz_j dz_j \ . \tag{7.2}$$

Now

$$ds^2 = \frac{\partial z_j}{\partial x_i}\frac{\partial z_j}{\partial x_k}\, dx_i dx_k \ ,$$

$$= g_{ik} dx_i dx_k \ , \tag{7.3}$$

where $$g_{ik} = \frac{\partial z_j}{\partial x_i}\frac{\partial z_j}{\partial x_k} . \tag{7.4}$$

The components g_{ik} are the components of the **Lagrangian deformation tensor g**.

From (7.1) and (7.3) we have:

$$\begin{aligned} ds^2 - ds_0^2 &= (g_{ik} - \delta_{ik})dx_i dx_k , \\ &= 2\epsilon_{ik} dx_i dx_k , \end{aligned}$$

where $$2\epsilon_{ik} = g_{ik} - \delta_{ik} . \tag{7.5}$$

The components ϵ_{ik} are known as the **Lagrangian strain components**. The tensor $\boldsymbol{\epsilon}$ is the **Langrangian Strain Tensor.**

Now $$z_j = x_j + r_j , \tag{7.6}$$

where the r_j are the components of the **displacement vector r**.

Therefore $$\frac{\partial z_j}{\partial x_i} = \delta_{ij} + \frac{\partial r_j}{\partial x_i} .$$

Hence, from (7.5) and (7.4):

$$\begin{aligned} 2\epsilon_{ik} &= \left(\delta_{ij} + \frac{\partial r_j}{\partial x_i}\right)\left(\delta_{jk} + \frac{\partial r_j}{\partial x_k}\right) - \delta_{ik} , \\ &= \frac{\partial r_k}{\partial x_i} + \frac{\partial r_i}{\partial x_k} + \frac{\partial r_j}{\partial x_i}\frac{\partial r_j}{\partial x_k} . \end{aligned} \tag{7.7}$$

We shall consider initially cases in which the product term in (7.7) may be ignored.

Then $$2\epsilon_{ik} = \frac{\partial r_k}{\partial x_i} + \frac{\partial r_i}{\partial x_k} . \tag{7.8}$$

The necessary and sufficient condition for (7.8) to be valid is that squares and products of the displacement gradients $\partial r_i/\partial x_k$ are negligible. It is therefore necessary (but not sufficient) that the strains are small. It is, however, sufficient (but not necessary) that the displacements are small (see Examples 7.1, No. 4). In suffix-free notation (7.8) becomes:

$$2\boldsymbol{\epsilon} = \nabla \mathbf{r} + (\nabla \mathbf{r})' . \tag{7.9}$$

The tensor $\boldsymbol{\epsilon}$ is called the **infinitesimal strain tensor** or simply the **strain tensor**. There is no longer any need to refer to it as the *Langrangian* strain tensor as there is now no distinction between it and the Eulerian strain tensor to be introduced later (section 7.6.2).

The quantity usually referred to simply as strain is a vector. The **strain** of a line element, for example, parallel to the unit vector **n**, is defined to be $\boldsymbol{\epsilon}.\mathbf{n}$. The component of the strain in the direction of the line element is therefore given by $\mathbf{n}.\boldsymbol{\epsilon}.\mathbf{n}$. Now $\boldsymbol{\epsilon}$ is symmetric and given by (7.9).

Hence $$\mathbf{n}.\boldsymbol{\epsilon}.\mathbf{n} = \mathbf{n}.(\nabla\mathbf{r}).\mathbf{n} \quad . \tag{7.10}$$

7.1.2 The Contribution of the Strain Tensor to the Deformation

We will now follow an analysis similar to that of section 6.2.1, and suppose that when the point $\mathbf{x}$ suffers a displacement $\mathbf{r}$, the neighbouring point to $\mathbf{x}$, with position vector $\mathbf{x} + \delta\mathbf{x}$, suffers a displacement $\mathbf{r} + \delta\mathbf{r}$, so that

$$\begin{aligned} \mathbf{r} + \delta\mathbf{r} &= \mathbf{r}(\mathbf{x} + \delta\mathbf{x}) \ , \\ &= \mathbf{r}(\mathbf{x}) + \delta\mathbf{x}.\nabla\mathbf{x} + \tfrac{1}{2}(\delta\mathbf{x}.\nabla)^2\mathbf{r} + \ldots \ . \end{aligned}$$

Thus $$\delta\mathbf{r} \approx \delta\mathbf{x}.\nabla\mathbf{r} = (\nabla\mathbf{r})'.\delta\mathbf{x} \ , \tag{7.11}$$

correct to the 1st order in $|\delta\mathbf{x}|$. Equation (7.11) may be written:

$$\delta\mathbf{r} = (\boldsymbol{\epsilon} + \boldsymbol{\zeta}).\delta\mathbf{x} \ , \tag{7.12}$$

where $\boldsymbol{\epsilon}$ and $\boldsymbol{\zeta}$ are the symmetric and antisymmetric parts of $(\nabla\mathbf{r})'$ respectively. $\boldsymbol{\epsilon}$ is the strain tensor defined by (7.9).

Following the method of section 6.2.2, we may write:

$$\delta\mathbf{r}^{(s)} = \boldsymbol{\epsilon}.\delta\mathbf{x} \ ,$$

which, when $\boldsymbol{\epsilon}$ is referred to principal axes, becomes:

$$\delta\mathbf{r}^{(s)} = \left(\frac{\partial r_1}{\partial x_1}\,\delta x_1, \frac{\partial r_2}{\partial x_2}\,\delta x_2, \frac{\partial r_3}{\partial x_3}\,\delta x_3 \right) , \tag{7.13}$$

[cf. (6.64)]. The kind of deformation represented by (7.13) is known as **pure strain**. In such a deformation a line element parallel to the x_1-axis would be stretched parallel to that axis. An element of material in the shape of a sphere would be distorted into an ellipsoid whose principal axes coincided with those of the strain tensor.

Consider a rectangular volume element with edges of length $\delta x_1, \delta x_2, \delta x_3$ parallel to the coordinate axes. The volume $V = \delta x_1\, \delta x_2\, \delta x_3$. Let the element suffer a pure strain so that the edge of length δx_1 becomes, from (7.13):

$$\delta x_1 + \delta r_1 = \delta x_1 + \frac{\partial r_1}{\partial x_1}\,\delta x_1 \ ,$$

with the other edges similarly extended.
The volume is now

$$V + \delta V = V[1 + \epsilon_{ii} + 0(\epsilon_{11}^2)] \, .$$

Hence, on ignoring higher powers of the strain components, we have:

$$\delta V/V = \epsilon_{ii} = \operatorname{div}\mathbf{r} \ .$$

The invariant ϵ_{ii} is known as the **dilatation**. It is the fractional increase in volume.

7.1.3 The Contribution of the Antisymmetric Tensor $\boldsymbol{\zeta}$ to the Deformation

The tensor $\boldsymbol{\zeta}$ mentioned in section 7.1.2 is the antisymmetric part of $(\nabla\mathbf{r})'$ given explicitly by:

$$\boldsymbol{\zeta} = \tfrac{1}{2}[(\nabla\mathbf{r})' - (\nabla\mathbf{r})] \; . \tag{7.14}$$

In component form (7.14) becomes:

$$\zeta_{ik} = \frac{1}{2}\left(\frac{\partial r_i}{\partial x_k} - \frac{\partial r_k}{\partial x_i}\right), \tag{7.15}$$

or, written in full:

$$\boldsymbol{\zeta} = \frac{1}{2}\begin{bmatrix} 0 & \partial r_1/\partial x_2 - \partial r_2/\partial x_1 & \partial r_1/\partial x_3 - \partial r_3/\partial x_1 \\ \partial r_2/\partial x_1 - \partial r_1/\partial x_2 & 0 & \partial r_2/\partial x_3 - \partial r_3/\partial x_2 \\ \partial r_3/\partial x_1 - \partial r_1/\partial x_3 & \partial r_3/\partial x_2 - \partial r_2/\partial x_3 & 0 \end{bmatrix} .$$

It contributes the 'antisymmetric' part to the displacement vector given by:

$$\delta\mathbf{r}^{(a)} = \boldsymbol{\zeta}.\delta\mathbf{x} \; .$$

We shall proceed along the lines of section 6.2.4 and define a vector $\boldsymbol{\theta}$ by:

$$\boldsymbol{\theta} = -\operatorname{vec}\boldsymbol{\zeta} = \tfrac{1}{2}\operatorname{curl}\mathbf{r}. \tag{7.16}$$

Then
$$\delta\mathbf{r}^{(a)} = \boldsymbol{\theta}\times\delta\mathbf{x} \; , \tag{7.17}$$

where $\boldsymbol{\theta}$ corresponds to $\frac{1}{2}\boldsymbol{\omega}$ in (6.74) and (6.75).

We shall show in Examples 7.1, No. 1 that the condition $\boldsymbol{\epsilon} = \mathbf{0}$ is necessary and sufficient for the displacement vector $\delta\mathbf{r}$ to represent a pure rotation. It follows therefore that $\delta\mathbf{r}^{(a)}$ represents a pure rotation, and reference to the result of Problems 3.4, No. 9 shows that $\boldsymbol{\theta}$ is the **vector angle** through which the element rotates, i.e. a vector whose magnitude is the angle of rotation and whose direction is along the axis of rotation.

7.1.4 Compatibility Equations

Since the six strain components ϵ_{ik} are related to r_1, r_2, r_3, they are not all independent.

We have from (7.8):

$$\epsilon_{ik} = \frac{1}{2}\left(\frac{\partial r_k}{\partial x_i} + \frac{\partial r_i}{\partial x_k}\right).$$

Therefore
$$\frac{\partial^2 \epsilon_{ik}}{\partial x_p \partial x_q} = \frac{1}{2}\left(\frac{\partial^3 r_k}{\partial x_i \partial x_p \partial x_q} + \frac{\partial^3 r_i}{\partial x_k \partial x_p \partial x_q}\right),$$

or
$$\frac{\partial^2 \epsilon_{ik}}{\partial x_i \partial x_k} = \frac{1}{2}\left(\frac{\partial^2 \epsilon_{kk}}{\partial x_i \partial x_i} + \frac{\partial^2 \epsilon_{ii}}{\partial x_k \partial x_k}\right), \tag{7.18}$$

where no summation is implied over i or k.

Also:
$$\frac{\partial^2 \epsilon_{ii}}{\partial x_j \partial x_k} = \frac{\partial}{\partial x_i}\left(\frac{\partial^2 r_i}{\partial x_j \partial x_k}\right),$$

i.e.
$$\frac{\partial^2 \epsilon_{ii}}{\partial x_j \partial x_k} = \frac{\partial}{\partial x_i}\left(\frac{\partial \epsilon_{ij}}{\partial x_k} + \frac{\partial \epsilon_{ik}}{\partial x_j} - \frac{\partial \epsilon_{jk}}{\partial x_i}\right), \tag{7.19}$$

where, again, no summation is implied. (7.18) and (7.19) each give rise to three relations known as **compatibility equations**. These equations must be satisfied in any real system.

Examples 7.1

(1) Prove that the condition $\boldsymbol{\epsilon} = \mathbf{0}$ throughout is necessary and sufficient for a pure rotation.

We have, from (7.5)

$$ds^2 - ds_0{}^2 = 2\epsilon_{ik} dx_i dx_k .$$

Hence lengths are preserved and the condition $\boldsymbol{\epsilon} = \mathbf{0}$ is sufficient for a pure rotation.

Now let us suppose we have a pure rotation. From (3.48) the displacement due to a rotation θ about an axis in the direction of the unit vector $\mathbf{n}$ is given by:

$$\mathbf{r} = [(\mathbf{x.n})\mathbf{n} - \mathbf{x}](1 - \cos\theta) + \mathbf{n} \times \mathbf{x} \sin\theta$$

or
$$r_i = (x_j n_j n_i - x_i)(1 - \cos\theta) + \epsilon_{ijm} n_j x_m \sin\theta .$$

Hence
$$\frac{\partial r_i}{\partial x_k} + \frac{\partial r_k}{\partial x_i} = 2(n_i n_k - \delta_{ik})(1 - \cos\theta),$$

and
$$\frac{\partial r_j}{\partial x_i}\frac{\partial r_j}{\partial x_k} = -2(n_i n_k - \delta_{ik})(1 - \cos\theta) .$$

Thus, from (7.7), $\boldsymbol{\epsilon} = \mathbf{0}$ for a pure rotation.

For a small rotation the expression for the displacement becomes:

$$r_i = \epsilon_{ijm} n_j x_m \sin\theta ,$$

whence
$$\partial r_i / \partial x_k = \epsilon_{ijk} n_j \sin\theta .$$

Thus the tensor $\nabla\mathbf{r}$ is antisymmetric and the infinitesimal strain tensor given by (7.8) vanishes.

(2) In a certain deformation of an elastic material, the particle whose position vector was originally $\mathbf{x}$ is displaced to the point with position vector $\mathbf{x} + \mathbf{r}$,

where $\mathbf{r} = \alpha(3x_2 - 7x_3, -3x_1 + 2x_3, 7x_1 - 2x_2)$.

Show that the displacement is a pure rotation and find the axis about which the rotation takes place.

We have $$\nabla\mathbf{r} = \alpha\begin{bmatrix} 0 & -3 & 7 \\ 3 & 0 & -2 \\ -7 & 2 & 0 \end{bmatrix}.$$

Hence $\boldsymbol{\epsilon} = \mathbf{0}$; $\boldsymbol{\zeta} = (\nabla\mathbf{r})'$.

$\boldsymbol{\epsilon} = \mathbf{0}$ is, as we have seen, a necessary and sufficient condition for a pure rotation. Now $\boldsymbol{\theta} = \frac{1}{2}$ curl $\mathbf{r} = \alpha(-2, -7, -3)$. Hence the axis of rotation is the line $\mathbf{r} = \lambda(2, 7, 3)$.

(3) Write out the compatibility equations explicitly and show that they are satisfied by the elements of the strain tensor

$$\boldsymbol{\epsilon} = k\begin{bmatrix} x_2^2x_3^2 & 2x_1x_2x_3^2 & 2x_1x_2^2x_3 \\ 2x_1x_2x_3^2 & x_1^2x_3^2 & 2x_1^2x_2x_3 \\ 2x_1x_2^2x_3 & 2x_1^2x_2x_3 & x_1^2x_2^2 \end{bmatrix},$$

where k is a small constant. Find the corresponding displacement vector.

The required equations are:

$$\frac{\partial^2 \epsilon_{23}}{\partial x_2 \partial x_3} = \frac{1}{2}\left[\frac{\partial^2 \epsilon_{22}}{\partial x_3^2} + \frac{\partial^2 \epsilon_{33}}{\partial x_2^2}\right],$$

and two similar equations:

$$\frac{\partial^2 \epsilon_{11}}{\partial x_2 \partial x_3} = \frac{\partial}{\partial x_1}\left[\frac{\partial \epsilon_{12}}{\partial x_3} + \frac{\partial \epsilon_{31}}{\partial x_2} - \frac{\partial \epsilon_{23}}{\partial x_1}\right],$$

and two similar equations.

We have: $$\frac{\partial^2 \epsilon_{23}}{\partial x_2 \partial x_3} = \frac{\partial^2 \epsilon_{22}}{\partial x_3^2} = \frac{\partial^2 \epsilon_{33}}{\partial x_2^2} = 2kx_1^2 ,$$

and the first equation is satisfied.

Also $$\frac{\partial^2 \epsilon_{11}}{\partial x_2 \partial x_3} = \frac{\partial^2 \epsilon_{12}}{\partial x_1 \partial x_3} = \frac{\partial^2 \epsilon_{31}}{\partial x_1 \partial x_2} = \frac{\partial^2 \epsilon_{23}}{\partial x_1^2} = 4kx_2x_3 ,$$

and the second equation is satisfied.

Integration of equation (7.8) yields for the displacement vector:

$$\mathbf{r} = k(x_1x_2^2x_3^2, x_1^2x_2x_3^2, x_1^2x_2^2x_3) .$$

(4) Examine the validity of equations (7.7) and (7.8) in relation to the displacement given by

$$\mathbf{r} = (0, x_3 - x_2, -x_2 - x_3) .$$

The configuration of the body after displacement is given by

$$\mathbf{z} = \mathbf{x} + \mathbf{r} = (x_1, x_3, -x_2) .$$

Hence the given displacement corresponds to a rigid body rotation of $\pi/2$ in the positive sense about the x_1-axis. Equation (7.7) yields $\boldsymbol{\epsilon} = \mathbf{0}$. Hence the strain is zero but the displacement is finite. Equation (7.8) yields $\boldsymbol{\epsilon} = -\mathbf{e}_2 \otimes \mathbf{e}_2 - \mathbf{e}_3 \otimes \mathbf{e}_3$, which is incorrect. The displacement gradients are, of course, not small. In fact the displacement gradient tensor $\overleftarrow{\nabla}\mathbf{r}$ is given by

$$\nabla\mathbf{r} = \begin{bmatrix} 0 & 0 & 0 \\ 0 & -1 & -1 \\ 0 & 1 & -1 \end{bmatrix} .$$

Problems 7.1

(1) Prove that the components of the deformation tensor $\mathbf{g}$ and the strain tensor $\boldsymbol{\epsilon}$ obey the transformation law for second-order tensors.

(2) In Example (4) the strain tensor was zero but the displacement was finite. Reference to section 7.1.1. will show that these two conditions taken together are not sufficient to establish the validity or otherwise of equation (7.8) and, in fact, in this example the equation was shown not to hold. Can you suggest a simple example in which the strain tensor is small or zero, the displacement is finite and equation (7.8) is valid?

(3) Let the line element of length l, parallel to the unit vector $\mathbf{n}$ suffer a strain in which the point $\mathbf{x}$ is moved to the point $\mathbf{x} + \mathbf{r}$. Prove that the strain of a line element of length l, defined by $\delta l/l$, is equivalent to that given by $\mathbf{n}.\nabla\mathbf{r}.\mathbf{n}$, where $\mathbf{n}$ is a unit vector in the direction of the line element.

(4) Show that the component of strain of a line element parallel to the unit vector $\mathbf{n}$, in the direction of the unit vector $\mathbf{p}$, is given by:

$$\mathbf{n}.\boldsymbol{\epsilon}.\mathbf{p} = \tfrac{1}{2}[\mathbf{n}.(\mathbf{p}.\nabla)\mathbf{r} + \mathbf{p}.(\mathbf{n}.\nabla)\mathbf{r}] .$$

(5) Consider a plane element of elastic material defined by $0 \leqslant x_1 \leqslant l$; $0 \leqslant x_2 \leqslant m$, deformed in its own plane such that the point $\mathbf{x}$ is moved to $\mathbf{x} + \mathbf{r}$.

the area of the deformed element is given approximately by:

$$lm\left(1+\frac{\partial r_1}{\partial x_1}+\frac{\partial r_2}{\partial x_2}\right) = lm(1+\epsilon_{11}+\epsilon_{22}) \ .$$

(6) A rectangular block shaped element of elastic material defined by $0 \leqslant x_1 \leqslant l$; $0 \leqslant x_2 \leqslant m; 0 \leqslant x_3 \leqslant n$, is deformed such that the point $\mathbf{x}$ is moved to $\mathbf{x}+\mathbf{r}$. Prove that, to a first order of approximation, the increase of volume is $lmn\epsilon_{ii}$.

(7) A small sphere centered on the origin is deformed so that the point $\mathbf{x}$ is moved to the point $\mathbf{x}+\mathbf{r}$,

where $$\mathbf{r} = k(10x_1+4x_2, 4x_1-5x_2, 11x_3) \ .$$

Show that the sphere is deformed initially into an ellipsoid of revolution, and find its axis of symmetry.

(8) A small sphere centred on the origin is deformed so that the point $\mathbf{x}$ is moved to the point $\mathbf{x}+\mathbf{r}$,

where $$\mathbf{r} = k(x_1+2x_2, x_2, x_3) \ .$$

Show that the sphere is deformed initially into an ellipsoid and rotated about a principal axis.

(9) Show that the components of the strain tensor

$$\boldsymbol{\epsilon} = k\begin{bmatrix} x_2^2+x_3^2 & 2x_1x_2 & 2x_1x_3 \\ 2x_1x_2 & x_1^2+x_3^2 & 2x_2x_3 \\ 2x_1x_3 & 2x_2x_3 & x_1^2+x_2^2 \end{bmatrix}$$

satisfy the compatibility equations, and find the corresponding displacement vector.

(10) Write down the only non-trivial compatibility equation applicable to a strain distribution in the (x_1, x_2)-plane, and show that it is satisfied by the distribution corresponding to the strain tensor $\boldsymbol{\epsilon}$ where

$$\begin{aligned} \epsilon_{11} &= \alpha_{11}+2\beta_{11}(x_1^2+x_2^2)+x_1^4+x_2^4 \ , \\ \epsilon_{22} &= \alpha_{22}+2\beta_{22}(x_1^2+x_2^2)+x_1^4+x_2^4 \ , \\ \epsilon_{12} &= \alpha_{12}+2x_1x_2(x_1^2+x_2^2+\beta_{11}+\beta_{22}) \ , \end{aligned}$$

and $\alpha_{11}, \alpha_{22}, \alpha_{12}, \beta_{11}, \beta_{22}$ are constants. Find the corresponding displacement vector.

7.2 THE RELATION BETWEEN STRESS AND STRAIN FOR ELASTIC SOLIDS

7.2.1 Hooke's Law

In school physics one encounters the elementary relation that the extension of a metal wire is directly proportional to the force producing it. This is **Hooke's Law**, and it is generally written:

$$T = \lambda(x/l) , \tag{7.20}$$

where T is the tension, x the extension, l the natural length, and λ a parameter of the material. In other words the stress (T) is directly proportional to the strain (x/l).

If the x_1-axis is taken to be along the wire, with the origin a fixed point, and the reduction in the cross-sectional area of the wire, is ignored, the strain tensor is given by:

$$\boldsymbol{\epsilon} = \begin{bmatrix} x/l & 0 & 0 \\ 0 & 0 & 0 \\ 0 & 0 & 0 \end{bmatrix} . \tag{7.21}$$

Similarly the stress tensor is given by:

$$\boldsymbol{\sigma} = \begin{bmatrix} T/A & 0 & 0 \\ 0 & 0 & 0 \\ 0 & 0 & 0 \end{bmatrix} , \tag{7.22}$$

where A is the cross-sectional area of the wire. Thus there is a linear relation between the stress and strain tensors, which is supported by experimental evidence.

The off-diagonal terms of the stress tensor arise from shearing forces.

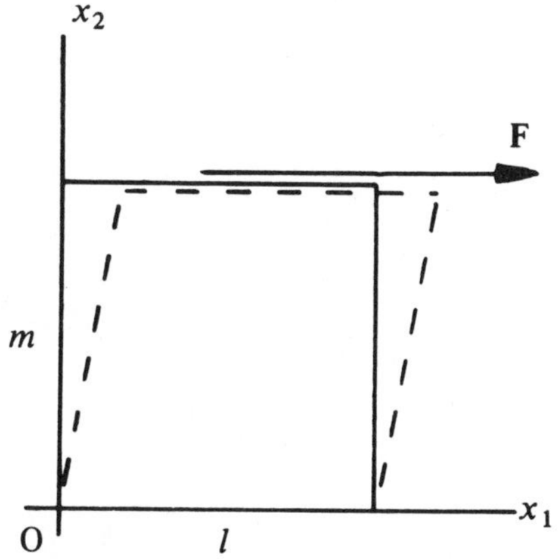

Fig. 7.1

Consider a rectangular block of unit thickness, defined by $0 \leqslant x_1 \leqslant l$, $0 \leqslant x_2 \leqslant m$, subjected to shearing force **F** per unit area (Fig. 7.1). It is found that the distortion is proportional to the force producing it.

Thus
$$\mathbf{r} = (kx_2, 0, 0),$$

$$\nabla\mathbf{r} = \begin{bmatrix} 0 & 0 & 0 \\ k & 0 & 0 \\ 0 & 0 & 0 \end{bmatrix},$$

$$\boldsymbol{\epsilon} = \frac{1}{2}\begin{bmatrix} 0 & k & 0 \\ k & 0 & 0 \\ 0 & 0 & 0 \end{bmatrix}, \tag{7.23}$$

$$\boldsymbol{\sigma} = \begin{bmatrix} 0 & F & 0 \\ F & 0 & 0 \\ 0 & 0 & 0 \end{bmatrix}. \tag{7.24}$$

There is again a linear relationship between the stress and strain tensors.

7.2.2 The Generalized Hooke's Law

Experimental evidence has tended to indicate that, at least over a limited range, stress and strain are linearly related in materials which are isotropic. This is true for both normal and shearing stress, although the constants of proportionality are not the same. It is reasonable to suppose, therefore, that for isotropic materials the stress tensor is a linear isotropic function of the strain tensor. Using the Representation Theorem as in (6.81) and (6.82), but without the product term, we have:

$$\boldsymbol{\sigma} = 2\mu\boldsymbol{\epsilon} + \beta\mathbf{1},$$

where $\boldsymbol{\sigma} = \mathbf{0}$ when $\boldsymbol{\epsilon} = \mathbf{0}$. Since $\boldsymbol{\sigma}$ is a linear fuction of $\boldsymbol{\epsilon}$ it follows that $\beta \propto \epsilon_{ii}$.

Hence
$$\boldsymbol{\sigma} = 2\mu\boldsymbol{\epsilon} + \lambda\epsilon_{jj}\mathbf{1}, \tag{7.25}$$

or
$$\sigma_{ik} = 2\mu\epsilon_{ik} + \lambda\epsilon_{jj}\delta_{ik}, \tag{7.26}$$

where λ and μ are parameters of the material. Equations (7.25) and (7.26) represent what is known as the **Generalized Hooke's Law**.

We note from (7.26) that

$$\sigma_{jj} = (2\mu + 3\lambda)\epsilon_{jj}. \tag{7.27}$$

Hence
$$\boldsymbol{\epsilon} = \frac{1}{2\mu}\left(\boldsymbol{\sigma} - \frac{\lambda}{2\mu + 3\lambda}\sigma_{jj}\mathbf{1}\right). \tag{7.28}$$

The constants λ and μ are known as **Lamé constants**. We shall now look at equation (7.25) in some specific cases.

7.2.3 Simple Shear

In the case of **simple shear** already considered (equations (7.23) and (7.24)) we note that (7.25) becomes:

$$\boldsymbol{\sigma} = 2\mu\boldsymbol{\epsilon} . \tag{7.29}$$

The stress across a plane normal to $\mathbf{e}_1$ is given by:

$$\boldsymbol{\sigma}.\mathbf{e}_1 = k\mu\mathbf{e}_2 = F\mathbf{e}_2 .$$

Therefore $\mu = F/k = \sigma_{12}/k$. (7.30)

The parameter μ is known as the **shear modulus**.

7.2.4 Isotropic Expansion

Suppose the displacement $\mathbf{r}$ at any point $\mathbf{x}$ is given by $\mathbf{r} = k\mathbf{x}$. In this case a spherical element of material would retain its shape but not its size. The strain tensor $\boldsymbol{\epsilon}$ would be $k\mathbf{1}$ and (7.25) would become:

$$\boldsymbol{\sigma} = (2\mu + 3\lambda)k\mathbf{1} . \tag{7.31}$$

The stress is everywhere normal to the surface on which it acts. The ratio K given by:

$$K = \frac{\mathbf{n}.\boldsymbol{\sigma}.\mathbf{n}}{\epsilon_{jj}} = \lambda + \frac{2}{3}\mu , \tag{7.32}$$

is known as the **bulk modulus**.

Eliminating λ from (7.25) and (7.32), we obtain the relation:

$$\boldsymbol{\sigma} = 2\mu(\boldsymbol{\epsilon} - \frac{1}{3}\epsilon_{jj}\mathbf{1}) + K\epsilon_{jj}\mathbf{1} , \tag{7.33}$$

which should be compared with (6.84).

7.2.5 The Extension of a Wire

Suppose the displacement at (x_1, x_2, x_3) is given by:

$$\mathbf{r} = kx_1\mathbf{e}_1 - k'(x_2\mathbf{e}_2 + x_3\mathbf{e}_3) ,$$

where k, k' are positive. The strain tensor $\boldsymbol{\epsilon}$, calculated from (7.9), is given by:

$$\boldsymbol{\epsilon} = k\mathbf{e}_1 \otimes \mathbf{e}_1 - k'(\mathbf{e}_2 \otimes \mathbf{e}_2 + \mathbf{e}_3 \otimes \mathbf{e}_3) . \tag{7.34}$$

If we choose $k \gg k'$, (7.34) is a more accurate representation of the extension of a wire along the x_1-axis under tension than (7.21). The stress tensor calculated from (7.25) is given by:

$$\boldsymbol{\sigma} = (k - 2k')\lambda\mathbf{1} + 2\mu\boldsymbol{\epsilon} \ . \tag{7.35}$$

The stresses in the directions of the coordinate axes are given by:

$$\begin{aligned} \boldsymbol{\sigma}.\mathbf{e}_1 &= [(k - 2k')\lambda + 2\mu k]\,\mathbf{e}_1 \ , \\ \boldsymbol{\sigma}.\mathbf{e}_2 &= [(k - 2k')\lambda - 2\mu k']\,\mathbf{e}_2 \ , \\ \boldsymbol{\sigma}.\mathbf{e}_3 &= [(k - 2k')\lambda - 2\mu k']\,\mathbf{e}_3 \ . \end{aligned} \tag{7.36}$$

For a wire under tension the stress normal to the wire must vanish. Hence we must have: $\boldsymbol{\sigma}.\mathbf{e}_2 = \boldsymbol{\sigma}.\mathbf{e}_3 = \mathbf{0}$,

i.e. $$\nu = k'/k = \lambda/(2(\lambda + \mu)) \ . \tag{7.37}$$

The dimensionless constant ν is known as **Poisson's Ratio**. The one remaining non-zero component of (7.36) may now be written:

$$\boldsymbol{\sigma}.\mathbf{e}_1 = \frac{k\mu}{\lambda + \mu}(3\lambda + 2\mu)\mathbf{e}_1 \ . \tag{7.38}$$

The ratio $$E = \sigma_{11}/k = \frac{\mu(3\lambda + 2\mu)}{\lambda + \mu} \tag{7.39}$$

is known as **Young's Modulus**.

From (7.34) we note that $k = \epsilon_{11}$ is the longitudinal strain, and $k' = e_{22} = e_{33}$ is the lateral strain. Moreover σ_{11} is the longitudinal stress. We may therefore define Young's Modulus as the ratio of longitudinal stress to longitudinal strain, and Poisson's Ratio as minus the ratio of lateral strain to longitudinal strain. However, these ratios only apply to a rod which is free to contract laterally in response to longitudinal stress. We therefore define two further ratios. The **Effective Young's Modulus** E' is defined by:

$$E' = \frac{\sigma_{11}}{\epsilon_{11}} = \frac{\text{Longitudinal stress}}{\text{Longitudinal strain}} \ . \tag{7.40}$$

The **Effective Poisson's Ratio** ν'' is defined for lateral strain in the x_2-direction by:

$$\nu'' = -\frac{\epsilon_{22}}{\epsilon_{11}} = \frac{\text{Lateral strain}}{\text{Longitudinal strain}} \ . \tag{7.41a}$$

A similar definition holds for ν''', the effective Poisson's Ratio for lateral strain in the x_3-direction:

$$\nu''' = -\epsilon_{33}/\epsilon_{11} \ . \tag{7.41b}$$

The relations (7.40) and (7.41) hold whether or not the rod is free to contract laterally. Of course, when the rod is free to contract there is no distinction between E and E' or between ν, ν'' and ν'''.

Examples 7.2

(1) From (7.37) and (7.39) it follows that:

$$\frac{\lambda}{2\mu(2\lambda + 3\mu)} = \frac{\nu}{E}, \quad \frac{1}{2\mu} = \frac{1+\nu}{E} .$$

Hence we may rewrite (7.28) as:

$$\boldsymbol{\epsilon} = \frac{1}{E}\left[(1+\nu)\boldsymbol{\sigma} - \nu\sigma_{jj}\mathbf{1}\right] . \qquad (7.42)$$

Expressing $\boldsymbol{\sigma}$ in terms of $\boldsymbol{\epsilon}$ [(the equivalent of (7.25)] we have:

$$\boldsymbol{\sigma} = \frac{E}{1+\nu}\left(\boldsymbol{\epsilon} + \frac{\nu}{1-2\nu}\epsilon_{jj}\mathbf{1}\right) . \qquad (7.43)$$

(2) Calculate the displacement due to a uniform tension T at any point within a right circular cylinder (Fig. 7.2).

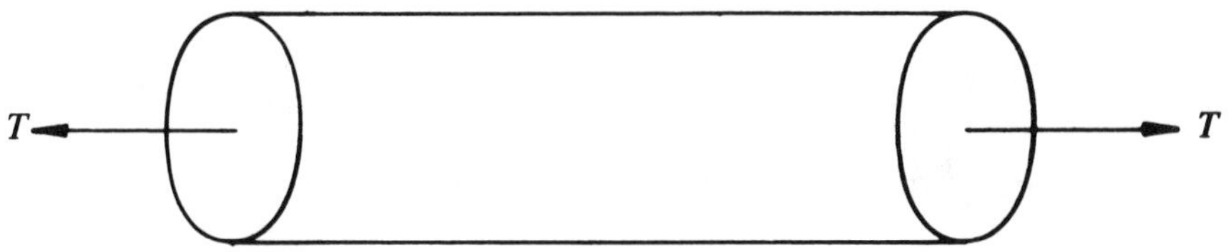

Fig. 7.2

This problem is the inverse of the one already considered under the heading 'The Extension of a Wire'. Using Cartesian axes with the x_1-axis along the axis of symmetry we have for the stress tensor:

$$\boldsymbol{\sigma} = \begin{bmatrix} T & 0 & 0 \\ 0 & 0 & 0 \\ 0 & 0 & 0 \end{bmatrix} .$$

Hence $\boldsymbol{\sigma}.\mathbf{n} = \mathbf{0}$ over the curved surface, $\boldsymbol{\sigma}.\mathbf{n} = T\mathbf{n}$ on the flat ends.

Equation (7.42) gives:

$$\boldsymbol{\epsilon} = \frac{T}{E}\begin{bmatrix} 1 & 0 & 0 \\ 0 & -\nu & 0 \\ 0 & 0 & -\nu \end{bmatrix}.$$

Therefore $\dfrac{\partial r_1}{\partial x_1} = \dfrac{T}{E}, \dfrac{\partial r_2}{\partial x_2} = \dfrac{\partial r_3}{\partial x_3} = -\dfrac{\nu T}{E}$,

$$\frac{\partial r_2}{\partial x_3} + \frac{\partial r_3}{\partial x_2} = \frac{\partial r_3}{\partial x_1} + \frac{\partial r_1}{\partial x_3} = \frac{\partial r_1}{\partial x_2} + \frac{\partial r_2}{\partial x_1} = 0 .$$

The solution of these equations is:

$$\mathbf{r} - \mathbf{r}_0 = \frac{T}{E}(x_1, -\nu x_2, -\nu x_3) ,$$

where $\mathbf{r}_0$ is an arbitrary constant vector.

(3) Rework the previous example using cylindrical polar coordinates.

The problem may be expressed naturally in cylindrical polar coordinates. We should therefore expect the solution to be simpler. Choose the axis of symmetry to be the z-axis and the stress tensor becomes:

$$\boldsymbol{\sigma} = \begin{bmatrix} 0 & 0 & 0 \\ 0 & 0 & 0 \\ 0 & 0 & T \end{bmatrix}.$$

Hence, from (7.42):

$$\boldsymbol{\epsilon} = \frac{T}{E}\begin{bmatrix} -\nu & 0 & 0 \\ 0 & -\nu & 0 \\ 0 & 0 & 1 \end{bmatrix}.$$

We now employ the expressions in (3.115) for the components of the rate of strain tensor, and adapt them for the strain tensor by replacing $\mathbf{v}$ by $\mathbf{r}$ to obtain:

$$\frac{\partial r_1}{\partial r} = \epsilon_{11} = \frac{-\nu T}{E} , \quad \frac{r_1}{r} = \epsilon_{22} = \frac{-\nu T}{E} ,$$

$$\frac{\partial r_3}{\partial z} = \epsilon_{33} = \frac{T}{E}, \quad \frac{\partial r_2}{\partial z} = 0 ,$$

$$\frac{\partial r_3}{\partial r} = -\frac{\partial r_1}{\partial z} = 0 , \quad \frac{\partial}{\partial r}\left(\frac{r_2}{r}\right) = 0 .$$

The solution of these equations is

$$\mathbf{r} - \mathbf{r}_0 = \frac{T}{E}(-\nu r, 0, z) \ ,$$

where $\mathbf{r}_0$ is an arbitrary vector in the direction of the axis of symmetry.
(4) A body is immersed in a fluid of the same density. Find the displacement at any point.

The hydrostatic pressure in still water is $(p + \rho g h)$ at a depth h below the point at which the pressure is p. Choose Cartesian axes with the x_3-axis vertically upwards. Then the stress tensor is given by:

$$\boldsymbol{\sigma} = -(p - \rho g x_3)\mathbf{1} \ .$$

Hence from (7.42) the strain tensor is given by:

$$\boldsymbol{\epsilon} = \frac{1}{E}(1 - 2\nu)(\rho g x_3 - p)\mathbf{1} \ .$$

Therefore $$\frac{\partial r_1}{\partial x_1} = \frac{\partial r_2}{\partial x_2} = \frac{\partial r_3}{\partial x_3} = \frac{1}{E}(1 - 2\nu)(\rho g x_3 - p)\mathbf{1} \ ,$$

and $$\frac{\partial r_2}{\partial x_3} + \frac{\partial r_3}{\partial x_2} = \text{etc.} = 0 \ .$$

Hence $$\mathbf{r} = \frac{1}{E}(1 - 2\nu)\left[(\rho g x_3 - p)\mathbf{x} - \tfrac{1}{2}\rho g x_3^2 \mathbf{e}_3\right] \ .$$

Problems 7.2

(1) Show that $E = 3K(1 - 2\nu)$, where E is Young's Modulus, K is the bulk modulus, and ν is Poisson's Ratio defined respectively by (7.39), (7.32) and (7.37). A body is in a state of uniform tension p_0 in all directions. Show using (i) Cartesian coordinates and (ii) spherical polar coordinates, that the displacement at any point $\mathbf{x}$ is given by $\mathbf{r} = (p_0/3K)\mathbf{x}$.
(2) Consider the cylinder of Examples 7.2 No. 2.
(i) Suppose it is constrained so that no lateral contraction takes place. Show that the Effective Young's Modulus is given by:

$$E' = \frac{E(1 - \nu)}{(1 + \nu)(1 - 2\nu)}$$

What is the Effective Poisson's Ratio?
(ii) Find the Effective Young's Modulus and Poisson's Ratio if the cylinder is constrained so that contraction takes place in the x_3-direction only.
(3) A uniform right circular cylinder of length l is suspended from the centre of one end and hangs under its own weight. Determine the displacement at any point.

7.3 STRAIN ENERGY

7.3.1 The Strain Energy Function

Consider an arbitrary body of elastic material of volume τ and surface area S. We propose to calculate the **Strain Energy Function** U, that is the internal energy per unit volume due to deformation from the equilibrium configuration. To do this we shall find the total work done by body and surface forces when changing the displacement at the point $\mathbf{x}$ from $\mathbf{r}$ to $\mathbf{r} + \delta\mathbf{r}$.

The work done by the body force $\mathbf{F}$ is

$$\int_\tau (\rho\mathbf{F}.\delta\mathbf{r})\mathrm{d}\tau \ .$$

The work done by the surface forces is

$$\int_S (\boldsymbol{\sigma}.\mathbf{n}).\delta\mathbf{r}\mathrm{d}S \ .$$

Hence, by use of the Divergence Theorem (3.67), the total work done is given by:

$$\int_\tau \delta W\mathrm{d}\tau = \int_\tau \left[\rho\mathbf{F}.\delta\mathbf{r} + \mathrm{div}(\boldsymbol{\sigma}.\delta\mathbf{r})\right]\mathrm{d}\tau \ ,$$

$$= \int_\tau \left[(\rho\mathbf{F} + \mathrm{div}\,\boldsymbol{\sigma}).\delta\mathbf{r} + \boldsymbol{\sigma}.\nabla(\delta\mathbf{r})\right]\mathrm{d}\tau \ .$$

Now $(\rho\mathbf{F} + \mathrm{div}\,\boldsymbol{\sigma})$ is the rate of change of momentum per unit mass (see (6.28)), and therefore contributes nothing to the strain energy. Indeed in the next section we shall equate this quantity to zero, the body being presumed to be at rest before and after the deformation takes place. We are left with a single term which represents the work done in deforming the body. We have then:

$$\int_\tau \delta U\mathrm{d}\tau = \int_\tau \delta W\mathrm{d}\tau = \int_\tau \boldsymbol{\sigma}.\nabla(\delta\mathbf{r})\mathrm{d}\tau \ .$$

Now the body of material was stated to be arbitrary. We may therefore write:

$$\delta U = \boldsymbol{\sigma}\ \nabla(\delta\mathbf{r}) = \sigma_{ki}\frac{\partial}{\partial x_i}(\delta r_k) \ . \tag{7.45}$$

We now introduce the assumption that the strain energy function U depends only upon the strain components ϵ_{ik}.

Then remembering that $\boldsymbol{\sigma}$ is a symmetric tensor,

$$\delta U = \tfrac{1}{2}\left[\sigma_{ik}\frac{\partial}{\partial x_i}(\delta r_k) + \sigma_{ki}\frac{\partial}{\partial x_k}(\delta r_i)\right]$$

$$= \sigma_{ik}\delta\epsilon_{ik} \ .$$

Since σ_{ik} is linearly dependent on the elements of $\boldsymbol{\epsilon}$ by (7.26) we deduce that the strain energy function U is given by:

$$U = \tfrac{1}{2}\sigma_{ik}\epsilon_{ik} = \tfrac{1}{2}\boldsymbol{\sigma}:\boldsymbol{\epsilon} \ . \tag{7.46}$$

7.3.2 Energy Stored in Dilatation and Distortion

By making use of the relation (7.33), which splits the stress tensor into two terms involving respectively the shear modulus μ and the bulk modulus K, we may write:

$$\boldsymbol{\sigma} = \mathbf{d} - p\mathbf{1} \ ,$$

where $\mathbf{d} = 2\mu(\boldsymbol{\epsilon} - \tfrac{1}{3}\epsilon_{jj}\mathbf{1})$ is the deviatoric stress tensor, (see section 6.1.5) and $p = -K\epsilon_{jj} = -\tfrac{1}{3}\sigma_{jj}$ is the pressure. Hence from (7.46):

$$U = \tfrac{1}{2}\mathbf{d}:\boldsymbol{\epsilon} - \tfrac{1}{2}p\mathbf{1}:\boldsymbol{\epsilon} = U' + U'' \ , \tag{7.47}$$

where
$$U' = \tfrac{1}{2}\mathbf{d}:\boldsymbol{\epsilon} = \tfrac{1}{2}\mathbf{d}:\left(\frac{1}{2\mu}\mathbf{d} - \frac{p}{3K}\mathbf{1}\right) ,$$

$$= \frac{1}{4\mu}\mathbf{d}:\mathbf{d} \ ,$$

and
$$U'' = \tfrac{1}{2}p\mathbf{1}:\boldsymbol{\epsilon} = -\tfrac{1}{2}p\epsilon_{jj} = \frac{p^2}{2K} \ .$$

We now recall that the scalar invariants of $\boldsymbol{\sigma}$ defined by (2.50), (2.51) and (2.52) are:

$$\left.\begin{aligned} I_1 &= \sigma_{ii} = \boldsymbol{\sigma}:\mathbf{1} = -3p \ , \\ I_2 &= \tfrac{1}{2}(\sigma_{ii}\sigma_{jj} - \sigma_{ij}\sigma_{ji}) = \tfrac{1}{2}\,(I_1^2 - \boldsymbol{\sigma}:\boldsymbol{\sigma}) \ , \\ I_3 &= \det \boldsymbol{\sigma} \ , \end{aligned}\right\} \tag{7.48}$$

(see Examples 2.6 No. 2). For simplicity we shall assume that $\boldsymbol{\sigma}$ is referred to principal axes.

Then
$$\left.\begin{aligned} I_1 &= \sigma_{11} + \sigma_{22} + \sigma_{33} \ , \\ I_2 &= \sigma_{22}\sigma_{33} + \sigma_{33}\sigma_{11} + \sigma_{11}\sigma_{22} \ , \\ I_3 &= \sigma_{11}\sigma_{22}\sigma_{33} \ . \end{aligned}\right\} \tag{7.49}$$

Hence
$$U' = \frac{1}{4\mu}(\boldsymbol{\sigma} + p\mathbf{1}):(\boldsymbol{\sigma} + p\mathbf{1}) \ ,$$

$$= \frac{1}{6\mu}(I_1^2 - 3I_2) \ , \tag{7.50}$$

and
$$U'' = \frac{I_1^2}{18K} \ . \tag{7.51}$$

U' is the energy due to distortion. It depends entirely on the shear modulus μ. U'' is the energy due to dilatation. It depends entirely on the bulk modulus K.

Example 7.3

(1) Find the strain energy U per unit volume due to the pure strain $\mathbf{r} = (k_1x_1, k_2x_2, k_3x_3)$. Find the strain energies U', U'' due respectively to distortion and dilatation, and show that

$$U = U' + U'' .$$

The strain tensor is given by (7.9) and the stress tensor by (7.25).

Thus
$$\boldsymbol{\epsilon} = \begin{bmatrix} k_1 & 0 & 0 \\ 0 & k_2 & 0 \\ 0 & 0 & k_3 \end{bmatrix} ,$$

$$\boldsymbol{\sigma} = 2\mu \begin{bmatrix} k_1 & 0 & 0 \\ 0 & k_2 & 0 \\ 0 & 0 & k_2 \end{bmatrix} + \lambda(k_1 + k_2 + k_3)\mathbf{1} .$$

The total strain energy is given by (7.46):

$$U = \mu(k_1^2 + k_2^2 + k_3^2) + \tfrac{1}{2}\lambda(k_1 + k_2 + k_3)^2 .$$

The scalar invariants of $\boldsymbol{\sigma}$ are, from (7.48):

$$\begin{aligned} I_1 &= \sigma_{ii} = (2\mu + 3\lambda)(k_1 + k_2 + k_3) , \\ I_2 &= \tfrac{1}{2}(\sigma_{ii}\sigma_{jj} - \sigma_{ij}\sigma_{ji}) , \\ &= \tfrac{1}{2}I_1^2 - 2\mu^2(k_1^2 + k_2^2 + k_3^2) - (\tfrac{3}{2}\lambda^2 + 2\mu\lambda)(k_1 + k_2 + k_3)^2 . \end{aligned}$$

The energy due to distortion is, from (7.50):

$$U' = \mu(k_1^2 + k_2^2 + k_3^2) - \tfrac{1}{3}\mu(k_1 + k_2 + k_3)^2 .$$

The energy due to dilatation is, from (7.51):

$$U'' = \tfrac{1}{2}K(k_1 + k_2 + k_3)^2 ,$$

where K is the bulk modulus given by (7.32). It is now a simple matter to show that $U = U' + U''$.

Problems 7.3

(1) Find the strain energy per unit volume due to (i) a pure tension T, (ii) an all-round tension T, and (iii) a simple shear $\mathbf{r} = ke_2\mathbf{e}_1$. Find in each case the energy due to distortion and the energy due to dilatation.

(2) A rectangular block, with its edges parallel to the coordinate directions, is

subject to a tension T in the x_1-direction and a compression of equal magnitude in the x_2-direction. Show that there is no change of volume, and find the energy of distortion.

(3) A uniform right circular cylinder of length l and radius a is subject to a turning couple at one end, which rotates through an angle θ relative to the other end which is fixed. Find the total strain energy.

7.4 THE DISPLACEMENT EQUATION

The problems discussed in section 7.2 were readily solved because the stress distribution throughout the body was either specified beforehand or could be determined from the given data in a straightforward manner. This is not always the case, and there is one further relation which is sometimes useful, known as the **Displacement Equation** or the **Equation of Equilibrium**. It is just the equation of Fluid Motion (6.28) with the acceleration term, $\mathrm{D}\mathbf{u}/\mathrm{D}t$, put equal to zero. Thus we have:

$$\rho\mathbf{F} + \mathrm{div}\,\boldsymbol{\sigma} = \mathbf{0} \ , \tag{7.52}$$

where $\mathbf{F}$ is the body force per unit mass. In the absence of body force (as in the case of most of the problems so far considered), the equation reduces to

$$\mathrm{div}\,\boldsymbol{\sigma} = \mathbf{0} \ . \tag{7.53}$$

Now, from (7.25): $\boldsymbol{\sigma} = 2\mu\boldsymbol{\epsilon} + \lambda\epsilon_{jj}\mathbf{1}$,

and from (7.9): $2\boldsymbol{\epsilon} = (\nabla\mathbf{r})' + \nabla\mathbf{r}$.

Hence, after some algebra similar to that used to derive equation (6.86):

$$\mathrm{div}\,\boldsymbol{\sigma} = \mu\nabla^2\mathbf{r} + (\lambda + \mu)\mathrm{grad\ div}\mathbf{r} \ , \tag{7.54}$$

or, using (3.26):

$$\mathrm{div}\,\boldsymbol{\sigma} = (\lambda + 2\mu)\mathrm{grad\ div}\mathbf{r} - \mu\mathrm{curl\ curl}\mathbf{r} \ . \tag{7.55}$$

The Displacement Equation (7.52) therefore becomes:

$$\rho\mathbf{F} + (\lambda + 2\mu)\mathrm{grad\ div}\mathbf{r} - \mu\mathrm{curl\ curl}\mathbf{r} = \mathbf{0} \ . \tag{7.56}$$

Equation (7.56) relates the body force (if any) to the displacement $\mathbf{r}$. Unfortunately, as it is a second-order partial differential equation, it is not usually possible to write down a solution for $\mathbf{r}$ in closed form. The chief exceptions to this rule are certain problems in which the displacement depends on one coordinate only, in which case the partial differential equation reduces to an ordinary differential equation.

In suffix notation the Displacement Equation is:

$$\rho F_i + (\lambda + \mu)\frac{\partial^2 r_j}{\partial x_i \partial x_j} + \mu\frac{\partial^2 r_i}{\partial x_j \partial x_j} = 0 \ . \tag{7.57}$$

We now give some examples in which the displacement depends upon a single coordinate, and in which the Displacement Equation is a useful tool in the solution of the problem.

Examples 7.4

(1) Show that, in the special case of spherical symmetry, the Displacement Equation, written in spherical polar coordinates, becomes:

$$\rho F_r + (\lambda + 2\mu)\left(\frac{\mathrm{d}^2 v}{\mathrm{d}r^2} + \frac{2}{r}\frac{\mathrm{d}v}{\mathrm{d}r} - \frac{2v}{r^2}\right) = 0 \ , \tag{7.58}$$

where v is the displacement in the radial direction. Hence rework Problems 7.2 No. 1 by solving the Displacement Equation.

Let $\mathbf{v}$ be the displacement in which $v_r = v, v_\theta = v_\phi = 0$. Then, from (3.105):

$$\operatorname{div} \mathbf{v} = \frac{1}{r^2}\frac{\mathrm{d}}{\mathrm{d}r}(vr^2) \ ,$$

$$(\operatorname{grad} \operatorname{div} \mathbf{v})_1 = \frac{\mathrm{d}}{\mathrm{d}r}\left(\frac{1}{r^2}\frac{\mathrm{d}}{\mathrm{d}r}(vr^2)\right) .$$

Also, from (3.109), we have:

$$(\operatorname{curl} \operatorname{curl} \mathbf{v})_1 = 0 \ .$$

Hence, from (7.56):

$$\rho F_r + (\lambda + 2\mu)\frac{\mathrm{d}}{\mathrm{d}r}\left(\frac{1}{r^2}\frac{\mathrm{d}}{\mathrm{d}r}(vr^2)\right) = 0 \ . \tag{7.59}$$

Alternatively, from (3.110) and (3.107):

$$(\nabla^2 \mathbf{v})_1 = \nabla^2 v_1 - \frac{2v}{r^2} = \frac{1}{r^2}\frac{\mathrm{d}}{\mathrm{d}r}\left(r^2\frac{\mathrm{d}v}{\mathrm{d}r}\right) - \frac{2v}{r^2} \ ,$$

and we again recover (7.58). Equation (7.59) is in the most convenient form for solution. In the special case of zero body force ($F_r = 0$), we obtain:

$$\frac{1}{r^2}\frac{\mathrm{d}}{\mathrm{d}r}(vr^2) = 3A \ ,$$

where A is a constant. Hence, by integration

$$v = Ar + B/r^2 \ , \tag{7.60}$$

where B is another constant. The strain tensor is, from (3.103), given by:

$$\boldsymbol{\epsilon} = \nabla \mathbf{v} = \begin{bmatrix} A - 2B/r^3 & 0 & 0 \\ 0 & A + B/r^3 & 0 \\ 0 & 0 & A + B/r^3 \end{bmatrix}.$$

Equation (7.60) represents the general solution for the displacement in the case of spherical symmetry. We now use the stress-strain relation in the form (7.26), with $\boldsymbol{\sigma} = p_0\mathbf{1}$ (since the body is in a state of uniform tension in all directions) to obtain:

$$p_0 = 2\mu(A - 2B/r^3) + 3A\lambda\ ,$$

$$p_0 = 2\mu(A + B/r^3) + 3A\lambda\ .$$

$$B = 0, \quad A = \frac{p_0}{2\mu + 3\lambda}\ , \quad v = \frac{p_0 r}{2\mu + 3\lambda}\ .$$

But, from (7.32), $K = \lambda + \dfrac{2}{3}\mu$.

Hence $$\mathbf{v} = \frac{p_0}{3K}\mathbf{r}\ .$$

(2) Find the displacement in a thick right circular cylinder of internal radius a and pressure p_1, and external radius b and pressure p_2.

Proceeding as in Example (1), we let $\mathbf{v}$ be the displacement in which $v_r = v$, $v_z = v_\theta = 0$, where r, θ, z are cylindrical polar coordinates. It is by no means obvious that v_z is necessarily zero. Indeed we shall consider an alternative condition later on (see Problem (1)). For the moment, however, we shall take $v_z = 0$, which is the case when the cylinder is rigidly confined between two parallel plates, perpendicular to its axis.

From (3.117) we have:

$$\operatorname{div} \mathbf{v} = \frac{1}{r}\frac{\mathrm{d}}{\mathrm{d}r}(rv)\ .$$

Therefore $$(\operatorname{grad}(\operatorname{div} \mathbf{v}))_1 = \frac{\mathrm{d}}{\mathrm{d}r}\left(\frac{1}{r}\frac{\mathrm{d}}{\mathrm{d}r}(rv)\right)\ .$$

Also, from (3.118) we have:

$$\operatorname{curl} \mathbf{v} = \mathbf{0}\ ,$$

therefore $$\operatorname{curl}\operatorname{curl} \mathbf{v} = \mathbf{0}\ .$$

Hence from (7.56), with **F** put equal to zero:

$$(\lambda + 2\mu)\frac{\mathrm{d}}{\mathrm{d}r}\left(\frac{1}{r}\frac{\mathrm{d}}{\mathrm{d}r}(rv)\right) = 0 . \tag{7.61}$$

The solution of (7.61) is:

$$v = Ar + B/r , \tag{7.62}$$

where A and B are constants. The strain tensor is, from (3.115), given by:

$$\boldsymbol{\epsilon} = \nabla\mathbf{v} = \begin{bmatrix} A - B/r^2 & 0 & 0 \\ 0 & A + B/r^2 & 0 \\ 0 & 0 & 0 \end{bmatrix},$$

Equation (7.62) represents the general solution for the displacement in the case of axial symmetry, when the longitudinal strain is zero. The stress-strain relation (7.26) yields:

$$\sigma_{11} = (\lambda + 2\mu)\epsilon_{11} + \lambda(\epsilon_{22} + \epsilon_{33}) = 2A\lambda + 2\mu(A - B/r^2) ,$$

$$\sigma_{22} = (\lambda + 2\mu)\epsilon_{22} + \lambda(\epsilon_{11} + \epsilon_{33}) = 2A\lambda + 2\mu(A + B/r^2) ,$$

$$\sigma_{33} = (\lambda + 2\mu)\epsilon_{33} + \lambda(\epsilon_{11} + \epsilon_{22}) = 2A\lambda .$$

The boundary conditions are:

$$\sigma_{11} = -p_1 \qquad (r = a) ,$$

$$\sigma_{11} = -p_2 \qquad (r = b) .$$

Hence $$-2(\lambda + \mu)A = \frac{p_1a^2 - p_2b^2}{a^2 - b^2} , \quad -2\mu B = a^2b^2\left(\frac{p_1 - p_2}{a^2 - b^2}\right),$$

$$r^2(a^2 - b^2)\sigma_{11} = p_1a^2(b^2 - r^2) - p_2b^2(a^2 - r^2) ,$$

$$r^2(a^2 - b^2)\sigma_{22} = p_2b^2(a^2 + r^2) - p_1a^2(b^2 + r^2) .$$

The radial displacement v is therefore given by:

$$2(a^2 - b^2)v = -\frac{r(p_1a^2 - p_2b^2)}{\lambda + \mu} - \frac{a^2b^2(p_1 - p_2)}{\mu r} .$$

(3) A thick right circular cylinder of internal radius b and external radius c is shrunk onto a thick cylinder of internal radius a and external radius $b + \delta(\delta \ll b)$. Assuming that the two cylinders are made from the same material (of Young's Modulus E and Poisson's Ratio $\nu \ll 1$), and that no longitudinal stresses are induced, find an approximate expression for the radial pressure between the two

cylinders. Find also the tangential stress in each cylinder.

We have, for the inner cylinder, using the result of Problem (1):

$$(b^2 - a^2)Evr = r^2(1-\nu)(p_1a^2 - p_2b^2) + a^2b^2(1+\nu)(p_1 - p_2) \ ,$$

where v is the radial displacement.

Now let $\nu = 0$ and $r = b$:

$$E(b^2 - a^2)v = 2p_1a^2b - p_2b(a^2 + b^2) \ ,$$

where p_1 is the internal pressure and p_2 is the radial pressure between the two cylinders.

Similarly, for the outer cylinder, we have:

$$E(c^2 - b^2)u = p_2b(b^2 + c^2) - 2p_3bc^2 \ ,$$

where u is the radial displacement in the outer cylinder.

$$\text{Now } \delta = u - v = \frac{b}{E}\left[\frac{-2p_1a^2}{b^2 - a^2} + p_2\left(\frac{c^2 + b^2}{c^2 - b^2}\right) + p_2\left(\frac{b^2 + a^2}{b^2 - a^2}\right) - \frac{2p_3c^2}{c^2 - b^2}\right]$$

$$= \frac{2b}{E}\left[\frac{-p_1a^2(c^2 - b^2) - p_3c^2(b^2 - a^2) + p_2b^2(c^2 - a^2)}{(c^2 - b^2)(b^2 - a^2)}\right].$$

Therefore

$$p_2b^2(c^2 - a^2) = p_1a^2(c^2 - b^2) + p_3c^2(b^2 - a^2) + \frac{\delta E}{2b}(c^2 - b^2)(b^2 - a^2) \ .$$

In the special case when $p_1, p_3 \ll p_2$, we have:

$$p_2 \approx \frac{\delta E(c^2 - b^2)(b^2 - a^2)}{2b^3(c^2 - a^2)} \ .$$

Again from the result of Problem (1) we have for the inner cylinder:

$$\sigma_{11} = \frac{a^2p_1(r^2 - c^2)}{r^2(c^2 - a^2)} + \frac{c^2p_3(a^2 - r^2)}{r^2(c^2 - a^2)} - \frac{\delta E}{2b}\,\frac{(r^2 - a^2)(c^2 - b^2)}{(c^2 - a^2)r^2} \ ,$$

$$\sigma_{22} = \frac{a^2p_1(r^2 + c^2)}{r^2(c^2 - a^2)} - \frac{c^2p_3(a^2 + r^2)}{r^2(c^2 - a^2)} - \frac{\delta E}{2b}\,\frac{(r^2 + a^2)(c^2 - b^2)}{(c^2 - a^2)r^2} \ .$$

And for the outer cylinder:

$$\sigma_{11} = \frac{a^2p_1(r^2 - c^2)}{r^2(c^2 - a^2)} + \frac{c^2p_3(a^2 - r^2)}{r^2(c^2 - a^2)} + \frac{\delta E}{2b}\,\frac{(r^2 - c^2)(b^2 - a^2)}{(c^2 - a^2)r^2} \ ,$$

$$\sigma_{22} = \frac{a^2p_1(r^2 - c^2)}{r^2(c^2 - a^2)} - \frac{c^2p_3(a^2 + r^2)}{r^2(c^2 - a^2)} + \frac{\delta E}{2b}\,\frac{(r^2 + c^2)(b^2 - a^2)}{r^2(c^2 - a^2)} \ .$$

Problems 7.4

(1) Rework Example No. (2) with the condition that the cylinder is allowed to expand or contract freely in a direction parallel to the axis of symmetry. Show that the non-zero components of the stress tensor are given by:

$$r^2(b^2 - a^2)\sigma_{11} = p_1 a^2(r^2 - b^2) + p_2 b^2(a^2 - r^2) ,$$

$$r^2(b^2 - a^2)\sigma_{22} = p_1 a^2(r^2 + b^2) - p_2 b^2(a^2 + r^2) .$$

Show also that the radial displacement, v, is given by:

$$(b^2 - a^2)Evr = r^2(1 - \nu)(p_1 a^2 - p_2 b^2) + a^2 b^2(1 + \nu)(p_1 - p_2) ,$$

where E is Young's Modulus and ν is Poissons Ratio.

(2) Consider a thick spherical shell of internal radius a and external radius b. If the internal pressure is p_1 and the external pressure is p_2, show that the displacement at any point is given by: $v = Ar + B/r^2$,

where $$3KA(b^3 - a^3) = p_1 a^3 - p_2 b^3 ,$$

$$4\mu B(b^3 - a^3) = (p_1 - p_2)a^3 b^3 ,$$

and $$3K = 3\lambda + 2\mu .$$

Show also that the radial and tangential stresses are given by:

$$\sigma_{11} = 3KA - 4\mu B/r^3 ,$$

$$\sigma_{22} = 3KA + 2\mu B/r^3 .$$

Suppose the spherical shell is shrunk onto a rigid sphere of radius a. Write down the new boundary conditions and calculate the displacement and the components of the stress tensor at any point within the spherical shell.

(3) A spherical shell of elastic material $b \leqslant r \leqslant a$ is enclosed by a rigid spherical envelope $r = a$, and is under constant pressure p inside the cavity $r = b$. Show that the principal stresses $\sigma_{11}, \sigma_{22}. \sigma_{33}$ are given by:

$$\sigma_{11} = -\frac{p(kr^3 + a^3)b^3}{(kb^3 + a^3)r^3} , \quad \sigma_{22} = \sigma_{33} = -\frac{p(2kr^3 - a^3)b^3}{2(kb^3 + a^3)} ,$$

where $k = \dfrac{1 + \nu}{2(1 - 2\nu)}$, and ν is Poisson's ratio.

(4) A hollow sphere of external radius $2a$ contains a concentric cavity of radius a, and is initially free from stress. Show that when the gas pressure inside the cavity is increased from zero to P the external radius increases by $\dfrac{3(\lambda + 2\mu)Pa}{14\mu(3\lambda + 2\mu)}$.

(Liverpool University, June 1979)

7.5 THE ROTATING DISC PROBLEM

As a further example of the use of the Displacement Equation, we shall consider the problem of a uniform circular disc of radius a rotating steadily about its axis of symmetry, with angular velocity $\boldsymbol{\omega}$. We shall choose cylindrical polar coordinates, with the z-axis along the axis of symmetry (Fig. 7.3).

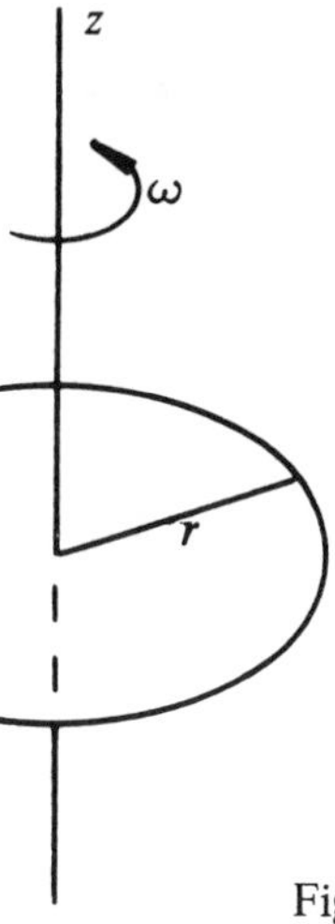

Fig. 7.3

The displacement vector $\mathbf{v}$ will have only one non-zero component, v_1 the radial displacement. As the problem is axially symmetric, v_1 will depend or r alone. Let us agree to drop the subscript and write v for v_1. We shall then be in a position to write the Displacement Equation as an ordinary differential equation. Before doing so, however, a word of caution seems appropriate.

It is not convenient to use the form of the Displacement Equation given by (7.56) as it contains the unknown strain component $\epsilon_{33}(\partial v_3/\partial z)$. This component cannot be assumed zero on the grounds that v_3 is zero. Indeed ϵ_{33} cannot be zero if σ_{33} is zero, and it is the latter condition which is correct.† There cannot be any normal stress on the flat faces of the disc. It is logical therefore to take σ_{33} as being identically zero everywhere.

Now the non-zero components of the strain tensor are given by:

$$\epsilon_{11} = \frac{\partial v_1}{\partial r}, \quad \epsilon_{22} = \frac{v_1}{r}, \tag{7.63}$$

†But see remarks at the end of this section.

with ϵ_{33} unknown. We may eliminate ϵ_{33}, however, with the aid of the stress-strain relations (7.26).

Thus

$$\left.\begin{aligned} \sigma_{11} &= 2\mu\epsilon_{11} + \lambda\epsilon_{jj} = (2\mu+\lambda)\frac{\partial v_1}{\partial r} + \lambda\frac{v_1}{r} + \lambda\epsilon_{33} \ , \\ \sigma_{22} &= 2\mu\epsilon_{22} + \lambda\epsilon_{jj} = (2\mu+\lambda)\frac{v_1}{r} + \lambda\frac{\partial v_1}{\partial r} + \lambda\epsilon_{33} \ , \\ 0 &= 2\mu\epsilon_{33} + \lambda\epsilon_{jj} \ . \end{aligned}\right\} \tag{7.64}$$

Eliminating the unknown ϵ_{33} from (7.64) we have:

$$\left.\begin{aligned} \sigma_{11} &= \frac{4\mu(\lambda+\mu)}{(\lambda+2\mu)}\frac{\partial v_1}{\partial r} + \frac{2\mu\lambda}{(\lambda+2\mu)}\frac{v_1}{r} = \frac{E}{1-\nu^2}\left(\frac{\partial v_1}{\partial r} + \nu\frac{v_1}{r}\right) , \\ \sigma_{22} &= \frac{4\mu(\lambda+\mu)}{(\lambda+2\mu)}\frac{v_1}{r} + \frac{2\mu\lambda}{(\lambda+2\mu)}\frac{\partial v_1}{\partial r} = \frac{E}{1-\nu^2}\left(\frac{v_1}{r} + \nu\frac{\partial v_1}{\partial r}\right) , \end{aligned}\right\} \tag{7.65}$$

where the values of E and ν given by (7.39) and (7.37) have been used. We may now use the basic form of the Displacement Equation given by (7.52) noting from Example 3.7 No. 1 that, $(\operatorname{div} \boldsymbol{\sigma})_1$, is given by:

$$(\operatorname{div}\boldsymbol{\sigma})_1 = \frac{\partial\sigma_{11}}{\partial r} + \frac{\sigma_{11} - \sigma_{22}}{r} \ . \tag{7.66}$$

Substituting this value into (7.52), we obtain for the Displacement Equation:

$$\frac{\partial\sigma_{11}}{\partial r} + \frac{\sigma_{11} - \sigma_{22}}{r} + \rho\omega^2 r = 0 \ , \tag{7.67}$$

where the final term represents the radial component of the body force per unit mass. We now substitute the expressions for σ_{11}, σ_{22} from (7.65) into (7.67) to obtain:

$$\frac{E}{1-\nu^2}\frac{\mathrm{d}}{\mathrm{d}r}\left[\frac{1}{r}\frac{\mathrm{d}}{\mathrm{d}r}(rv)\right] + \rho\omega^2 r = 0 \ . \tag{7.68}$$

The general solution of (7.68) is:

$$v = Ar + B/r - \frac{(1-\nu^2)\rho\omega^2 r^3}{8E} \ , \tag{7.69}$$

with the principal stresses given by:

$$\sigma_{11} = \frac{AE}{1-\nu} - \frac{BE}{r^2(1+\nu)} - \tfrac{1}{8}(3+\nu)\rho\omega^2 r^2 \ , \qquad (7.70)$$

$$\sigma_{22} = \frac{AE}{1-\nu} + \frac{BE}{r^2(1+\nu)} - \tfrac{1}{8}(1+3\nu)\rho\omega^2 r^2 \ . \qquad (7.71)$$

There are two cases to consider according as the point $r = 0$ is included within the body or not.

Case 1. $r = 0$ included
Clearly $B = 0$. In order to find A we may use the boundary condition $\sigma_{11} = 0$ at $r = a$.

Thus $$A = \frac{1}{8E}(1-\nu)(3+\nu)\rho\omega^2 a^2 \ .$$

Hence $$v = \frac{1}{8E}(1-\nu)\rho r\omega^2\left[(3+\nu)a^2 - (1+\nu)r^2\right] \ ,$$

$$\sigma_{11} = \tfrac{1}{8}\left[(3+\nu)(a^2 - r^2)\rho\omega^2\right] \ , \qquad (7.72)$$

$$\sigma_{22} = \tfrac{1}{8}\left[(3+\nu)a^2 - (1+3\nu)r^2\right]\rho\omega^2 \ . \qquad (7.73)$$

At the boundary $r = a$ we have:

$$v = \frac{1}{4E}(1-\nu)\rho a^3\omega^2 \ , \qquad (7.74)$$

$$\sigma_{22} = \tfrac{1}{4}(1-\nu)\rho a^2\omega^2 \ . \qquad (7.75)$$

At $r = 0$: $$\sigma_{11} = \tfrac{1}{8}(3+\nu)\rho a^2\omega^2 \ , \qquad (7.76)$$

$$\sigma_{22} = \tfrac{1}{8}(3+\nu)\rho a^2\omega^2 \ . \qquad (7.77)$$

The maximum radial and tangential stresses are attained at the centre of the disc.

Case 2. $r = 0$ not included
We now have a disc with a concentric hole of radius (say) b. The boundary conditions are now $\sigma_{11} = 0$ at $r = a,b$.

Thus, from (7.70): $$\frac{AE}{1-\nu} = \tfrac{1}{8}(a^2 + b^2)(3+\nu)\rho\omega^2 \ ,$$

$$\frac{BE}{1+\nu} = \tfrac{1}{8}a^2 b^2(3+\nu)\rho\omega^2 \ .$$

Hence
$$v = \frac{\rho\omega^2}{8Er}[(3+\nu)\{(a^2+b^2)r^2(1-\nu)+a^2b^2(1+\nu)\}-(1-\nu^2)r^4] \;, \tag{7.78}$$

$$\sigma_{11} = \frac{\rho\omega^2}{8r^2}(3+\nu)[(a^2+b^2)r^2-a^2b^2-r^4] \;, \tag{7.79}$$

therefore
$$\sigma_{22} = \frac{\rho\omega^2}{8r^2}[(3+\nu)\{(a^2+b^2)r^2+a^2b^2\}-(1+3\nu)r^4] \;. \tag{7.80}$$

At the outer boundary ($r = a$):

$$v = \frac{1}{4E}[(1-\nu)a^2+(3+\nu)b^2]\rho\omega^2 a \;, \tag{7.81}$$

$$\sigma_{22} = \tfrac{1}{4}[(1-\nu)a^2+(3+\nu)b^2]\rho\omega^2 \;. \tag{7.82}$$

At the inner boundary ($r = b$):

$$v = \frac{1}{4E}[(1-\nu)b^2+(3+\nu)a^2]\rho\omega^2 b \;, \tag{7.83}$$

$$\sigma_{22} = \tfrac{1}{4}[(1-\nu)b^2+(3+\nu)a^2]\rho\omega^2 \;. \tag{7.84}$$

If we now let $b \to 0$ in (7.84) we obtain:

$$\sigma_{22} = \tfrac{1}{4}(3+\nu)a^2\rho\omega^2 \;, \tag{7.85}$$

which is twice that given by (7.77).

The foregoing solution is given by Southwell (1941) and elsewhere. However, the condition $v_3 = 0, \partial v_3/\partial z \neq 0$ is not entirely satisfactory, and Hunter (1976) argues that it is not possible to satisfy the condition $\sigma_{33} = 0$ with the type of solution given here.

Hunter looks for a solution of the form

$$\mathbf{v} = v_1(r)\mathbf{e}_1 + v_3(z)\mathbf{e}_3 \;,$$

with (7.64) replaced by:

$$\sigma_{33} = (2\mu+\lambda)\epsilon_{33} + \lambda\left(\frac{v_1}{r}+\frac{\partial v_1}{\partial r}\right) .$$

The third component of the Displacement Equation yields:

$$(\mathrm{div}\,\boldsymbol{\sigma})_3 = \frac{\partial\sigma_{33}}{\partial z} = 0 \;,$$

whence ϵ_{33} is constant.

Equation (7.68) is replaced by

$$\frac{E(1-\nu)}{(1+\nu)(1-2\nu)}\frac{\mathrm{d}}{\mathrm{d}r}\left[\frac{1}{r}\frac{\mathrm{d}}{\mathrm{d}r}(rv_1)\right]+\rho\omega^2 r = 0 \ ,$$

with solution:

$$v_1 = Ar + B/r - \frac{(1+\nu)(1-2\nu)\rho\omega^2 r^3}{8(1-\nu)E}$$

in place of (7.69).

In the case of a solid disc (7.72) and (7.73) are replaced by:

$$\sigma_{11} = \frac{(3-2\nu)}{8(1-\nu)}\rho\omega^2(a^2-r^2) \ ,$$

$$\sigma_{22} = \frac{\rho\omega^2}{8(1-\nu)}[(3-2\nu)a^2-(1+2\nu)r^2] \ .$$

There is in addition an expression for σ_{33}:

$$\sigma_{33} = E\epsilon_{33} + [(3-2\nu)a^2 - 2r^2]\frac{\rho\omega^2\nu}{4(1-\nu)} \ .$$

Thus, since ϵ_{33} is constant, it is not possible to satisfy the condition $\sigma_{33} = 0$ exactly. It is, however, possible to find a solution for which the total force exerted on the disc is zero, i.e. $\int_0^a \sigma_{33} r dr = 0$. This condition yields:

$$\epsilon_{33} = -\frac{1}{2E}\rho\nu\omega^2 a^2 \ ,$$

$$\sigma_{33} = \frac{\rho\omega^2\nu}{4(1-\nu)}(a^2-2r^2) \ .$$

7.6 FINITE DEFORMATION

7.6.1 Lagrangian Finite Strain Components

We have in this chapter been concerned with the theory of infinitesimal strains, a theory which is essentially linear and therefore mathematically straightforward. We shall now consider briefly the theory of finite deformations and take as our starting point equation (7.7) for the strain components in which the product term will no longer be ignored. Thus we have:

$$2\epsilon_{ik} = \frac{\partial r_k}{\partial x_i} + \frac{\partial r_i}{\partial x_k} + \frac{\partial r_j}{\partial x_i}\frac{\partial r_j}{\partial x_k} \ . \qquad (7.86)$$

or $$2\boldsymbol{\epsilon} = (\nabla\mathbf{r}) + (\nabla\mathbf{r})' + (\nabla\mathbf{r}).(\nabla\mathbf{r})' \;. \tag{7.87}$$

The tensor $\boldsymbol{\epsilon}$ is the **Lagrangian finite strain tensor.**

7.6.2 The Eulerian Finite Strain Components

The Lagrangian approach, which we considered in the previous section, took for independent variables the coordinates x_i of a material point in the undeformed state. The Eulerian approach, on the other hand, is to take for independent variables the coordinates z_i of a material point in the deformed state. From (7.1) we have:

$$\mathrm{d}s_0^2 = \frac{\partial x_j}{\partial z_i}\frac{\partial x_j}{\partial z_k}\mathrm{d}z_i\mathrm{d}z_k = G_{ik}\mathrm{d}z_i\mathrm{d}z_k \;, \tag{7.88}$$

where $$G_{ik} = \frac{\partial x_j}{\partial z_i}\frac{\partial x_j}{\partial z_k} \;. \tag{7.89}$$

The components G_{ik} are the components of the **Eulerian Deformation Tensor.** From (7.2) we have:

$$\mathrm{d}s^2 = \delta_{ik}\mathrm{d}z_i\mathrm{d}z_k \;,$$

therefore $$\mathrm{d}s^2 - \mathrm{d}s_0^2 = (\delta_{ik} - G_{ik})\mathrm{d}z_i\mathrm{d}z_k \;,$$

$$= 2\eta_{ik}\mathrm{d}z_i\mathrm{d}z_k \;,$$

where $$2\eta_{ik} = \delta_{ik} - G_{ik} \;. \tag{7.90}$$

The components η_{ik} are known as the **Eulerian Strain Components**. The tensor $\boldsymbol{\eta}$ is the **Eulerian Strain Tensor**.

Now $$x_j = z_j - r_j \;, \tag{7.91}$$

therefore $$\frac{\partial x_j}{\partial z_i} = \delta_{ij} - \frac{\partial r_j}{\partial z_i} \;.$$

Hence, from (7.90) and (7.89):

$$2\eta_{ik} = \delta_{ik} - \left(\delta_{ij} - \frac{\partial r_j}{\partial z_i}\right)\left(\delta_{jk} - \frac{\partial r_j}{\partial z_k}\right),$$

$$= \frac{\partial r_k}{\partial z_i} + \frac{\partial r_i}{\partial z_k} - \frac{\partial r_j}{\partial z_i}\frac{\partial r_j}{\partial z_k} \;, \tag{7.92}$$

or $$2\boldsymbol{\eta} = \nabla\mathbf{r} + (\nabla\mathbf{r})' - (\nabla\mathbf{r}).(\nabla\mathbf{r})' \;. \tag{7.93}$$

This expression reduces to that for the infinitesimal strain tensor when the product term is omitted. There is then no distinction between the Eulerian and Lagrangian specifications.

7.6.3 The Relation between the Eulerian Finite Strain Tensor and the Rate - of - Strain Tensor

Consider a time-dependent deformation characterized by the Eulerian finite strain tensor $\boldsymbol{\eta}$. We may pose the question as to what is the relation between the rate of change of this tensor and the rate-of-strain tensor defined in Chapter 6, equation (6.58). To answer this question it is necessary to introduce the convected derivative defined by Oldroyd (1950). Before doing so, however, it must be pointed out that here, for the first time, it is necessary to distinguish between the covariant and contravariant components of a tensor, a distinction which is explained in the Appendix. This distinction is necessary even when working in Cartesian coordinates.

For a tensor $\boldsymbol{\eta}$ with covariant components, which transform according to (A.17), the **convected derivative** is given by:

$$\frac{\mathscr{D}\boldsymbol{\eta}}{\mathscr{D}t} = \frac{\mathrm{D}\boldsymbol{\eta}}{\mathrm{D}t} + (\nabla\mathbf{v}).\boldsymbol{\eta} + \boldsymbol{\eta}.(\nabla\mathbf{v})' \,, \tag{7.94}$$

where
$$\frac{\mathrm{D}\boldsymbol{\eta}}{\mathrm{D}t} = \frac{\partial\boldsymbol{\eta}}{\partial t} + \mathbf{v}.\nabla\boldsymbol{\eta} \,. \tag{7.95}$$

The quantity $\frac{\mathscr{D}\boldsymbol{\eta}}{\mathscr{D}t}$ is the **total time derivative** for a coordinate system embedded in a deformable continuum and deforming with it. In other words the tensor whose components are $\mathrm{D}\eta_{jl}/\mathrm{D}t$ in the convected system has components $\mathscr{D}\eta_{ik}/\mathscr{D}t$ in the fixed system. In suffix notation (7.94) becomes:

$$\frac{\mathscr{D}\eta_{ik}}{\mathscr{D}t} = \frac{\mathrm{D}\eta_{ik}}{\mathrm{D}t} + \frac{\partial v_j}{\partial z_i}\eta_{jk} + \frac{\partial v_j}{\partial z_k}\eta_{ij} \,, \tag{7.96}$$

where
$$\frac{\mathrm{D}\eta_{ik}}{\mathrm{D}t} = \frac{\partial\eta_{ik}}{\partial t} + v_j\frac{\partial\eta_{ik}}{\partial z_j} \,. \tag{7.97}$$

The proof of (7.96) is as follows. From (A.17) we have:

$$\bar{\eta}_{jl} = \frac{\partial z_i}{\partial \bar{z}_j}\frac{\partial z_k}{\partial \bar{z}_l}\eta_{ik} \,, \tag{7.98}$$

where $\bar{\eta}_{jl}$ are the components of $\boldsymbol{\eta}$ with respect to a convected coordinate system $(\bar{z}_1, \bar{z}_2, \bar{z}_3)$. We operate on both sides of (7.98) with D/Dt to obtain:

$$\frac{\mathrm{D}\bar{\eta}_{jl}}{\mathrm{D}t} = \frac{\mathrm{D}}{\mathrm{D}t}\left(\frac{\partial z_i}{\partial \bar{z}_j}\frac{\partial z_k}{\partial \bar{z}_l}\eta_{ik}\right),$$

$$= \eta_{ik}\frac{\mathrm{D}}{\mathrm{D}t}\left(\frac{\partial z_i}{\partial \bar{z}_j}\frac{\partial z_k}{\partial \bar{z}_l}\right) + \frac{\partial z_i}{\partial \bar{z}_j}\frac{\partial z_k}{\partial \bar{z}_l}\frac{\mathrm{D}\eta_{ik}}{\mathrm{D}t},$$

$$= \eta_{ik}\left[\frac{\partial z_i}{\partial \bar{z}_j}\frac{\mathrm{D}}{\mathrm{D}t}\left(\frac{\partial z_k}{\partial \bar{z}_l}\right) + \frac{\partial z_k}{\partial \bar{z}_l}\frac{\mathrm{D}}{\mathrm{D}t}\left(\frac{\partial z_i}{\partial \bar{z}_j}\right)\right] + \frac{\partial z_i}{\partial \bar{z}_j}\frac{\partial z_k}{\partial \bar{z}_l}\frac{\mathrm{D}\eta_{ik}}{\mathrm{D}t},$$

$$= \eta_{ik}\left(\frac{\partial z_i}{\partial \bar{z}_j}\frac{\partial v_k}{\partial \bar{z}_l} + \frac{\partial z_k}{\partial \bar{z}_l}\frac{\partial v_i}{\partial \bar{z}_j}\right) + \frac{\partial z_i}{\partial \bar{z}_j}\frac{\partial z_k}{\partial \bar{z}_l}\frac{\mathrm{D}\eta_{ik}}{\mathrm{D}t}. \tag{7.99}$$

Equation (7.99) holds for tensors with covariant components whatever system of convected coordinates $\bar{z}_j$ has been chosen. Let us now suppose that $\bar{z}_j$ is instantaneously coincident with the Cartesian system z_i. Then $\bar{z}_i = z_i$, $\partial z_i/\partial \bar{z}_j = \delta_{ij}$, and $\mathrm{D}\bar{\eta}_{jl}/\mathrm{D}t = \mathscr{D}\eta_{jl}/\mathscr{D}t$. Thus we obtain (7.96).

Equation (7.96) gives an expression for the convected derivative of the covariant tensor $\boldsymbol{\eta}$. We now obtain the important result that the covariant rate-of-strain tensor is the convected derivative of the covariant finite strain tensor.

Using (7.90) and (7.89) for the components η_{ik} we have, from (7.96):

$$\frac{\mathscr{D}\eta_{ik}}{\mathscr{D}t} = -\frac{1}{2}\frac{\partial}{\partial t}\left(\frac{\partial x_j}{\partial z_i}\frac{\partial x_j}{\partial z_k}\right) - \frac{1}{2}v_j\frac{\partial}{\partial z_j}\left(\frac{\partial x_m}{\partial z_i}\frac{\partial x_m}{\partial z_k}\right)$$

$$= \frac{1}{2}\frac{\partial v_j}{\partial z_i}\left(\delta_{jk} - \frac{\partial x_m}{\partial z_j}\frac{\partial x_m}{\partial z_k}\right) + \frac{1}{2}\frac{\partial v_j}{\partial z_k}\left(\delta_{ij} - \frac{\partial x_m}{\partial z_i}\frac{\partial x_m}{\partial z_j}\right),$$

$$= \frac{1}{2}\left(\frac{\partial v_k}{\partial z_i} + \frac{\partial v_i}{\partial z_k}\right) - \frac{1}{2}\frac{\partial x_m}{\partial z_i}\frac{\partial}{\partial z_k}\left(\frac{\partial x_m}{\partial t} + v_j\frac{\partial x_m}{\partial z_j}\right)$$

$$- \frac{1}{2}\frac{\partial x_m}{\partial z_k}\frac{\partial}{\partial z_i}\left(\frac{\partial x_m}{\partial t} + v_j\frac{\partial x_m}{\partial z_j}\right),$$

$$= e_{ik} - \frac{1}{2}\frac{\partial x_m}{\partial z_i}\frac{\partial}{\partial z_k}\left(\frac{\mathrm{D}x_m}{\mathrm{D}t}\right) - \frac{1}{2}\frac{\partial x_m}{\partial z_k}\frac{\partial}{\partial z_i}\left(\frac{\mathrm{D}x_m}{\mathrm{D}t}\right),$$

where e_{ik} are the components of the rate-of-strain tensor defined by (6.58).

Now x_m are the coordinates of a point in the material in its undeformed state. Hence, by definition $\mathrm{D}x_m/\mathrm{D}t = 0$. Thus we finally obtain:

$$\mathscr{D}\eta_{ik}/\mathscr{D}t \equiv e_{ik}, \tag{7.100}$$

or

$$\mathscr{D}\boldsymbol{\eta}/\mathscr{D}t \equiv \mathbf{e}. \tag{7.101}$$

The relation expressed by (7.100) or (7.101) is, of course, true in general. We are therefore justified in writing them as identities.

Examples 7.6

(1) Find the Lagrangian finite strain tensors for

(i) the shear strain $\mathbf{z} = (x_1 + kx_3, x_2, x_3)$;

(ii) the uniform compression $\mathbf{z} = (1 + k)\mathbf{x}$, $(k < 0)$;

(iii) the simple torsion and cross-sectional contraction of a rod defined by:

$$\mathbf{z} = (x_1 - \sigma kx_1, x_2 - \sigma kx_2, x_3 + kx_3) \; .$$

(i) The displacement vector $\mathbf{r}$ is given by $\mathbf{r} = (kx_3, 0, 0)$, hence the only non-zero component of the tensor $\overline{\nabla}\mathbf{r}$ is $\partial r_1/\partial x_3 = k$. The strain tensor is given by (7.86), (7.87):

$$\boldsymbol{\epsilon} = \frac{1}{2}\begin{bmatrix} 0 & 0 & k \\ 0 & 0 & 0 \\ k & 0 & k^2 \end{bmatrix}$$

(ii) The displacement vector is $\mathbf{r} = k\mathbf{x}$. Hence $\nabla\mathbf{r} = k\mathbf{1}$; $\boldsymbol{\epsilon} = k(1 + \tfrac{1}{2}k)\mathbf{1}$.

(ii) The displacement vector is $\mathbf{r} = (-\sigma kx_1, -\sigma kx_2, kx_3)$.

Hence $$\nabla\mathbf{r} = k\begin{bmatrix} -\sigma & 0 & 0 \\ 0 & -\sigma & 0 \\ 0 & 0 & 1 \end{bmatrix} ;$$

$$\boldsymbol{\epsilon} = \frac{1}{2}\begin{bmatrix} \sigma^2k^2 - 2\sigma k & 0 & 0 \\ 0 & \sigma^2k^2 - 2\sigma k & 0 \\ 0 & 0 & k^2 + 2k \end{bmatrix} .$$

(2) Find the Eulerian finite strain tensor for the time-dependent shear flow specified by the velocity vector $\mathbf{u} = kx_2\mathbf{e}_1$. Calculate the rate of strain tensor and show that the identity (7.100) is satisfied.

Let the time $t = 0$ correspond to the undeformed state, then the displacement vector is given by

$$\mathbf{r} = (kx_2t, 0, 0) = (kz_2t, 0, 0) \; .$$

Therefore $\nabla\mathbf{r} = kt\mathbf{e}_2 \otimes \mathbf{e}_1$.

The Eulerian strain tensor is given by (7.92), (7.93):

$$\boldsymbol{\eta} = \frac{1}{2}\begin{bmatrix} 0 & kt & 0 \\ kt & -(kt)^2 & 0 \\ 0 & 0 & 0 \end{bmatrix}.$$

The rate-of-strain tensor $\mathbf{e}$ is given by (6.58):

$$\mathbf{e} = \frac{1}{2}\begin{bmatrix} 0 & k & 0 \\ k & 0 & 0 \\ 0 & 0 & 0 \end{bmatrix}.$$

Now
$$\frac{\partial \boldsymbol{\eta}}{\partial t} = \frac{1}{2}\begin{bmatrix} 0 & k & 0 \\ k & -2k^2 t & 0 \\ 0 & 0 & 0 \end{bmatrix},$$

$$(\nabla \mathbf{v}).\boldsymbol{\eta} = \boldsymbol{\eta}.(\nabla \mathbf{v})' = \frac{1}{2}\begin{bmatrix} 0 & 0 & 0 \\ 0 & k^2 t & 0 \\ 0 & 0 & 0 \end{bmatrix},$$

$$\mathbf{v}.\nabla\boldsymbol{\eta} = \mathbf{0}.$$

Hence (7.100) and (7.101) are satisfied.

(3) Consider the two-dimensional flow given by $\mathbf{v} = \nabla\phi$, $\phi = \frac{1}{2}\alpha(z_1^2 - z_2^2)$ where $\mathbf{v}$ is the velocity vector. Find the Eulerian finite strain tensor and the rate of strain tensor, and show that the identity (7.100) is satisfied.

We have $\mathbf{v} = \nabla\phi = (\alpha z_1, -\alpha z_2)$.

Hence $\quad dz_1/dt = \alpha z_1, \quad dz_2/dt = -\alpha z_2$.

Therefore $\mathbf{z} = (Ae^{\alpha t}, Be^{-\alpha t})$, where A and B are arbitrary constants.

Consider a particle at (x_1, x_2) at time $t = 0$, and at $(z_1, z_2) = (x_1 + r_1, x_2 + r_2)$ at time t.

Then $\quad \mathbf{x} = (A, B) = (z_1 e^{-\alpha t}, z_2 e^{\alpha t})$,

therefore $\quad \mathbf{r} = \mathbf{z} - \mathbf{x} = [(1 - e^{-\alpha t})z_1, (1 - e^{\alpha t})z_2]$.

Proceeding now as in Example (2) we have:

$$\boldsymbol{\eta} = \tfrac{1}{2}\begin{bmatrix} 1 - e^{-2\alpha t} & 0 & 0 \\ 0 & 1 - e^{2\alpha t} & 0 \\ 0 & 0 & 0 \end{bmatrix},$$

$$\mathbf{e} = \begin{bmatrix} \alpha & 0 & 0 \\ 0 & -\alpha & 0 \\ 0 & 0 & 0 \end{bmatrix}, \quad \frac{\partial \boldsymbol{\eta}}{\partial t} = \begin{bmatrix} \alpha e^{-2\alpha t} & 0 & 0 \\ 0 & -\alpha e^{2\alpha t} & 0 \\ 0 & 0 & 0 \end{bmatrix},$$

$$\nabla \mathbf{v}.\boldsymbol{\eta} = \boldsymbol{\eta}.(\nabla \mathbf{v})' = \tfrac{1}{2}\alpha \begin{bmatrix} 1 - e^{-2\alpha t} & 0 & 0 \\ 0 & -1 + e^{2\alpha t} & 0 \\ 0 & 0 & 0 \end{bmatrix},$$

$$\mathbf{v}.\nabla \boldsymbol{\eta} = \mathbf{0}\ .$$

Hence (7.100) and (7.101) are satisfied.

Problems 7.6

(1) Calculate the components of the Lagrangian finite strain tensor for the torsion

$$\mathbf{z} = [r\cos(\phi + kz), r\sin(\phi + kz), z]\ ,$$

where (r, ϕ, z) are cylindrical coordinates given by $r\cos\phi = x_1, r\sin\phi = x_2, z = x_3$.

(2) Calculate, at time $t = 0$, the components of the Lagrangian finite strain tensor for the time-dependent displacement given by:

$$\mathbf{z} = \mathbf{x} - [R\cos(\omega t - \delta), \alpha_1 x_1, R\sin(\omega t - \delta)]\ ,$$

where $R\cos\delta = \alpha_2 x_2 + \beta_2, R\sin\delta = \alpha_3 x_3 + \beta_3$.

(3) Find the Eulerian finite strain tensors for the deformations specified in Example (1).

(4) Consider the fluid flows given by the following velocity vectors:
(i) $\mathbf{v} = \nabla\phi$ where $\phi = (z_1 - t)(z_2 - t)$;
(ii) $\mathbf{v} = (az_2 - z_2^2)\mathbf{e}_1$ (flow of a viscous fluid between parallel planes);
(iii) $\mathbf{v} = (a^2 - r^2)\mathbf{e}_3$ where (r, ϕ, z) are cylindrical polar coordinates (flow of a viscous fluid in a pipe of circular cross-section and radius a).

Find in each case the Eulerian finite strain tensor and the rate-of-strain tensor. Show that the identities (7.100), (7.101) are satisfied.

(5) The positions of the particles of a deformable continuum are referred to an arbitrarily chosen fixed system of Cartesian coordinates z_i. The displacement at z_i varies with the time t in the following way:

$$r_1 = z_1 + a_1 + \sigma_3 z_3 \ ,$$

$$r_2 = z_2 + a_2 - (\sigma_1 z_1 + b_1)\cos\omega t + (\sigma_2 z_2 + b_2)\sin\omega t \ ,$$

$$r_3 = z_3 + a_3 + (\sigma_1 z_1 + b_1)\sin\omega t + (\sigma_2 z_2 + b_2)\cos\omega t \ .$$

where a_i, b_i, σ_i and ω are constants and $\sigma_i > 0$. Evaluate the Eulerian strain tensor $\boldsymbol{\eta}$ at $\mathbf{z}$ at time t.

Show that the only non-vanishing components of the rate-of-strain tensor at $\mathbf{z}$ at time t are:

$$e_{12} = e_{21} = \tfrac{1}{2}\omega\left(\frac{\sigma_2}{\sigma_1} - \frac{\sigma_1}{\sigma_2}\right) .$$

Show also that

$$\mathbf{e} = \frac{\mathscr{D}\boldsymbol{\eta}}{\mathscr{D}t} = \frac{\partial\boldsymbol{\eta}}{\partial t} + \mathbf{v}.\nabla\boldsymbol{\eta} + (\nabla\mathbf{v}).\boldsymbol{\eta} + \boldsymbol{\eta}.(\nabla\mathbf{v})' \ .$$

(Liverpool University, June 1971).

(6) Complete information about the kinematics of a moving continuum is given (with reference to a fixed Cartesian coordinate system) in the following statement: the particle that is at $\mathbf{z}$ at time t is at

$$\mathbf{x} = \mathbf{z} - (t - \tau)\phi(z_2, z_3)\mathbf{e}_1$$

at any time τ. Evaluate the following tensors at $\mathbf{z}$:
(i) the deformation tensor at time t (taking the configuration at an arbitrarily selected instant τ as reference configuration);

(ii) the corresponding strain tensor $\boldsymbol{\eta}$ at time t;
(iii) the velocity vector $\mathbf{v}$ at time t;
(iv) the rate-of-strain tensor $\mathbf{e}$ at time t.

Verify by direct calculation that

$$\mathbf{e} = \frac{\partial \boldsymbol{\eta}}{\partial t} + \mathbf{v}.\nabla \boldsymbol{\eta} + (\nabla \mathbf{v}).\boldsymbol{\eta} + \boldsymbol{\eta}.(\nabla \mathbf{v})' \;.$$

(Liverpool University, June 1970).

(7) Let $\boldsymbol{\eta}$ be a Cartesian tensor whose components transform according to (A.15). Show that the convected derivative is given by

$$\frac{\mathscr{D}\boldsymbol{\eta}}{\mathscr{D}t} = \frac{\mathrm{D}\boldsymbol{\eta}}{\mathrm{D}t} - \boldsymbol{\eta}.\nabla \mathbf{v} - (\nabla \mathbf{v})'.\boldsymbol{\eta} \;.$$

Appendix

A.1 GENERAL TENSORS

A.1.1 Contravariant and covariant components

In section 3.6.2 we remarked that the physical components of vectors and tensors referred to curvilinear coordinates no longer obeyed the transformation laws. We shall now look briefly at the modifications necessary to restore tensorial properties referred to curvilinear coordinates.

Let us consider a curvilinear coordinate system (u^1, u^2, u^3) where, in agreement with the accepted notation for non-Cartesian tensors, we have use superscripts rather than subscripts to distinguish the individual curvilinear coordinates. It is of course important that the superscripts are not confused with powers.

In such a system the physical components of the vector **v** are the projections of the vector on tangents to the coordinates lines. Thus if the physical components of **v** are $(v_\alpha, v_\beta, v_\gamma)$,

then $$v_\alpha = \mathbf{v}.\mathbf{e}'_\alpha \ , \tag{A.1}$$

where the $\mathbf{e}'_\alpha$ are unit vectors along the coordinate lines (or, in the case of orthogonal curvilinear coordinates, perpendicular to the coordinate surfaces).

We now define the **contravariant components** v^i, of the vector **v** by:

$$\mathbf{v} = v^i \frac{\partial \mathbf{r}}{\partial u^i} \ , \tag{A.2}$$

and the **covariant components** v_i, of **v** by:

$$\mathbf{v} = v_i \nabla u^i \ . \tag{A.3}$$

Note that the vectors $\partial\mathbf{r}/\partial u^i$, ∇u^i are not unit vectors. In fact, by definition $h_i = |\partial\mathbf{r}/\partial u^i|$, where the h_i are the scale factors defined in section 3.6.1. It is true, however, that $(\partial\mathbf{r}/\partial u^i).\nabla u^j = \delta_{ij}$,

for $$\frac{\partial \mathbf{r}}{\partial u^i}.\nabla u^j = \frac{\partial}{\partial u^i}(x_k \mathbf{e}_k).\left(\frac{\partial u^j}{\partial x_m}\mathbf{e}_m\right) , \tag{A.4}$$

where the x_i are Cartesian coordinates and the $\mathbf{e}_i$ form a Cartesian basis.

The transformation law for the contravariant components of a vector may be obtained from (A.2). Suppose $(\bar{v}^1, \bar{v}^2, \bar{v}^3)$ are the contravariant components of $\mathbf{v}$ with respect to the curvilinear coordinates $(\bar{u}^1, \bar{u}^2, \bar{u}^3)$.

Then $$\mathbf{v} = \bar{v}^i \frac{\partial \mathbf{r}}{\partial \bar{u}^i} = v^i \frac{\partial \mathbf{r}}{\partial u^i} . \tag{A.5}$$

We form the scalar product with $\nabla \bar{u}^j$ to obtain:

$$\bar{v}^i \frac{\partial \mathbf{r}}{\partial \bar{u}^i} . \nabla \bar{u}^j = v^i \frac{\partial \bar{u}^k}{\partial u^i} \frac{\partial \mathbf{r}}{\partial \bar{u}^k} . \nabla \bar{u}^j , \tag{A.6}$$

hence, using (A.4), $$\bar{v}^j = \frac{\partial \bar{u}^j}{\partial u^i} v^i , \tag{A.7}$$

similarly $$v^j = \frac{\partial u^j}{\partial \bar{u}^i} \bar{v}^i . \tag{A.8}$$

A similar argument may be applied to (A.3) to obtain the transformation law for covariant components:

$$\bar{v}_j = \frac{\partial u^i}{\partial \bar{u}^j} v_i , \tag{A.9}$$

$$v_j = \frac{\partial \bar{u}^i}{\partial u^j} \bar{v}_i . \tag{A.10}$$

The **contravariant components** T^{ik}, of the tensor $\mathbf{T}$ are given by:

$$\mathbf{T} = T^{ik} \frac{\partial \mathbf{r}}{\partial u^i} \otimes \frac{\partial \mathbf{r}}{\partial u^k} , \tag{A.11}$$

and the **covariant components** T_{ik} by:

$$\mathbf{T} = T_{ik} \nabla u^i \otimes \nabla u^k . \tag{A.12}$$

The transformation law for the contravariant components of a tensor may be obtained from (A.11).

Thus $$\mathbf{T} = \bar{T}^{ik} \frac{\partial \mathbf{r}}{\partial \bar{u}^i} \otimes \frac{\partial \mathbf{r}}{\partial \bar{u}^k} = T^{ik} \frac{\partial \mathbf{r}}{\partial u^i} \otimes \frac{\partial \mathbf{r}}{\partial u^k} . \tag{A.13}$$

We form the double inner product with $\nabla \bar{u}^l \otimes \nabla \bar{u}^j$ to obtain:

$$\bar{T}^{ik} \frac{\partial \mathbf{r}}{\partial \bar{u}^i} \cdot \nabla \bar{u}^j \frac{\partial \mathbf{r}}{\partial \bar{u}^k} \cdot \nabla \bar{u}^l = T^{ik} \frac{\partial \bar{u}^m}{\partial u^i} \frac{\partial \bar{u}^n}{\partial u^k} \frac{\partial \mathbf{r}}{\partial \bar{u}^m} \cdot \nabla \bar{u}^j \frac{\partial \mathbf{r}}{\partial \bar{u}^n} \cdot \nabla \bar{u}^l , \tag{A.14}$$

or $$\bar{T}^{jl} = \frac{\partial \bar{u}^j}{\partial u^i} \frac{\partial \bar{u}^l}{\partial u^k} T^{ik} . \tag{A.15}$$

Similarly $$T^{jl} = \frac{\partial u^j}{\partial \bar{u}^i}\frac{\partial u^l}{\partial \bar{u}^k}\bar{T}^{ik} \,. \tag{A.16}$$

A similar argument may be applied to (A.12) to obtain the transformation law for covariant components:

$$\bar{T}_{jl} = \frac{\partial u^i}{\partial \bar{u}^j}\frac{\partial u^k}{\partial \bar{u}^l}T_{ik} \,, \tag{A.17}$$

$$T_{jl} = \frac{\partial \bar{u}^i}{\partial u^j}\frac{\partial \bar{u}^k}{\partial u^l}\bar{T}_{ik} \,. \tag{A.18}$$

Tensors whose components transform according to (A.15), (A.16) or (A.17), (A.18) are called respectively **contravariant** and **covariant tensors of the second order**. In the same way a tensor whose components transform according to the law:

$$\bar{T}^j_l = \frac{\partial \bar{u}^j}{\partial u^i}\frac{\partial u^k}{\partial \bar{u}^l}T^i_k \,, \tag{A.19}$$

$$T^j_l = \frac{\partial u^j}{\partial \bar{u}^i}\frac{\partial \bar{u}^k}{\partial u^l}\bar{T}^i_k \,, \tag{A.20}$$

is called a **mixed tensor of the second order.**

A.1.2 The Metric Tensor

Suppose in the system of curvilinear coordinates (u^1, u^2, u^3) the square of the line element is given by:

$$\delta s^2 = \delta \mathbf{r}.\delta \mathbf{r} = g_{ik}\mathrm{d}u^i\mathrm{d}u^k \,.$$

Then $$g_{ik} = \frac{\partial \mathbf{r}}{\partial u^i}\cdot\frac{\partial \mathbf{r}}{\partial u^k} \,. \tag{A.21}$$

Also $$g_{ik} = \delta_{ik}h_i h_k \,, \tag{A.22}$$

if the curvilinear coordinate system is orthogonal. The quantities g_{ik} are the components of a symmetric covariant tensor called the **metric tensor**;

for $$\bar{g}_{jl} = \frac{\partial \mathbf{r}}{\partial \bar{u}^j}\cdot\frac{\partial \mathbf{r}}{\partial \bar{u}^l} \,,$$

$$= \frac{\partial \mathbf{r}}{\partial u^i}\cdot\frac{\partial \mathbf{r}}{\partial u^k}\frac{\partial u^i}{\partial \bar{u}^j}\frac{\partial u^k}{\partial \bar{u}^l} \,,$$

$$= \frac{\partial u^i}{\partial \bar{u}^j}\frac{\partial u^k}{\partial \bar{u}^l}g_{ik} \,. \tag{A.23}$$

The tensor whose components are g^{ik} where

$$g^{ik} = \frac{\text{cofactor of } g_{ik}}{\det g_{ik}} , \tag{A.24}$$

is called the **associate** or **reciprocal** tensor to g_{ik}. It may be shown that

$$g_{ij}g^{il} = \delta_j^l , \tag{A.25}$$

where the δ_j^l are the components of the mixed tensor

$$\delta_j^l = \frac{\partial x^l}{\partial x^j} . \tag{A.26}$$

The metric tensor has the property of raising and lowering suffices. From (A.2) and (A.3) we have:

$$v^k \frac{\partial \mathbf{r}}{\partial u^k} = v_k \nabla u^k ,$$

therefore $$v^k \frac{\partial \mathbf{r}}{\partial u^k}\cdot\frac{\partial \mathbf{r}}{\partial u^i} = v_k \nabla u^k \cdot \frac{\partial \mathbf{r}}{\partial u^i} ,$$

i.e. $$v^k g_{ik} = v_k \delta_{ik} = v_i ,$$

or $$v_i = g_{ik} v^k .$$

Similarly $$v^i = g^{ik} v_k .$$

A.1.3 The Quotient Theorem

We were able to show that the quantities g_{ik} were the components of a tensor by use of the transformation law (A.23). It is not, however, convenient to repeat the process for the quantities g^{ik}.

In such a case the **Quotient Theorem** is helpful. The theorem states that a collection of quantities form the components of a tensor provided an inner product of these quantities with an *arbitrary* tensor is itself a tensor.

The Quotient Theorem may not be applied directly to (A.25) because g_{ij} are not the components of an *arbitrary* tensor. Consider, however, an arbitrary contravariant vector **c** then $d_i = g_{ij}c^j$ are the components of an arbitrary covariant vector **d**.

But $$d_i g^{il} = g_{ij} g^{il} c^j = \delta_j^l c^j = c^l .$$

The Quotient Theorem now tells us that the g^{ik} are the components of a contravariant tensor.

A.2 DIFFERENTIATION OF GENERAL TENSORS

A.2.1 The Christoffel Symbols

By defining the covariant and contravariant components of tensors we have been able to obtain quantities which transform according to laws which are formally the same as the laws for the transformation of components of Cartesian tensors. We encounter further difficulties, however, when finding derivatives. In the case of Cartesian tensors their derivatives are also tensors. The same is not true in the general case. A modification of the derivative has therefore to be found such that (i) it reduces to the derivative in the Cartesian case and (ii) the modified derivatives are themselves tensors. It is to this end that we introduce the Christoffel symbols.

The Christoffel symbols of the first and second kind are defined respectively by:

$$[ij, k] = \frac{1}{2}\left(\frac{\partial g_{ik}}{\partial u^j} + \frac{\partial g_{jk}}{\partial u^i} - \frac{\partial g_{ij}}{\partial u^k}\right), \tag{A.27}$$

$$\begin{Bmatrix} l \\ ij \end{Bmatrix} = g^{lk}[ij, k] . \tag{A.28}$$

From (A.25) and (A.28) we obtain at once:

$$[ij, m] = g_{lm}\begin{Bmatrix} l \\ ij \end{Bmatrix} . \tag{A.29}$$

Also, from (A.27):

$$[ij, k] + [kj, i] = \frac{\partial g_{ik}}{\partial u^j} . \tag{A.30}$$

From (A.25) we obtain by differentiation:

$$\frac{\partial g^{ik}}{\partial u^l} g_{ij} + \frac{\partial g_{ij}}{\partial u^l} g^{ik} = 0 , \tag{A.31}$$

$$\frac{\partial g^{mk}}{\partial u^l} + g^{jm} g^{ik} \frac{\partial g_{ij}}{\partial u^l} = 0 . \tag{A.32}$$

Then, from (A.30) and (A.32), we have:

$$-\frac{\partial g^{mk}}{\partial u^l} = +g^{ik}\begin{Bmatrix} m \\ il \end{Bmatrix} + g^{jm}\begin{Bmatrix} k \\ jl \end{Bmatrix} . \tag{A.33}$$

A.2.2 The Transformation Law for Christoffel Symbols

By using the transformation law for the components of the metric tensor (A.23) it may be shown that the Christoffel symbol of the first kind transforms according to the law:

$$[\overline{lm, n}] = [ij, k]\frac{\partial u^i}{\partial \bar{u}^l}\frac{\partial u^j}{\partial \bar{u}^m}\frac{\partial u^k}{\partial \bar{u}^n} + g_{ij}\frac{\partial u^i}{\partial \bar{u}^n}\frac{\partial^2 u^j}{\partial \bar{u}^l \partial \bar{u}^m} . \tag{A.34}$$

The presence of the second term means that these symbols do not in general behave like the components of third-order tensors. (A.23), (A.28) and (A.34) may then be used to deduce the transformation law for Christoffel symbols of the second kind:

$$\overline{\left\{ {p \atop lm} \right\}} = \left\{ {s \atop ij} \right\} \frac{\partial \bar{u}^p}{\partial u^s} \frac{\partial u^i}{\partial \bar{u}^l} \frac{\partial u^j}{\partial \bar{u}^m} + \frac{\partial \bar{u}^p}{\partial u^j} \frac{\partial^2 u^j}{\partial \bar{u}^l \partial \bar{u}^m} . \tag{A.35}$$

From (A.35) an explicit expression may be obtained for $\dfrac{\partial^2 u_r}{\partial \bar{u}^l \partial \bar{u}^m}$:

$$\frac{\partial^2 u^r}{\partial \bar{u}^l \partial \bar{u}^m} = \overline{\left\{ {p \atop lm} \right\}} \frac{\partial u^r}{\partial \bar{u}^p} - \left\{ {r \atop ij} \right\} \frac{\partial u^i}{\partial \bar{u}^l} \frac{\partial u^j}{\partial \bar{u}^m} . \tag{A.36}$$

A.2.3 The Covariant Differentiation of Vectors

Consider a vector whose contravariant components transform according to the law (A.8):

$$v^j = \frac{\partial u^j}{\partial \bar{u}^i} \bar{v}^i .$$

On differentiation we obtain:

$$\frac{\partial v^j}{\partial u^k} = \frac{\partial^2 u^j}{\partial \bar{u}^i \partial \bar{u}^m} \frac{\partial \bar{u}^m}{\partial u^k} \bar{v}^i + \frac{\partial u^j}{\partial \bar{u}^i} \frac{\partial \bar{v}^i}{\partial \bar{u}^m} \frac{\partial \bar{u}^m}{\partial u^k} ,$$

$$= \overline{\left\{ {p \atop im} \right\}} \frac{\partial u^j}{\partial \bar{u}^p} \frac{\partial \bar{u}^m}{\partial u^k} \bar{v}^i - \left\{ {j \atop qr} \right\} \frac{\partial u^q}{\partial \bar{u}^i} \frac{\partial u^r}{\partial \bar{u}^m} \frac{\partial \bar{u}^m}{\partial u^k} \bar{v}^i + \frac{\partial u^j}{\partial \bar{u}^i} \frac{\partial \bar{v}^i}{\partial \bar{u}^m} \frac{\partial \bar{u}^m}{\partial u^k} .$$

$$\frac{\partial v^j}{\partial u^k} + \left\{ {j \atop qk} \right\} v^q = \left[\frac{\partial \bar{v}^p}{\partial \bar{u}^m} + \overline{\left\{ {p \atop im} \right\}} \bar{v}^i \right] \frac{\partial \bar{u}^m}{\partial u^k} \frac{\partial u^j}{\partial \bar{u}^p} . \tag{A.37}$$

We introduce the notation

$$v^j{}_{,k} = \frac{\partial v^j}{\partial u^k} + \left\{ {j \atop qk} \right\} v^q \tag{A.38}$$

for the **covariant differentiation** of v^j, and (A.37) becomes:

$$v^j{}_{,k} = \frac{\partial \bar{u}^m}{\partial u^k} \frac{\partial u^j}{\partial \bar{u}^p} \bar{v}^p{}_{,m} . \tag{A.39}$$

For covariant vectors we have:

$$v_{j,n} = \frac{\partial v^j}{\partial u^n} - \left\{ {r \atop jn} \right\} v_r , \tag{A.40}$$

$$\bar{v}_{i,l} = v_{j,n} \frac{\partial u^j}{\partial \bar{u}^i} \frac{\partial u^n}{\partial \bar{u}^l} . \tag{A.41}$$

The covariant derivatives $v^j{}_{,k}$ and $v_{j,n}$ fulfil the criteria, set out in section A.2.1, for a modification of the derivative. From (A.38) and (A.40) we see that they reduce to ordinary derivatives when referred to Cartesian coordinates since the Christoffel symbols are in that case identically zero. From (A.39) and (A.41) we see that the covariant derivatives are indeed the components of tensors. The reader requiring further information on general tensors is referred to the text-books on the subject, for example the book by Spain mentioned in the Bibliography.

Examples A.1

(1) Consider the case of cylindrical polar coordinates, (r, θ, z). The physical components of displacement, velocity and acceleration are:

$$\begin{aligned} \delta\mathbf{r} &= (\delta r, r\delta\theta, \delta z) \ , \\ \mathbf{v} &= (\dot{r}, r\dot{\theta}, \dot{z}) \ , \\ \mathbf{a} &= (\ddot{r} - r\dot{\theta}^2, r\ddot{\theta} + 2\dot{r}\dot{\theta}, \ddot{z}) \ . \end{aligned}$$

The contravariant components, calculated from (A.2) are:

$$\begin{aligned} \delta\mathbf{r} &= (\delta r, \delta\theta, \delta z) \ , \\ \mathbf{v} &= (\dot{r}, \dot{\theta}, \dot{z}) \ , \\ \mathbf{a} &= (\ddot{r} - r\dot{\theta}^2, \ddot{\theta} + 2\dot{r}\dot{\theta}/r, \ddot{z}) \ . \end{aligned}$$

The covariant components, calculated from (A.6) are:

$$\begin{aligned} \delta\mathbf{r} &= (\delta r, r^2\delta\theta, \delta z) \\ \mathbf{v} &= (\dot{r}, r^2\dot{\theta}, \dot{z}) \\ \mathbf{a} &= (\ddot{r} - r\dot{\theta}^2, r^2\ddot{\theta} + 2r\dot{r}\dot{\theta}, \ddot{z}) \end{aligned}$$

(2) The covariant components of the gradient vector $\nabla\Phi$ in cylindrical polar coordinates are

$$\nabla\Phi = \left(\frac{\partial\Phi}{\partial r}, \frac{\partial\Phi}{\partial\theta}, \frac{\partial\Phi}{\partial z} \right) ,$$

(calculated from (3.88) and (A.3)).

(3) In the case of orthogonal curvilinear coordinates, in which case the components of the metric tensor are given by (A.22), the expressions (A.27) for the Christoffel symbols of the first kind are considerably simplified. Thus we have

$$[ij, k] = 0 \ ,$$

$$[ii, i] = \frac{1}{2}\frac{\partial g_{ii}}{\partial u_i} \ ,$$

$$[ii, j] = -\frac{1}{2}\frac{\partial g_{ii}}{\partial u_j},$$

$$[ij, i] = \frac{1}{2}\frac{\partial g_{ii}}{\partial u_j},$$

$$[ij, j] = \frac{1}{2}\frac{\partial g_{ij}}{\partial u_i},$$

where i, j, k are supposed to be all different and no summation is implied over the indices i or j.

(4) Consider the case of cylindrical polar coordinates (r, θ, z) in which $h_1 = 1$, $h_2 = r, h = 1$. The metric tensor (g_{ik}) is given by (A.22):

$$(g_{ik}) = \begin{bmatrix} 1 & 0 & 0 \\ 0 & r^2 & 0 \\ 0 & 0 & 1 \end{bmatrix}.$$

The associate tensor (g^{ik}) is given by (A.24):

$$(g^{ik}) = \begin{bmatrix} 1 & 0 & 0 \\ 0 & 1/r^2 & 0 \\ 0 & 0 & 1 \end{bmatrix}.$$

Using the formulae of Example (3), the only non-zero Christoffel symbols of the first kind are found to be:

$$[22, 1] = -r, \quad [12, 2] = [21, 2] = r .$$

From (A.28) we find that the only non-zero Christoffel symbols of the second kind are:

$$\left\{ {1 \atop 22} \right\} = -r, \quad \left\{ {2 \atop 12} \right\} = \left\{ {2 \atop 21} \right\} = \frac{1}{r} .$$

(5) In the case of spherical polar coordinates (r, θ, ϕ), in which $h_1 = 1, h_2 = r$, $h_3 = r \sin \theta$, the metric tensor (g_{ik}) is given by:

$$(g_{ik}) = \begin{bmatrix} 1 & 0 & 0 \\ 0 & r^2 & 0 \\ 0 & 0 & r^2 \sin^2 \theta \end{bmatrix}.$$

The associate tensor (g^{ik}) is given by:

$$(g^{ik}) = \begin{bmatrix} 1 & 0 & 0 \\ 0 & 1/r^2 & 0 \\ 0 & 0 & 1/(r^2 \sin^2\theta) \end{bmatrix} .$$

The non-zero Christoffel symbols of the first kind are:

$$[22, 1] = -r, \quad [33, 1] = -r \sin^2 \theta ,$$

$$[33, 2] = -r^2 \sin\theta \cos\theta ,$$

$$[12, 2] = [21, 2] = r, \quad [13, 3] = [31, 3] = r \sin^2\theta ,$$

$$[23, 3] = [32, 3] = r^2 \sin\theta \cos\theta .$$

The non-zero Christoffel symbols of the second kind are:

$$\begin{Bmatrix} 1 \\ 22 \end{Bmatrix} = -r, \quad \begin{Bmatrix} 2 \\ 12 \end{Bmatrix} = \begin{Bmatrix} 2 \\ 21 \end{Bmatrix} = \frac{1}{r} ,$$

$$\begin{Bmatrix} 1 \\ 33 \end{Bmatrix} = -r \sin^2\theta, \begin{Bmatrix} 2 \\ 33 \end{Bmatrix} = -\sin\theta \cos\theta ,$$

$$\begin{Bmatrix} 3 \\ 31 \end{Bmatrix} = \begin{Bmatrix} 3 \\ 13 \end{Bmatrix} = \frac{1}{r} ,$$

$$\begin{Bmatrix} 3 \\ 23 \end{Bmatrix} = \begin{Bmatrix} 3 \\ 32 \end{Bmatrix} = \cot\theta .$$

(6) The covariant derivatives of the covariant vector **v** are given by (A.40). In the case of cylindrical polar coordinates, we have, using the Christoffel symbols of the second kind, calculated in Example (4):

$$v_{1,1} = \frac{\partial v_1}{\partial r}, \quad v_{2,2} = \frac{\partial v_2}{\partial \theta} + r v_1 , v_{3,3} = \frac{\partial v_3}{\partial z} ,$$

$$v_{2,3} = \frac{\partial v_2}{\partial z}, \quad v_{3,2} = \frac{\partial v_3}{\partial \theta}, \, v_{3,1} = \frac{\partial v_3}{\partial r} ,$$

$$v_{1,3} = \frac{\partial v_1}{\partial z}, \quad v_{1,2} = \frac{\partial v_1}{\partial \theta} - \frac{v_2}{r}, \, v_{2,1} = \frac{\partial v_2}{\partial r} - \frac{v_2}{r} .$$

These are the covariant components of the tensor $\nabla\mathbf{v}$, calculated according to the relation $(\nabla\mathbf{v})_{ik} = v_{k,i}$.

Thus
$$\nabla\mathbf{v} = \begin{bmatrix} \partial v_1/\partial r & r\dfrac{\partial}{\partial r}\left(\dfrac{v_2}{r}\right) & \partial v_3/\partial r \\ \partial v_1/\partial\theta - v_2/r & \partial v_2/\partial\theta + r v_1 & \partial v_3/\partial\theta \\ \partial v_1/\partial z & \partial v_2/\partial z & \partial v_3/\partial z \end{bmatrix}.$$

This result should be compared with the physical components of $\nabla\mathbf{v}$ given by (3.114).

(7) The Christoffel symbols of the second kind, calculated in Example (5), may be used to calculate the covariant derivatives of a covariant vector expressed in spherical polar coordinates.

Thus
$$v_{1,1} = \frac{\partial v_1}{\partial r}, \quad v_{2,2} = \frac{\partial v_2}{\partial\theta} + r v_1 ,$$

$$v_{3,3} = \frac{\partial v_3}{\partial\phi} + r\sin^2\theta\, v_1 + \sin\theta\cos\theta\, v_2 ,$$

$$v_{2,3} = \frac{\partial v_2}{\partial\phi} - \cot\theta\, v_3, \quad v_{3,2} = \frac{\partial v_3}{\partial\theta} - \cot\theta\, v_3 ,$$

$$v_{3,1} = \frac{\partial v_3}{\partial r} - \frac{v_3}{r}, \quad v_{1,3} = \frac{\partial v_1}{\partial\phi} - \frac{v_3}{r},$$

$$v_{1,2} = \frac{\partial v_1}{\partial\theta} - \frac{v_2}{r}, \quad v_{2,1} = \frac{\partial v_2}{\partial r} - \frac{v_2}{r}.$$

Hence

$$\nabla\mathbf{v} = \begin{bmatrix} \partial v_1/\partial r & (r\partial/\partial r)(v_2/r) & (r\partial/\partial r)(v_3/r) \\ \partial v_1/\partial\theta - v_2/r & \partial v_2/\partial\theta + r v_1 & (\sin\theta\,\partial/\partial\theta)(v_3/\sin\theta) \\ \partial v_1/\partial\phi - v_3/r & \partial v_2/\partial\phi - \cot\theta\, v_3 & \partial v_3/\partial\phi + r\sin^2\theta\, v_1 + \sin\theta\cos\theta\, v_2 \end{bmatrix}.$$

This result should be compared with the physical components of $\nabla\mathbf{v}$ given by (3.102).

Answers to Problems

1.1

(1)2; (2)–10; (3)8; (4)7; (5)–48; (6)0; (7)–49; (8)15; (9)2172; (10) $4a^2b^2c^2$.

1.2

(1) (i) $\begin{bmatrix} a_1 & b_1 & c_1 \\ a_2 & b_2 & c_2 \\ a_3 & b_3 & c_3 \end{bmatrix}\begin{bmatrix} x \\ y \\ z \end{bmatrix} = \begin{bmatrix} d_1 \\ d_2 \\ d_3 \end{bmatrix}$; (ii) $\begin{bmatrix} 1 & 2 & -1 \\ 2 & -1 & 3 \\ 1 & 2 & 0 \end{bmatrix}\begin{bmatrix} x \\ y \\ z \end{bmatrix} = \begin{bmatrix} -4 \\ 0 \\ 3 \end{bmatrix}$.

(2) $A(2 \times 2)$; $B(3 \times 2)$; $C(2 \times 3)$; $D(2 \times 3)$.
(i), (ii), (v), (vii) not possible;

(iii) $\begin{bmatrix} 0 & 3 & 3 \\ 3 & -10 & -4 \end{bmatrix}$; (iv) $\begin{bmatrix} -2 & 5 & -1 \\ 3 & -2 & -10 \end{bmatrix}$; (vi) $\begin{bmatrix} 3 & 0 & -11 \\ 11 & -20 & -27 \\ 6 & -12 & -14 \end{bmatrix}$;

(viii) $\begin{bmatrix} 1 & 16 \\ 3 & -32 \end{bmatrix}$; (ix) $\begin{bmatrix} 8 & 7 \\ 6 & 9 \\ 2 & 4 \end{bmatrix}$.

(3)

(i) $AC = \begin{bmatrix} 1 & 2 & -5 \\ 5 & -8 & -13 \end{bmatrix}$, $AD = \begin{bmatrix} 2 & -6 & 7 \\ 1 & -9 & 8 \end{bmatrix}$

$$\text{(ii) } B(AC) = (BA)C = \begin{bmatrix} +13 & -10 & -41 \\ 21 & -30 & -57 \\ 10 & -16 & -26 \end{bmatrix};$$

$$\text{(iii) } BD = \begin{bmatrix} 3 & -11 & 12 \\ 1 & -17 & 14 \\ 0 & -8 & 6 \end{bmatrix}, \quad DB = \begin{bmatrix} 2 & 2 \\ -4 & -10 \end{bmatrix}.$$

(4) $x'' = x\cos\theta + y\sin\theta$,

$y'' = -x\cos\phi\sin\theta + y\cos\phi\cos\theta + z\sin\phi$,

$z'' = x\sin\phi\sin\theta - y\sin\theta\cos\theta + z\cos\phi$.

(6) $A = \begin{bmatrix} 1 & -2 \\ 1 & -2 \end{bmatrix}, \quad B = \begin{bmatrix} 4 & -4 \\ 2 & -2 \end{bmatrix}$

$$\frac{b_{22}}{b_{12}} = \frac{b_{21}}{b_{11}} = -\frac{a_{11}}{a_{12}} = -\frac{a_{21}}{a_{22}}.$$

(7) (i) $\begin{bmatrix} 2 & 1 \\ 0 & -3 \end{bmatrix} = \begin{bmatrix} 2 & 0.5 \\ 0.5 & -3 \end{bmatrix} + \begin{bmatrix} 0 & 0.5 \\ -0.5 & 0 \end{bmatrix};$

$$\text{(ii) } \begin{bmatrix} 3 & 2 & -1 \\ 6 & -3 & -2 \\ -4 & 5 & 0 \end{bmatrix} = \begin{bmatrix} 3 & 4 & -2.5 \\ 4 & -3 & 1.5 \\ -2.5 & 1.5 & 0 \end{bmatrix} + \begin{bmatrix} 0 & -2 & 1.5 \\ 2 & 0 & -3.5 \\ -1.5 & 3.5 & 0 \end{bmatrix}.$$

(8) 9, $R = 2$.

(9) (i)2; (ii)2; (iii)1; (iv)3; (v)2; (vi)2; (vii)3; (viii)1.

1.3

(1) (i)0; (ii)1; (iii)1.

(2) (i)1; (ii)21; (iii)$39 - 2\sqrt{13}$.

1.4

(2) Replace the real numbers α, β in the definition of a real vector space, by complex numbers, to obtain the definition of a complex vector space.

1.5

(1) (i)1, 3, −4; (ii)$-\mathbf{e}_2, -2\mathbf{e}_1 - 2\mathbf{e}_3, -6\mathbf{e}_1 + 4\mathbf{e}_2$.

(3) $y = \sqrt{17}, z = 2\sqrt{17}$.

1.6

(2) (i)$\cos^{-1}(2/\sqrt{7})$; (ii)$\cos^{-1}(1/\sqrt{7})$; (iii)$\cos^{-1}(-2/7)$; (vi)$\cos^{-1}(-1/7)$.

(4) $\breve{\mathbf{a}} = \frac{1}{5\sqrt{2}}(3, 4, 5)$. The components of $\breve{\mathbf{a}}$ are the direction cosines.

1.7

(4) $[\mathbf{x}_1, \mathbf{x}_2, \mathbf{x}_3]\mathbf{x}_4 = [\mathbf{x}_4, \mathbf{x}_2, \mathbf{x}_3]\mathbf{x}_1 + [\mathbf{x}_1, \mathbf{x}_4, \mathbf{x}_3]\mathbf{x}_2 + [\mathbf{x}_1, \mathbf{x}_2, \mathbf{x}_4]\mathbf{x}_3$.
(5) Let the vectors be $\mathbf{x}_1, \mathbf{x}_2$, etc.
(i) $2\mathbf{x}_1 - \mathbf{x}_2 = \mathbf{0}$;
(ii) $3\mathbf{x}_1 - 2\mathbf{x}_2 - \mathbf{x}_3 = \mathbf{0}$;
(iii) $\mathbf{x}_1, \mathbf{x}_2, \mathbf{x}_3$ are linearly independent ;
(iv) $7\mathbf{x}_1 + 5\mathbf{x}_2 - 11\mathbf{x}_3 - 10\mathbf{x}_4 = \mathbf{0}$.

1.8

(2) (i) $(2x_1, 2, 0)$; (ii) $(8r^2 + 9r + 8 + \frac{1}{r})\mathbf{r}$; (iii) $\frac{-1}{r^3}\mathbf{r}$; (iv) $nr^{n-2}\mathbf{r}$; (v) $\frac{1}{r^2}\mathbf{r}$;

(vi) $\frac{1}{r}f'(r)\mathbf{r}$; (vii) $\frac{1}{r}\mathbf{e}_1 - \frac{x_1}{r^3}\mathbf{r}$.
(3) $(1/\sqrt{6}, 1/\sqrt{6}, 2/\sqrt{6})$.
(4) 16/3.

1.9

(1) (i)$\mathbf{a}$; (ii)$\mathbf{a}.\mathbf{b}$;
(3) (ii)$a_1c_1 + a_2c_2 + a_3c_3$; (iii)$a_1(b_1 + c_1) + a_2(b_2 + c_2) + a_3(b_3 + c_3)$.
(4) (i)3; (ii)δ_{km}.

1.10

(3) (i) Reflection in the plane $x_1 = x_2$;
(ii) Reflection in the plane $x_3 = 0$;
(iii) Rotation of $\pi/2$ about x_1-axis, followed by a similar rotation about the new x_2-axis, both in the positive sense. Alternatively it is a rotation of $2\pi/3$ in the positive sense about the line $x_1 = x_2 = x_3$ (see equations (3.52) and (3.54) of Chapter 3).
(iv) Rotation of α in the positive sense about the x_1-axis.
(v) Rotation of α about x_3-axis, followed by a rotation of β about the new x_1-axis, both in the positive sense.

(4)

$$\begin{bmatrix} \cos\beta & \sin\alpha\sin\beta & -\cos\alpha\sin\beta \\ 0 & \cos\alpha & \sin\alpha \\ \sin\beta & -\sin\alpha\cos\beta & \cos\alpha\cos\beta \end{bmatrix},$$

The components of **a** in the new system are:

$$a_1^* = (\sqrt{3}/2)a_1 + \tfrac{1}{4}a_2 - (\sqrt{3}/4)a_3,$$
$$a_2^* = (\sqrt{3}/2)a_2 + \tfrac{1}{2}a_3,$$
$$a_3^* = \tfrac{1}{2}a_1 - (\sqrt{3}/4)a_2 + (3/4)a_3.$$

2.1

(1) (i)Yes; (ii)Yes; (iii)Not linear in **v**; (iv)Not linear in **v**; (v)Yes; (vi)Not a vector; (vii)Yes; (viii)Yes; (ix)Not a vector; (x)Not linear in **v**; (xi)Not a vector.

(2) (i)0; (ii)$\mathbf{e}_1$; (iii)0; (iv)0; (v)0.

(3) (i)(6, 8, 2); (ii)(10, 0, 10,); (iii)(8, 4, 0); (iv)(10, 0, 10).

2.2

(1) (i) $\begin{bmatrix} 1 & 0 & 0 \\ 0 & 0 & 0 \\ 0 & 0 & 0 \end{bmatrix}$; (ii) $\begin{bmatrix} 0 & 0 & 1 \\ 0 & 0 & 0 \\ 0 & 0 & 0 \end{bmatrix}$; (iii) $\begin{bmatrix} 0 & 0 & 0 \\ 0 & 1 & 1 \\ 0 & 0 & 0 \end{bmatrix}$.

(3) δ_{ik}, $\begin{bmatrix} 1 & 0 & 0 \\ 0 & 1 & 0 \\ 0 & 0 & 1 \end{bmatrix}$. (5) $\begin{bmatrix} 3 & 0 & 3 \\ 4 & 0 & 4 \\ 1 & 0 & 1 \end{bmatrix}$.

(6) (23, 17, 7), $\begin{bmatrix} 3 & 1 & 2 \\ 4 & 0 & 1 \\ 1 & 1 & 0 \end{bmatrix}$. (7) (–33, 0, –22), $\begin{bmatrix} 0 & 3 & -9 \\ 0 & 0 & 0 \\ -2 & 1 & -4 \end{bmatrix}$,

(8) (9, –4, 13), $\begin{bmatrix} 3/2 & 3 & -3/2 \\ -2 & -2 & 2 \\ 5/2 & -3 & 7/2 \end{bmatrix}$.

(9) $$T_{jl} = \frac{(\mathbf{f}_1)_j\,[\mathbf{e}_l, \mathbf{b}_2, \mathbf{b}_3] + (\mathbf{f}_2)_j\,[\mathbf{b}_1, \mathbf{e}_l, \mathbf{b}_3] + (\mathbf{f}_3)_j\,[\mathbf{b}_1, \mathbf{b}_2, \mathbf{e}_l]}{[\mathbf{b}_1, \mathbf{b}_2, \mathbf{b}_3]}$$

from which the result follows. If $[\mathbf{b}_1, \mathbf{b}_2, \mathbf{b}_3] = 0$, the vectors $\mathbf{b}_1, \mathbf{b}_2, \mathbf{b}_3$ are linearly dependent and $\mathbf{T}$ is no longer defined uniquely.

2.3

(1) (i)(14, 20, –1); (ii)(18, 18, –5); (iii)(4, 0, 1); (iv)13; (v)13; (vi)13; (vii)26; (viii)20.
The fact that $\mathbf{a.T.b} = \mathbf{b.T.a}$ implies that $\mathbf{a.(T - T').b} = 0$. Check that this is the case.
(5) $\mathbf{a} = \mathbf{b}$.

2.4

(1) (i)Neither; (ii)Symmetric; (iii)Antisymmetric; (iv)Symmetric; (v)Neither; (vi)Symmetric; (vii)Symmetric; (viii)Antisymmetric; (ix)Neither; (x)Neither.
(2) (i)$\frac{1}{2}(\mathbf{e}_1 \otimes \mathbf{e}_2 + \mathbf{e}_2 \otimes \mathbf{e}_1) + \frac{1}{2}(\mathbf{e}_1 \otimes \mathbf{e}_2 - \mathbf{e}_2 \otimes \mathbf{e}_1)$;

(v) $$\begin{bmatrix} 3 & 0 & 0 \\ 0 & 1 & 0 \\ 0 & 0 & 5 \end{bmatrix} + \begin{bmatrix} 0 & 2 & -7 \\ -2 & 0 & 4 \\ 7 & -4 & 0 \end{bmatrix};$$

(ix) $$\mathbf{a} \otimes \mathbf{b} + \mathbf{b} \otimes \mathbf{a} + \tfrac{1}{2}[(\mathbf{a}+\mathbf{b}) \otimes \mathbf{c} + \mathbf{c} \otimes (\mathbf{a}+\mathbf{b})]$$
$$+\tfrac{1}{2}[(\mathbf{a}+\mathbf{b}) \otimes \mathbf{c} - \mathbf{c} \otimes (\mathbf{a}+\mathbf{b})];$$

(x) $$\frac{1}{2}\begin{bmatrix} 0 & 1 & 1 \\ 1 & 0 & 1 \\ 1 & 1 & 0 \end{bmatrix} + \frac{1}{2}\begin{bmatrix} 0 & 1 & -1 \\ -1 & 0 & 0 \\ 1 & -1 & 0 \end{bmatrix}.$$

(3) $$\begin{bmatrix} 2 & 2.5 & 3 \\ 2.5 & -3 & -5 \\ 3 & -5 & -8 \end{bmatrix} + \begin{bmatrix} 0 & -3.5 & -5 \\ 3.5 & 0 & -1 \\ 5 & 1 & 0 \end{bmatrix}.$$

2.5

(1) The components of the transformed tensor are

$$\mathbf{T}^* = \begin{bmatrix} 1 & 0 & 0 \\ \sin^2\theta & \cos^2\theta & -\sin\theta\cos\theta \\ \cos\theta & -\sin\theta\cos\theta & \sin^2\theta \end{bmatrix}$$

(2)

$$\mathbf{a}\otimes\mathbf{b} = \begin{bmatrix} -6 & -4 & 8 \\ -3 & -2 & 4 \\ -3 & -2 & 4 \end{bmatrix}, \quad \mathbf{a}^* = (3\sqrt{2}/2, 1, -\sqrt{2}/2),$$

$\mathbf{b}^* = (\sqrt{2}/2, -2, 7\sqrt{2}/2)$.

$$\mathbf{a}^*\otimes\mathbf{b}^* = \begin{bmatrix} 3/2 & -3\sqrt{2} & 21/2 \\ \sqrt{2}/2 & -2 & 7\sqrt{2}/2 \\ -1/2 & \sqrt{2} & -7/2 \end{bmatrix}.$$

(5)

$$\begin{bmatrix} \sin\theta\cos\phi & \sin\theta\sin\phi & \cos\theta \\ \cos\theta\cos\phi & \cos\theta\sin\phi & -\sin\theta \\ -\sin\phi & \cos\phi & 0 \end{bmatrix}$$

(i)
$$\begin{bmatrix} (5-2\sqrt{2})/4 & -3/4 & -(5\sqrt{2}+2)/4 \\ -3/4 & (5+2\sqrt{2})/4 & (2-5\sqrt{2})/4 \\ -(5\sqrt{2}+2)/4 & (2-5\sqrt{2})/4 & -3/2 \end{bmatrix}$$

(ii)
$$\begin{bmatrix} (1-2\sqrt{3})/8 & (2\sqrt{3}-15)/8 & -(4\sqrt{2}+5\sqrt{6})/8 \\ (2\sqrt{3}-15)/8 & (1+6\sqrt{3})/8 & -5\sqrt{6}/8 \\ -(4\sqrt{2}+5\sqrt{6})/8 & -5\sqrt{6}/8 & (3-2\sqrt{3})/4 \end{bmatrix}.$$

2.8

(1) $\mathbf{x}.\mathbf{y} = 1 - 3i$; $\mathbf{y}.\mathbf{x} = 1 + 3i$;
$\mathbf{x}\times\mathbf{y} = (1, -1-i, i-2)$; $|\mathbf{x}| = \sqrt{6}$; $\mathbf{x}.(\mathbf{x}\times\mathbf{y}) = (\mathbf{x}\times\mathbf{y}).\mathbf{y} = 0$;
$\mathbf{y}.(\mathbf{T}.\mathbf{x}) = (\mathbf{y}.\mathbf{T}).\mathbf{x} = 2i - 2$; $\mathbf{x}.\mathbf{T}^{\mathrm{H}}.\mathbf{y} = -2 - 2i$.

(2)

$$(\mathbf{x} \times \mathbf{y}).\mathbf{z} = \begin{vmatrix} x_1 & x_2 & x_3 \\ y_1 & y_2 & y_3 \\ z_1 & z_2 & z_3 \end{vmatrix}.$$

$\overline{(\mathbf{x} \times \mathbf{y}).\mathbf{z}} = \mathbf{x}.(\mathbf{y} \times \mathbf{z}) = 1 + 3i.$

2.9

(1) (i) $2, \sqrt{2}, -\sqrt{2}$; $(3+i, 1+i, 2), (1+\sqrt{2}, 1, 0), (1-\sqrt{2}, 1, 0)$;
$2, \sqrt{2}, -\sqrt{2}$; $(0, 0, 1), (\sqrt{2}, 2-\sqrt{2}, i-\sqrt{2}-1)$,
$(\sqrt{2}, -2-\sqrt{2}, -i-\sqrt{2}+1)$;
(ii) $1, 2+3i, 2-3i$; $(20+30i, 20-9i, 13), (0, 1, 1), (0, -1, 1)$;
$1, 2-3i, 2+3i$; $(1, 0, 0), (3, 3i-1, 3i-1), (1, 3i+1, -3i-1)$;
(iii) $0, 1, 2$; $(0, -1, -1), (1, 1+i, -3-i), (0, 0, 1)$; $0, 1, 2$; $(i-1, 1, 0)$,
$(1, 0, 0), (2, 1, 1)$;
(iv) $1+i, 3/2-i, -\frac{1}{2}-i$; $(72-56i, 65, -130i), (4, 2i-1, -4)$,
$(4, 3, -2i, -4)$; $1-i, 3/2+i, -\frac{1}{2}+i$; $(1, 0, 1)$,
$(17-6i, 24i-28, -16-2i), (22i-19, 24i+36, -16-2i)$.

(2) (i) $-1, 1, 2$; $(1, 1, -1), (0, 1, 0), (1, 0, 0)$; $-1, 1, 2$; $(0, 0, 1), (0, 1, 1)$,
$(1, 0, 1)$;
(ii) $4, 1+2i, 1-2i$; $(-10, -3, 1)$, $(1, -1, 1+i), (1, -1, 1-i)$;
$4, 1-2i, 1+2i$; $(1, 1, 0), (-1, 4-i, 2-3i), (-1, 4+i, 2+3i)$;
(iii) $1, 5, -5$; $(2, -1, -2), (7, 6, 3), (1, -2, -1)$; $1, 5, -5$; $(0, 1, -2)$,
$(1, 0, 1), (9, -20, 19)$.

(3) $\Delta = 1$, hence Θ is a rotation matrix. Eigenvalues: $\lambda = 1, i, -i$. Eigenvector corresponding to $\lambda = 1$ is $\mathbf{v} = (1, 0, 1)$. This vector is left unchanged by the rotation. [Check this statement by showing that $\Theta\mathbf{v} = \mathbf{v}$].

(4) Eigenvalues: $\lambda = -1, 1, 2$; Eigenvectors: $\mathbf{v} = (0, 0, 1), (0, 1, 1), (1, 0, 1)$.

(5) $(T_{23}, -T_{13}, T_{12})$.

(8)

$$\lambda = 3, \mathbf{X} = \begin{vmatrix} a & b \\ -a & -b \end{vmatrix}.$$

2.10

(1) (i) $1, 3, 3$; $(1, -1, 2), (1, 0, 1)$. Rank $= 2$. $1, 3, 3$; $(-1, 3, 1), (-1, 1, 1)$.
(ii) $i, 2, 2$; $(0, 1, 0), (0, -1, 1+2i)$. Rank $= 2, -i, 2, 2$;
$(17i-4, 5-10i, 5), (1, 0, 0)$.
(iii) $2, -1, -1$; $(1, 0, 2), (1, 0, -1)$. Rank $= 2$. $2, -1, -1$; $(3, -2i, 3)$,
$(0, 1, 0)$.
(iv) $6, -2, -2$; $(2, 1, -3), (\alpha, \beta, 2\alpha+\beta)$. Rank $= 1$. $6, -2, -2$; $(2, 1, -1)$,
$(\alpha, 3\beta-2\alpha, \beta)$. ($\alpha, \beta$ arbitrary).

(v) 1, 1, 1; Rank $= 1, \mathbf{u} = (1, 1, -4). \mathbf{u}' = (2, 2, 1)$,
$\mathbf{v} = \mathbf{u} \times \mathbf{u}' = (7, -7, 0)$.
(vi) 1, 1, 1; Rank $= 2, \mathbf{u} = (1, -1, 1), \mathbf{u}' = (0, 1, 1)$.
(vii) 2, 2, 2; Rank $= 2, \mathbf{u} = (1, -13, -7), \mathbf{u}' = (4, 3, -5)$.

2.11

(1) (i) $2, \sqrt{2}, -\sqrt{2}$; $(1, 0, 1), (1, -\sqrt{2}i, -1), (1, \sqrt{2}i, -1)$;
(ii) $0, 1 + \sqrt{2}, 1 - \sqrt{2}$; $(0, 0, 1), (1 + i, \sqrt{2}, 0), (1 + i, -\sqrt{2}, 0)$;
(iii) $0, 1, 2$; $(1, 0, i), (0, 1, 0), (1, 0, -i)$.

2.12

(1) (i) $1, 1 + \sqrt{11}, 1 - \sqrt{11}$; $(1, 1, -3), (3 + \sqrt{11}, 3 - \sqrt{11}, 2)$,
$(3 - \sqrt{11}, 3 + \sqrt{11}, 2)$;
(ii) $4, 2, 2$; $(1, 1, 0), (\alpha, -\alpha, \beta), (\alpha, \beta$ arbitrary);
(iii) $6, 6 + 2\sqrt{3}, 6 - 2\sqrt{3}$; $(1, 1, 0), (-1, 1, \sqrt{3} - 1), (1, -1, \sqrt{3} + 1)$;
(iv) $6, 2, 2$; $(-1, 1, \sqrt{2}), (\sqrt{2}\alpha, \sqrt{2}\beta, \alpha - \beta), (\alpha, \beta$ arbitrary).

2.13

(1) (i) $\mathbf{T}^3 + 3\mathbf{T}^2 + 2\mathbf{T} + 6\mathbf{1} = \mathbf{0}$,
(ii) $\mathbf{T}^3 - 16\mathbf{T} = \mathbf{0}$,
(iii) $\mathbf{T}^3 + 12\mathbf{T}^2 + 44\mathbf{T} + 48\mathbf{1} = \mathbf{0}$.

(2)
$$\mathbf{T}^{-1} = \begin{bmatrix} 1 & -6 & 2 \\ 0 & 5 & 0 \\ -2 & -3 & 1 \end{bmatrix}.$$

2.15

(1) (i) $(2x_1, 2x_3, 3x_3^2)$; (ii) $2x_1 + 2x_3 + 3x_3^2$.

3.1

(1) (i)
$$\begin{bmatrix} 30 & 0 & 30 \\ 40 & 0 & 40 \\ 10 & 0 & 10 \end{bmatrix};$$

(ii) Let $\mathbf{T} = \mathbf{a} \otimes \mathbf{b} \otimes \mathbf{c}$. $T_{111} = 6, T_{112} = 3, T_{131} = 6, T_{132} = 3$,
$T_{211} = 8, T_{212} = 4, T_{231} = 8, T_{232} = 4, T_{311} = 2, T_{312} = 1$,
$T_{331} = 2, T_{332} = 1$. The remaining 15 components are zero.

(iv) $\mathbf{S}:\mathbf{T} = 80\begin{bmatrix} 6 & 3 & 0 \\ 8 & 4 & 0 \\ 2 & 1 & 0 \end{bmatrix}.$

3.2

(1) (i) $A_{112} = A_{123} = A_{131} = +1, A_{121} = A_{132} = A_{113} = -1$. The remaining 21 components of $\mathbf{A}$ are zero.

(ii) $d_1 = -1, d_2 = 1, d_3 = 0$.

(3) $\mathbf{u} = -\frac{1}{2}\mathbf{E}:\mathbf{T}$.

3.3

(5) Tens (vec $\mathbf{T}$) $= \frac{1}{2}(\mathbf{T} - \mathbf{T}')$.

3.4

(1)

$$\Theta = \frac{1}{98}\begin{bmatrix} 8 + 45\sqrt{2} & 12 + 36\sqrt{2} & 24 - 33\sqrt{2} \\ 12 - 48\sqrt{2} & 18 + 40\sqrt{2} & 36 - 4\sqrt{2} \\ 24 + 9\sqrt{2} & 36 - 32\sqrt{2} & 72 + 13\sqrt{2} \end{bmatrix}.$$

(2)

$$\Theta = \frac{1}{162}\begin{bmatrix} 82 & 4 + 72\sqrt{3} & 8 - 36\sqrt{3} \\ 4 - 72\sqrt{3} & 97 & 32 + 9\sqrt{3} \\ 8 + 36\sqrt{3} & 32 - 9\sqrt{3} & 145 \end{bmatrix}.$$

(5)

$$\text{(i)} \begin{bmatrix} 1 & 0 & 0 \\ 0 & \sqrt{2}/2 & \sqrt{2}/2 \\ 0 & -\sqrt{2}/2 & \sqrt{2}/2 \end{bmatrix}; \quad \text{(ii)} \begin{bmatrix} \sqrt{2}/2 & 0 & -\sqrt{2}/2 \\ 0 & 1 & 0 \\ \sqrt{2}/2 & 0 & \sqrt{2}/2 \end{bmatrix};$$

$$\text{(iii)} \begin{bmatrix} \sqrt{2}/2 & \sqrt{2}/2 & 0 \\ -\sqrt{2}/2 & \sqrt{2}/2 & 0 \\ 0 & 0 & 0 \end{bmatrix}.$$

For a rotation of $\pi/4$ in the negative sense:

$$\Theta = \begin{bmatrix} 1 & 0 & 0 \\ 0 & \sqrt{2}/2 & -\sqrt{2}/2 \\ 0 & \sqrt{2}/2 & \sqrt{2}/2 \end{bmatrix}.$$

(6) For a rotation of $\pi/2$:

$$\text{(i)} \begin{bmatrix} 1 & 0 & 0 \\ 0 & 0 & 1 \\ 0 & -1 & 0 \end{bmatrix}; \quad \text{(ii)} \begin{bmatrix} 0 & 0 & -1 \\ 0 & 1 & 0 \\ 1 & 0 & 0 \end{bmatrix}; \quad \text{(iii)} \begin{bmatrix} 0 & 1 & 0 \\ -1 & 0 & 0 \\ 0 & 0 & 1 \end{bmatrix}.$$

For a rotation of π:

$$\text{(i)} \begin{bmatrix} 1 & 0 & 0 \\ 0 & -1 & 0 \\ 0 & 0 & -1 \end{bmatrix}; \quad \text{(ii)} \begin{bmatrix} -1 & 0 & 0 \\ 0 & 1 & 0 \\ 0 & 0 & -1 \end{bmatrix}; \quad \text{(iii)} \begin{bmatrix} -1 & 0 & 0 \\ 0 & -1 & 0 \\ 0 & 0 & 1 \end{bmatrix}.$$

(8) T^*_{11} is a maximum when $\sin 2\theta = 2AT_{12}$, $\cos 2\theta = A(T_{11} - T_{22})$ where $A = +\sqrt{[4T^2_{12} + (T_{11} - T_{22})^2]}$. T^*_{11} is a minimum when $\sin 2\theta = -2AT_{12}$, $\cos 2\theta = -A(T_{11} - T_{22})$.

3.6

(5) $h_1 = h_3 = 1, h_2 = r$.

4.2

(1) $\mathbf{S} = \dfrac{1}{\mu_2}\mathbf{T}' - \dfrac{\mu_1 T_{ii}}{\mu_2(3\mu_1 + \mu_2)}\mathbf{1}$.

(ii) First show that $T_{ii} = (3\mu_1 + \mu_2 + \mu_3)S_{ii}$,

also $\mu_3 T_{ik} - \mu_2 T_{ki} = \mu_1 S_{jj}\delta_{ik}(\mu_3 - \mu_2) + S_{ik}(\mu_3^2 - \mu_2^2)$.

$$\mathbf{S} = \frac{\mu_3\mathbf{T} - \mu_2\mathbf{T}'}{\mu_3^2 - \mu_2^2} - \frac{\mu_1 T_{ii}\mathbf{1}}{(\mu_3 + \mu_2)(3\mu_1 + \mu_2 + \mu_3)}.$$

(2)

$$\Theta = \begin{bmatrix} 2/3 & 2/3 & -1/3 \\ -1/3 & 2/3 & 2/3 \\ 2/3 & -1/3 & 2/3 \end{bmatrix}.$$

The complex eigenvalues are $\lambda = \frac{1}{2} \pm i\sqrt{3}/2$. The corresponding eigenvectors are $\mathbf{v} = (1 \pm \sqrt{3}i, -2, 1 \mp \sqrt{3}i)$.

(3) Each component is a linear combination of 105 terms of the form $\epsilon_{ijk}\delta_{lm}\delta_{np}$.

5.1

(1) $\mu = \frac{10}{9}$, $\mathbf{H}(O) = (4, 0, 2)$, $T = 5$.

The required lines are $x = y = 0$ and $y = (1 \mp \sqrt{2})x, z = 0$. The moments of inertia about these lines are $2, 2 \pm \sqrt{2}$.

(2) $\mu = \frac{1}{81}(1329 + 32\sqrt{6} - 56\sqrt{2} + 56\sqrt{3})$;

$\mathbf{H}(O) = (64 + 4\sqrt{6} - 7\sqrt{2}, 4\sqrt{6} + 60 + 7\sqrt{3}, -4\sqrt{2} + 4\sqrt{3} + 119)$;

The required lines are ($x = \sqrt{2}, y = -\sqrt{3}, z = 1$) and any line in the plane $z = \sqrt{3}y - \sqrt{2}x$. The moments of inertia about these lines are 12, 18, 18.

(3) $I_{11} = \mu_1, I_{22} = \mu_2, I_{33} = \mu_3, I_{23} = \mu_4 - \tfrac{1}{2}(\mu_2 + \mu_3)$,

$I_{31} = \mu_5, -\tfrac{1}{2}(\mu_3 + \mu_1), I_{12} = \mu_6 - \tfrac{1}{2}(\mu_1 + \mu_2)$.

(4) $I_{11} = \mu_1, I_{22} = \mu_2, I_{33} = \mu_3$,

$I_{23} = \tfrac{1}{2}(\mu_1 + \mu_2 + \mu_3) - 3(\mu_5 + \mu_6)/4$

$I_{31} = \tfrac{1}{2}(\mu_1 + \mu_2 + \mu_3) - 3(\mu_6 + \mu_4)/4$

$I_{12} = \tfrac{1}{2}(\mu_1 + \mu_2 + \mu_3) - 3(\mu_4 + \mu_5)/4$.

(5) $I_{11} = \frac{1}{10}(-12\mu_1 + 8\mu_2 + 8\mu_3 + 12\mu_4 - 3\mu_5 - 3\mu_6)$

$I_{23} = \frac{1}{20}(4\mu_1 + 4\mu_2 + 4\mu_3 + 6\mu_4 - 9\mu_5 - 9\mu_6)$,

and four similar expressions.

(7)
$$I = \begin{bmatrix} I_1 + M(a_2^2 + a_3^2) & -Ma_1a_2 & -Ma_1a_3 \\ -Ma_1a_2 & I_2 + M(a_1^2 + a_3^2) & -Ma_2a_3 \\ -Ma_1a_3 & -Ma_2a_3 & I_3 + M(a_1^2 + a_2^2) \end{bmatrix}.$$

The principal moments of inertia are given by $M(a_1^2 + a_2^2 + a_3^2 + \lambda_i)$, $i = 1, 2, 3$ where λ_i are the three solutions of

$$\frac{a_1^2}{(I_1/M) - \lambda} + \frac{a_2^2}{(I_2/M) - \lambda} + \frac{a_3^2}{(I_3/M) - \lambda} = 1 .$$

5.2

(1) $M \begin{bmatrix} 13 & -2 & -3 \\ -2 & 11 & -3 \\ -3 & -3 & 4 \end{bmatrix}$. (2) $\tfrac{1}{4}M(a^2 + b^2)$. (3) $M \begin{bmatrix} 2 & 0 & -1 \\ 0 & 2 & -2 \\ -1 & -2 & 2 \end{bmatrix}$

5.3

(1) $Ma^2(8/3 - \sin 2\theta)$.

(2) $(1/3)Ma^2\,[5\mathbf{1} - 3\mathbf{e}_3 \otimes \mathbf{e}_3]$;

$$\tfrac{1}{3}Ma^2 \begin{bmatrix} 23 & 0 & 0 \\ 0 & 17 & -6\sqrt{2} \\ 0 & -6\sqrt{2} & 8 \end{bmatrix}.$$

Principal moments of inertia (eigenvalues) are given by

$$\lambda_1 = (23/3)\,Ma^2\,,\ \lambda_{2,3} = \tfrac{1}{6}Ma^2(25 \pm 3\sqrt{41}).$$

Corresponding principal axes are given by

$$\mathbf{n}_1 = (1,0,0),\ \mathbf{n}_{2,3} = (0,3 \pm \sqrt{41}, 4\sqrt{2}).$$

(3) Three rods each of mass M, of lengths $2a$, $2b$, $2c$, and coincident with the corresponding diameters of the block, together with a 'mass' $-2M$ at the centroid.
(4) Let the x_1, x_2-axes be parallel respectively to sides of length $2a$, $2b$, the origin at the centroid of the lamina and P the point (a, b). Then:

$$\mathbf{I}(\mathrm{P}) = \tfrac{1}{3}M \begin{bmatrix} 4b^2 & -3ab & 0 \\ -3ab & 4a^2 & 0 \\ 0 & 0 & 4(a^2+b^2) \end{bmatrix}.$$

(5) Let the origin be at the centre of the cube with the masses $m, 2m, 3m$ on the positive x_1-, x_2- and x_3- axes respectively. Then $\mathbf{I}(\mathrm{O}) = 14ma^2\,\mathbf{1}$. The position of the centroid G is given by $\bar{\mathbf{r}} = \dfrac{-a}{21}(5, 3, 1)$.
(6) The eigenvalues are $\frac{23}{3}Ma^2$, $\frac{1}{6}Ma^2(25 \pm 3\sqrt{41})$ with corresponding eigenvectors $(1, 0, 0)$, $(0, 4\sqrt{2}, 3 \mp \sqrt{41})$.

5.4

(1) With origin at the centroid let the x_1-axis be parallel to one pair of sides and the x_2-axis in the plane of the rhombus. Principal moments of inertia are $1 \pm \sqrt{2}/2$, 2. Principal axes are $(1, \mp\sqrt{2}, -1, 0)$, $(0, 0, 1)$.
(4) $\frac{7}{10}Ma^2$.

5.5

(1) $\mathbf{I}(\mathrm{O}) = \tfrac{1}{4}Ma^2\{(\mathbf{1} - \mathbf{e}_1 \otimes \mathbf{e}_1) + (\mathbf{1} - \mathbf{e}_2 \otimes \mathbf{e}_2)\} + \tfrac{1}{2}Mh^2(\mathbf{1} - \mathbf{e}_3 \otimes \mathbf{e}_3)$.

(2) $M\left\{\dfrac{a^2}{4}(\mathbf{1} - \mathbf{e}_1 \otimes \mathbf{e}_1) + \dfrac{b^2}{4}(\mathbf{1} - \mathbf{e}_2 \otimes \mathbf{e}_2) + \dfrac{l^2}{3}(\mathbf{1} - \mathbf{e}_3 \otimes \mathbf{e}_3)\right\}$

(3) $M\{\frac{1}{4}a^2 + \frac{1}{3}l^2)\mathbf{1} + M(\frac{1}{4}a^2 - \frac{1}{3}l^2)\mathbf{e}_3 \otimes \mathbf{e}_3$.

(4) $\frac{3}{20}Ma^2\,\{(\mathbf{1} - \mathbf{e}_1 \otimes \mathbf{e}_1) + (\mathbf{1} - \mathbf{e}_2 \otimes \mathbf{e}_2)\} + \frac{3}{5}Mh^2(\mathbf{1} - \mathbf{e}_3 \otimes \mathbf{e}_3)$,

where $h^2 = l^2 - a^2$.

$$\frac{3M}{20}l^2 \sin^2\alpha(5\cos^2\alpha + 1).$$

(6) $\mathbf{I}(\mathrm{O}) = \frac{1}{5}M\{a^2(\mathbf{1} - \mathbf{e}_1 \otimes \mathbf{e}_1) + b^2(\mathbf{1} - \mathbf{e}_2 \otimes \mathbf{e}_2) + c^2(\mathbf{1} - \mathbf{e}_3 \otimes \mathbf{e}_3)\}$.

(7) $\frac{83}{320} Ma^2 \mathbf{1} + \frac{9}{64} Ma^2 \, \mathbf{e}_3 \otimes \mathbf{e}_3$

(8) $\mathbf{I}(O) = \frac{1}{4} Ma^2 \{(\mathbf{1} - \mathbf{e}_1 \otimes \mathbf{e}_1) + (\mathbf{1} - \mathbf{e}_2 \otimes \mathbf{e}_2)\}$

$$+ \frac{1}{8} Ma^2 \frac{\sin 2\phi}{\phi} \{(\mathbf{1} - \mathbf{e}_1 \otimes \mathbf{e}_1) - (\mathbf{1} - \mathbf{e}_2 \otimes \mathbf{e}_2)\} .$$

$$\mu = \frac{1}{4} Ma^2 \left(1 - \frac{\sin 4\phi}{4\phi}\right) .$$

(10) $\mathbf{I}(O) = \frac{3}{10} Ma^2 \{(\mathbf{1} - \mathbf{e}_1 \otimes \mathbf{e}_1) + (\mathbf{1} - \mathbf{e}_2 \otimes \mathbf{e}_2)\}$

$$- \frac{1}{10} Ma^2 \{(\mathbf{1} - \mathbf{e}_1 \otimes \mathbf{e}_1) + (\mathbf{1} - \mathbf{e}_2 \otimes \mathbf{e}_2) - 2(\mathbf{1} - \mathbf{e}_3 \otimes \mathbf{e}_3)\} \ (1 + \cos\phi + \cos^2\phi).$$

(12) The result is given in equation (5.113).

(13) $\frac{2}{5} Ma^2 \mathbf{1}$.

(14) Let O be the centre of one sphere and A the point of contact of the other two. The principal axes are in the directions of the unit vectors $\mathbf{e}_1, \mathbf{e}_2, \mathbf{e}_3$ where $\mathbf{e}_2$ is along OA, $\mathbf{e}_1$ is perpendicular to $\mathbf{e}_2$ and in the plane of the centres, $\mathbf{e}_3 = \mathbf{e}_1 \times \mathbf{e}_2$.

$$I_{11} = 14{\cdot}4 Ma^2 , I_{22} = 6{\cdot}4 Ma^2 , I_{33} = 18{\cdot}4 Ma^2 \cos^{-1}\left(\frac{21}{2\sqrt{129}}\right) .$$

(15)

(i) $$\mathbf{I}(G) = \frac{2Ma^2}{5} \begin{bmatrix} 8 & 0 & 0 \\ 0 & 8 & 0 \\ 0 & 0 & 13 \end{bmatrix} ;$$

(ii) $$\mathbf{I}(A) = \frac{Ma^2}{5} \begin{bmatrix} 21 & 0 & 0 \\ 0 & 16 & 0 \\ 0 & 0 & 31 \end{bmatrix} .$$

(16) Equation of momental ellipsoid is $\mathbf{I}(O): \mathbf{r} \otimes \mathbf{r} = 1$, i.e. $I_{ik} x_i x_k = 1$. Diagonalize $\mathbf{I}(O)$ to find the principal moments and axes.

Let the cube and sphere be centred on the origin. The principal axes at the corner $B(b, b, b)$ are BO together with any axis through B and perpendicular to BO. Principal moments at B are $A = (16b^5/3 - 8\pi a^5/15)\rho$, $C = (88b^5/3 - 8\pi a^5/15 - 4\pi a^3 b^2)\rho$.
Equation of momental ellipsoid is

$$\tfrac{1}{3} (64b^5 - \frac{8\pi}{5} a^5 - 8\pi a^3 b^2)(x_1^2 + x_2^2 + x_3^2)$$
$$+ \tfrac{2}{3}(-24b^5 + 4\pi a^3 b^2)(x_2 x_3 + x_3 x_1 + x_1 x_2) = 1.$$

5.6

(1) Roll the two spheres down an inclined plane and compare their accelerations. For either sphere $\dot{\omega} = \dfrac{Mga \sin\alpha}{\mu(G) + Ma^2}$ where $\dot{\omega}$ = angular acceleration, M = mass, a = radius, α = inclination of plane, $\mu(G)$ = moment of inertia about centroid G. For the solid sphere $\mu(G) = \frac{2}{5}Ma^2$, $\dot{v}_i = \frac{5}{7} g \sin\alpha$ where $\dot{v}_i = a\dot{\omega}_i$ is the linear acceleration of the solid sphere. For the hollow sphere $\mu(G) = \frac{2}{5} M\left(\dfrac{a^5 - b^5}{a^3 - b^3}\right)$

$$\dot{v}_2 = \frac{5ga^2 \sin\alpha(a^2 + ab + b^2)}{7a^2(a^2 + ab + b^2) + 2b^3(a + b)}$$

where b is the inner radius. Since $\dot{v}_2 < \dot{v}_1$ the acceleration of the hollow sphere is less than that of the solid sphere.

(2) The couple is $\mathbf{\Gamma} = \{(I_{33} - I_{22})\omega_2\omega_3, (I_{11} - I_{33})\omega_3\omega_1, 0\}$ where $\boldsymbol{\omega} = \omega(\cos\alpha, \sin\alpha\cos\theta, \sin\alpha\sin\theta)$ and $I_{11} = I_{22} = \frac{1}{4}Ma^2$, $I_{33} = \frac{1}{2}Ma^2$, $|\mathbf{\Gamma}| = \frac{1}{2}Ma^2\omega^2\ |\sin 2\alpha|$.

(3) $\dfrac{mgl}{Cn}$.

(4) The magnitude of the applied couple $|\mathbf{\Gamma}|$ is zero when the axis $(\cos\alpha, \cos\beta, \cos\gamma)$ is parallel to $\mathbf{e}_1$, $\mathbf{e}_2$ or $\mathbf{e}_3$. Hence any axis about which the body will rotate with no applied couple is a principal axis.

6.1

(1) $\frac{166}{61}$, $x + y = 0$.

(2)

$$\boldsymbol{\sigma} = \begin{bmatrix} 6 & 0 & 0 \\ 0 & 4 & 0 \\ 0 & 0 & 1 \end{bmatrix}. \quad \boldsymbol{\sigma}^* = 1/9 \begin{bmatrix} 32 & 14 & -16 \\ 14 & 41 & 2 \\ -16 & 2 & 26 \end{bmatrix}.$$

The components of $\boldsymbol{\sigma}^*$ are the required normal and shearing stresses.

(3) Eigenvalues of $\boldsymbol{\sigma}$ are $(1, -2, -2)$. Eigenvector corresponding to $\lambda = 1$ is $\mathbf{n} = (\sqrt{3}/2, 0, \frac{1}{2})$. If we let $\mathbf{n}_0 = \mathbf{n}$ the eigenvectors of $\boldsymbol{\sigma}$ are either parallel or perpendicular to $\mathbf{n}_0$.

(7)

$$\sigma = \begin{bmatrix} 0 & 3 & -1 \\ 3 & 6 & -2 \\ -1 & -2 & -7 \end{bmatrix}.$$

6.2

(3) At $(a, a, 0)$ $\quad \mathbf{e} = a^2(\mathbf{e}_2 \otimes \mathbf{e}_3 + \mathbf{e}_3 \otimes \mathbf{e}_2)$, $\boldsymbol{\omega} = -2a^2\mathbf{e}_1$.

Sphere is deformed into ellipsoid with major axis $x = 0, z = -y$,
minor axis $x = 0, z = +y$.

7.1

(1) The result is true whether $\boldsymbol{\epsilon}$ is taken to be the finite Lagrangian strain tensor defined by (7.7) or the infinitesimal strain tensor defined by (7.8).
(2) Consider a rod of length $2l$, diameter d with its axis lying along the x_1-axis between $(-l, 0, 0)$ and $(l, 0, 0)$. Suppose the rod is bent into a circular arc radius a, centre $(0, a, 0)$ such that $d/l \ll l/a \ll 1$. The displacement of the point $(x, 0, 0)$ in the rod is approximately represented by

$$\mathbf{r} = \left(\frac{-x_1^3}{6a^2}, \frac{x_1^2}{2a}\right), \quad |x_1| \leqslant 1 .$$

Equation (7.7) yields:

$$2\boldsymbol{\epsilon} = \begin{bmatrix} x_1^4/4a^4 & x_1/a & 0 \\ x_1/a & 0 & 0 \\ 0 & 0 & 0 \end{bmatrix}.$$

Equation (7.8) yields:

$$2\boldsymbol{\epsilon} = \begin{bmatrix} -x_1{}^2/a & x_1/a & 0 \\ x_1/a & 0 & 0 \\ 0 & 0 & 0 \end{bmatrix}.$$

Equation (7.8) is seen to be a good approximation to (7.7) but the displacement at $x_1 = \pm l$ is large compared with the diameter of the rod.

(7) The eigenvalues of the infinitesimal strain tensor $\boldsymbol{\epsilon}$ are $(-6, 11, 11)$. The axis corresponding to $\lambda = -6$ is $(1, -4, 0)$. This is the direction of the axis of symmetry.
(9) $\mathbf{r} = k[x_1(x_2^2 + x_3^2), x_2(x_3^2 + x_1^2), x_3(x_1^2 + x_2^2)]$.

(10) $$\frac{\partial^2 \epsilon_{12}}{\partial x_1 \partial x_2} = \frac{1}{2}\left(\frac{\partial^2 \epsilon_{22}}{\partial x_1^2} + \frac{\partial^2 \epsilon_{11}}{\partial x_2^2}\right)$$

$$r_1 = \alpha_{11}x_1 + Ax_2 + 2\beta_{11}x_1(x_1^2/3 + x_2^2) + x_1^5/5 + x_1x_2^4.$$

$$r_2 = (2\alpha_{12} - A)x_1 + \alpha_{22}x_2 + 2\beta_{22}x_2(x_1^2 + x_2^2/3) + x_2^5/5 + x_1^4x_2.$$

7.2

(2) (i) $\nu'' = 0;$

(ii) $E' = \dfrac{E}{1-\nu^2}, \nu''' = \dfrac{\nu}{1-\nu}$.

(3) Use cylindrical polar coordinates with the origin at the point of suspension and the z-axis along the axis of symmetry of the cylinder. Then

$$v_3 = \frac{\rho g}{2E}(\nu r^2 + z^2), v_1 = -\frac{\nu}{E}\rho g z r.$$

7.3

(1) (i) $U = \dfrac{T^2}{2E}, \; U' = \dfrac{T^2}{6\mu}, \; U'' = \dfrac{T^2}{18k};$

(ii) $U = \dfrac{T^2}{2k}, \; U' = 0, U'' = U;$

(iii) $U = U' = \tfrac{1}{2}\mu k^2, U'' = 0.$

(2) $$U = U' = \frac{T^2}{2\mu}, U'' = 0 .$$

(3) $\tfrac{1}{4}\pi\mu\theta^2 a^4/l$.

7.6

(1) $$\epsilon = \frac{1}{2}k \begin{bmatrix} 0 & 0 & -x_1 \\ 0 & 0 & x_2 \\ -x_2 & x_1 & k(x_1^2 + x_2^2) \end{bmatrix}.$$

(2) $$\epsilon = \frac{1}{2} \begin{bmatrix} \alpha_1^2 & \alpha_1 - \alpha_2 & 0 \\ -\alpha_1 - \alpha_2 & \alpha_2^2 & 0 \\ 0 & 0 & \alpha_3^2 + 2\alpha_3 \end{bmatrix}.$$

(3)

(i) $$\boldsymbol{\eta} = \tfrac{1}{2}\begin{bmatrix} 0 & 0 & k \\ 0 & 0 & 0 \\ k & 0 & -k^2 \end{bmatrix};$$

(ii) $$\boldsymbol{\eta} = \frac{k(k+2)}{2(k+1)^2}\mathbf{1}\ .$$

(iii) $$\boldsymbol{\eta} = \tfrac{1}{2}\begin{bmatrix} \dfrac{\sigma^2k^2 - 2\sigma k}{(1-\sigma k)^2} & 0 & 0 \\ 0 & \dfrac{\sigma^2k^2 - 2\sigma k}{(1-\sigma k)^2} & 0 \\ 0 & 0 & \dfrac{k(k+2)}{(1+k)^2} \end{bmatrix}.$$

(4)

(i) $$\boldsymbol{\eta} = \begin{bmatrix} -\sinh^2 t & \sinh t \cosh t & 0 \\ \sinh t \cosh t & -\sinh^2 t & 0 \\ 0 & 0 & 0 \end{bmatrix},$$

$$\mathbf{e} = \mathbf{e}_1 \otimes \mathbf{e}_2 + \mathbf{e}_2 \otimes \mathbf{e}_1 .$$

(ii) $$\boldsymbol{\eta} = \tfrac{1}{2}\begin{bmatrix} 0 & (a-2z_2)t & 0 \\ (a-2z_2)t & -(a-2z_2)^2t^2 & 0 \\ 0 & 0 & 0 \end{bmatrix},$$

$$\mathbf{e} = (\tfrac{1}{2}a - z_2)(\mathbf{e}_1 \otimes \mathbf{e}_2 + \mathbf{e}_2 \otimes \mathbf{e}_1).$$

(iii) Choose Cartesian coordinates (z_1, z_2, z_3).

$$\boldsymbol{\eta} = -\begin{bmatrix} 2z_1^2t^2 & 2z_1z_2t^2 & z_1t \\ 2z_1z_2t^2 & 2z_2^2t^2 & z_2t \\ z_1t & z_2t & 0 \end{bmatrix},$$

$$\mathbf{e} = -z_1(\mathbf{e}_1 \otimes \mathbf{e}_3 + \mathbf{e}_3 \otimes \mathbf{e}_1) - z_2(\mathbf{e}_2 \otimes \mathbf{e}_3 + \mathbf{e}_3 \otimes \mathbf{e}_2)\ .$$

(5)

$$\boldsymbol{\eta} = \frac{1}{2}\begin{bmatrix} 1-\sigma_1^2 & 0 & 0 \\ 0 & 1-\sigma_2^2 & 0 \\ 0 & 0 & 1-\sigma_3^2 \end{bmatrix},$$

$$\mathbf{e} = \tfrac{1}{2}\omega(\sigma_2^2 - \sigma_1^2)(\mathbf{e}_1 \otimes \mathbf{e}_2 + \mathbf{e}_2 \otimes \mathbf{e}_1)/(\sigma_1\sigma_2)$$

(6)

$$\text{(i) } \mathbf{G} = (\nabla\mathbf{r}).(\nabla\mathbf{r})' = (t-\tau)^2 \begin{bmatrix} 0 & 0 & 0 \\ 0 & (\partial\phi/\partial z_2)^2 & \partial\phi/\partial z_2\,\partial\phi/\partial z_3 \\ 0 & \partial\phi/\partial z_2\,\partial\phi/\partial z_3 & (\partial\phi/\partial z_3)^2 \end{bmatrix};$$

$$\text{(ii) } \boldsymbol{\eta} = \tfrac{1}{2}(t-\tau)\begin{bmatrix} 0 & \partial\phi/\partial z_2 & \partial\phi/\partial z_3 \\ \partial\phi/\partial z_2 & 0 & 0 \\ \partial\phi/\partial z_3 & 0 & 0 \end{bmatrix} - \mathbf{G}\ ;$$

(iii) $\mathbf{v} = -\phi\mathbf{e}_1$;

$$\text{(iv) } \mathbf{e} = -\frac{1}{2}\begin{bmatrix} 0 & \partial\phi/\partial z_2 & \partial\phi/\partial z_3 \\ \partial\phi/\partial z_2 & 0 & 0 \\ \partial\phi/\partial z_3 & 0 & 0 \end{bmatrix}.$$

BIBLIOGRAPHY

Batchelor, G. K., *An Introduction to Fluid Dynamics*, Cambridge University Press.

Bourne, D. W. and Kendall, P. C., *Vector Analysis and Cartesian Tensors*, Nelson.

Hunter, S. C., *Mechanics of Continuous Media*, Ellis Horwood.

Jeffreys, H., *Cartesian Tensors*, Cambridge University Press.

Kreyszig, E., *Advanced Engineering Mathematics*, John Wiley.

Milne, E. A., *Vectorial Mechanics*, Methuen.

Noble, B., *Applied Linear Algebra*, Prentice Hall Inc.

Oldroyd, J. G., *Proc. Roy. Soc.* A. **200**, 523 - 541.

Oldroyd, J. G., *Proc. Roy. Soc.* A. **202**, 345 - 358.

Oldroyd, J. G., *Proc. Roy. Soc.* A. **245**, 278 - 297.

Southwell, R. V., *An Introduction to the Theory of Elasticity*, Oxford University Press.

Spain, B., *Tensor Calculus*, Oliver and Boyd.

Index

A

adjoint matrix 31, 94
adjugate matrix 31, 94
alternate tensor 114
angular
    momentum 152
    velocity 63
        tensor 222
antisymmetric
    matrix 30
    tensor 74
associate tensor 269

B

basis 38
bulk modulus 239

C

Cayley-Hamilton theorem 103
characteristic
    equation 87
    values 86
Christoffel symbols 270
cofactor 23
compatibility equations 232
complete elliptic integrals 186
complex
    scalar product 84
    vector product 84
    isotropic vectors 149
components
    contravariant 266
    covariant 266
    of a dyad 69
    of the inertia tensor 159
    of the stress tensor 207
    of a tensor 68
    of a vector 35
conjugate
    tensor 85
    vector 85
continued vector product 116
contraction 82
contravariant components 266
convected derivative 259
coordinate line 133
covariant
    components 266
    differentiation 271
Cramer's rule 19
cross products of tensors and vectors 119
curl 116
curvilinear coordinates 132
cylindrical polar coordinates 140

D

defective
    matrix 91
    tensor 91
deformation 231
    finite 257
    tensor (Eulerian) 258
    tensor (Lagrangian) 230
derivative
    convected 259
    of a scalar field 55
    of a tensor field 143
    of a vector field 109
determinant 19
determinantal equation 87
deviatoric stress tensor 210
diagonal tensor 101
diagonalization (of a tensor) 101
dilatation 232
    rate of 228
dimensions (of a matrix) 26
direction cosines 44
directional derivative 56
displacement
    equation 247
    vector 230
Divergence theorem 130
double inner product 83
dummy suffix 57
dyad 66

E

effective
    Poisson's ratio 240

Young's modulus 240
eigenvalues
of a second-order tensor 86
of the Hermitian transpose tensor 88
of a Hermitian tensor 97
of a real symmetric tensor 101
eigenvector problem 86
degenerate cases 91
eigenvectors
of a second-order tensor 86
of the Hermitian transpose tensor 88
of a Hermitian tensor 97
of a real symmetric tensor 101
elliptic integrals 186
energy
kinetic 153
stored in dilatation and distortion 245
strain 244
equation
displacement 247
Navier-Stokes 227
of equilibrium 247
of fluid motion 205, 208, 227
equations
compatibility 232
simultaneous 19
equilibrium configuration 229
equimomental systems 167
Eulerian
deformation tensor 258
strain tensor 258
Euler's equation of motion 199
expansion of a determinant by cofactors 23
extension of a wire 239

F

finite
deformation 257
strain tensor 258
fluid
incompressible 227
Newtonian 227
pressure 210
Reiner-Rivlin 227
free suffix 57

G

generalised Hooke's law 238
gradient
of a scalar field 55
of a vector field 109
of a tensor field 143
Gram-Schmidt orthogonlization process 53
Green's theorem 130

H

Hermitian
tensor 97
transpose tensor 85
Hooke's Law 237

I

incompressible fluid 227
inertia
moment of 155
principal axes of 153
principal moments of 153
products of 159
tensor 152
components of 159
inertia tensor of
circular arc 195
circular cylinder (hollow) 182
circular cylinder (solid) 184
circular lamina 181
cone (hollow) 197
cone (solid) 197
disc 181
ellipsoid 197
elliptical lamina 189
elliptical ring 185
parallelepiped 174
parallelogram lamina 173
rectangular block 165, 168
rectangular lamina 166
ring 180
rod 166
sphere (hollow) 193
sphere (solid) 194
tetrahedron 177
torus (hollow) 190
torus (solid) 192
triangular lamina 175
inner
product of two tensors 82
suffices 82
infinitesimal strain 229
tensor 230
integral theorems 130
invariance of tensors 76
invariants of a second-order tensor 80
inverse matrix 31
isotropic
expansion 239
function 106
tensors 145
vectors 149

K

kinetic energy of a rotating body 153
Kronecker delta 57

L

Lagrangian
- deformation tensor 230
- finite strain tensor 258
- strain tensor 230

Lamé constants 239
Laplacian 110
latent
- roots 86
- vectors 86

leading diagonal of a matrix 27
Levi-Civita density function 115
linear
- dependence 51
- independence 51
- vector functions 66

longitudinal
- strain 240
- stress 240

lower triangular matrix 90

M

matrices 26
matrix
- adjoint 31
- adjugate 31
- antisymmetric 30
- inverse 31
- lower triangular 90
- null 29
- rank of 31
- rotation 60
- singular 31
- skew-symmetric 30
- square 26
- symmetric 30
- transpose 30
- unit 29
- upper triangular 90
- zero 29

metric tensor 268
minor 23
modulus of a vector 38
Mohr's circle of tress 217
moment of inertia 155
momental ellipsoid 156
multiplicity (of eigenvalue) 91

N

Navier-Stokes equation 227
Newtonian fluid 227
normal stress 207
null
- matrix 29
- vector 37

O

orthogonal unit triad 38

P

parallel axis theorem 159
parallelogram law of addition 40
perpendicular axis theorem 162
physical components of tensors 134
plane
- shear 223
- stress 216

Poisson's ratio 240
position vector 35
post-multiplication of a tensor and a vector 72
pre-multiplication of a tensor and a vector 72
pressure (fluid) 210
principal
- axes of inertia 153
- moments of inertia 153
- stresses 210
- values 86

products of inertia 159

Q

quotient theorem 269

R

radius of gyration 155
rank (of a matrix) 31
rate-of-strain 220
- quadric 222
- tensor 219, 259

real
- symmetric tensor 101
- vector space 36

reciprocal tensor 269
reference configuration 229
Reiner-Rivlin fluid 227
repeated suffix 57
representation theorem for second-order tensors 106
right-hand screw rule 46
rotating
- disc 253
- frame of reference 121

rotation matrix 60, 123
rotations 59, 123

about coordinate axes 126
small 128

S

scalar 34
field 34, 54
invariants of a second-order tensor 80
product of two vectors 37
scale factor 133
second-order tensors 66
secular equation 87
shear
simple 223, 239
modulus 239
shearing stress 207
simultaneous equations 19
singular matrix 31
skew-symmetric
matrix 30
tensor 75
spherical polar coordinates 137
spin tensor 222
Stokes's theorem 131
strain 229
components 257
energy 244
function 244
longitudinal 240
tensor 229, 258
stress 205
longitudinal 240
normal 207
plane 216
principal 216
shearing 207
tangential 207
tensor 204
symmetry of 209
deviatoric 210
submatrix 31
substitution operator 57
suffices 56
dummy 57
free 57
repeated 57
summation convention 56
symmetric
matrix 30
tensor 74
real 101

T

tangential stress 207
tensor
alternate 114
angular velocity 222
antisymmetric 74
associate 269
defective 91
deformation 230, 258
deviatoric stress 210
diagonal 101
Eulerian deformation 258
Eulerian strain 238
field
second-order 108
third-order 143
finite strain 258
first-order 35
fourth-order 145
general 266
Hermitian 97
inertia 152
infinitesimal strain 230
isotropic 145
Lagranigan deformation 230
Lagrangian strain 230
metric 268
of a vector 120
physical components of 134
product of
four vectors 145
three vectors 112
two vectors 67
rate-of-strain 219, 259
real symmetric 101
reciprocal 269
scalar invariants of 80
second-order 66
skew-symmetric 74
spin 222
strain 229, 258
stress 204
symmetric 74
third-order 112
trace of 81
transformation law for 76
transpose 72
unit 70
vorticity 222
zero 70
zero-order 34
top 202
trace of a tensor 80
transformation law for
second-order tensors 76
third-order tensors 113
vectors 58
transpose
matrix 30
tensor 72
treble inner product 113

triangle law of addition 41
triple
    scalar product 50
    vector product 50

U

unit
    matrix 29
    tensor 70
upper triangular matrix 90

V

vector 35
    angle 232
    complex isotropic 149
    field 54
    null 37
    of a tensor 120
    position 35
    product 38
        (continued) 116
    space 36
    zero 37
vorticity 222
    tensor 222

Y

Young's modulus 240

Z

zero
    matrix 29
    tensor 70
    vector 37